Studies in Systems, Decision and Control

Volume 140

Series editor

Janusz Kacprzyk, Polish Academy of Sciences, Warsaw, Poland
e-mail: kacprzyk@ibspan.waw.pl

The series "Studies in Systems, Decision and Control" (SSDC) covers both new developments and advances, as well as the state of the art, in the various areas of broadly perceived systems, decision making and control- quickly, up to date and with a high quality. The intent is to cover the theory, applications, and perspectives on the state of the art and future developments relevant to systems, decision making, control, complex processes and related areas, as embedded in the fields of engineering, computer science, physics, economics, social and life sciences, as well as the paradigms and methodologies behind them. The series contains monographs, textbooks, lecture notes and edited volumes in systems, decision making and control spanning the areas of Cyber-Physical Systems, Autonomous Systems, Sensor Networks, Control Systems, Energy Systems, Automotive Systems, Biological Systems, Vehicular Networking and Connected Vehicles, Aerospace Systems, Automation, Manufacturing, Smart Grids, Nonlinear Systems, Power Systems, Robotics, Social Systems, Economic Systems and other. Of particular value to both the contributors and the readership are the short publication timeframe and the world-wide distribution and exposure which enable both a wide and rapid dissemination of research output.

More information about this series at http://www.springer.com/series/13304

Vassil Sgurev · Vladimir Jotsov
Janusz Kacprzyk
Editors

Practical Issues of Intelligent Innovations

 Springer

Editors
Vassil Sgurev
Institute of Information and Communication
    Technologies
Bulgarian Academy of Sciences
Sofia
Bulgaria

Janusz Kacprzyk
Department of Computer Science,
    Systems Research Institute
Polish Academy of Sciences
Warsaw
Poland

Vladimir Jotsov
University of Library Studies
    and Information Technologies
Sofia
Bulgaria

ISSN 2198-4182   ISSN 2198-4190   (electronic)
Studies in Systems, Decision and Control
ISBN 978-3-030-08698-5   ISBN 978-3-319-78437-3   (eBook)
https://doi.org/10.1007/978-3-319-78437-3

This Springer imprint is published by the registered company Springer International Publishing AG
part of Springer Nature
The registered company address is: Gewerbestrasse 11, 6330 Cham, Switzerland

Preface

In this volume on "Practical Issues of Intelligent Innovations", some relevant selected scientific results and applications in the field of intelligent systems are presented. A wide scope of the presented research encompasses issues from industrial data science applications to scientific and applied ones. One can see that the research and especially its application parts are multifaceted and rather different. On the other hand, they are like parts of a large puzzle and altogether form a large wave appearing on the horizon. Will they completely change the main paradigms of Information Technologies and the whole world and when? The answer depends on too many factors and on the world economy as a whole. Will the consequences of these advanced applications be dangerous? Yes, if security issues are not explored in time. Contemporary complex intelligent systems are the main target for future attacks and the cyber war tools.

We hope that the wave of the progress is inevitable, and it will change the world for good, and that the proposed intelligent standards and tools will be used more efficiently for cyber defense and not for cyber wars. This book is aimed at a wide audience of advanced intelligent system users and developers. Its chapters are written by well-known experts and representatives of the young generation of talented researchers. The research groups are established and composed from many continents: from New Zealand through Asia to the European Union. This is the relevant key for a sustainable development of intelligent systems.

In Chapter "Non-conventional Control Design by Sigmoid Generated Fixed Point Transformation Using Fuzzy Approximation", the authors Adrienn Dineva, József Tar (Hungary), Annamária Várkonyi-Kóczy, János Tóth (Slovakia), and Vincenzo Piuri (Italy) present an original and innovative combination of methods and techniques for adaptive nonlinear control design. The fuzzy approximation is considered in order to handle errors occurring due to modeling imprecision. Iterative learning has been applied. The results have confirmed that by applying fuzzy modeling in the Sigmoid Generated Fixed Point Transformation (SGFPT) the control design the performance level can be significantly increased.

Chapter "From von Neumann Architecture and Atanasoff's ABC to Neuromorphic Computation and Kasabov's NeuCube. Part II: Applications" is written by a large international team: Maryam G. Doborjeh, Zohreh Gholami, Akshay Raj Gollahalli, Kaushalya Kumarasinghe, Vivienne Breen, Neelava Sengupta, Josafath Israel Espinosa Ramos, Reggio Hartono, Elisa Capecci (New Zealand), Hideaki Kawano (Japan), Muhaini Othman (Malaysia), Lei Zhou, Jie Yang (China), Pritam Bose (Italy), and Chenjie Ge (New Zealand). NeuCube is one of the first machine learning systems to analyze integrated space and time aspects of big data to deliver deeper insights. Inspired by the human brain, the most evolved learning system, NeuCube, does the same advanced pattern recognition of complex data streams in just seconds. Much like the brain, NeuCube uses a network of virtual neurons connecting to each other or disconnecting depending on the timing of signals encoded in incoming data streams. Its architecture is elaborated for different prominent applications of spatio/spectro-temporal data. Pattern recognition from satellite images, remote sensing, and other innovations is widely discussed in this chapter.

Chapter "Data-Driven Interval Type-2 Fuzzy Modelling for the Classification of Imbalanced Data" written by Rubio Solis, Ali Baraka, George Panoutsos, and Steve Thornton (UK) is one of the chapters from this book dedicated to data-driven applications. The Fuzzy Decision Engine developed is based on the Interval Type-2 RBF Neural Network and is used as the core facet of a modeling framework for imbalanced data classification. Here, the fuzzy methods are used for industrial applications at TATA factories in the UK. An original iterative information granulation is considered. Simulations are carried out using different number of fuzzy rules. The authors reveal that the associated accuracy is a compromise between the specificity and sensitivity. The overall framework is designed to specifically address the issue of imbalanced data in the data-driven modeling for manufacturing industrial processes.

In Chapter "Network Flows and Risks", written by Vassil Sgurev and Stanislav Drangajov (Bulgaria), some network flow models are considered aiming at the analysis of transportation of resources: material, financial, communication, commodities, services, etc., and risks related to these processes. As a result, two interrelated flows arise—that of resources and of risks. A method is proposed for coordinating the values of the capacities and the arc evaluations of both flows, respectively, in which, in the examples given, the parameters of one of the flows are taken into consideration in the parameters of the other one and vice versa. The Intelligent Combined Resource and Network Risk Flow Model is proposed and discussed. Numerical examples for optimal transportation schemes have been considered.

Data science applications in intelligent measurement systems have been discussed in Chapter "Decreasing Influence of the Error due to Acquired Inhomogeneity of Sensors by the Means of Artificial Intelligence" written by Vladimir Jotsov (Bulgaria), Orest Kochan (Ukraine), and Su Jun (China). Different types of constraints are developed or applied; some of them are the different classical forms while the others are used for logical bounds between different pieces of knowledge. The applied original puzzle standards revealed the strength of semantic knowledge extraction and processing methods which could directly influence the practical applications which, in this chapter, concern the measurement and analysis of sensory data. While investigating the influence of the proposed new types of constraints, namely the binding ones, on linear constraints or other classical types, the authors reveal the fact that some of the introduced new types of constraints may influence and fuzzify the classical ones and make their applications more universal and/or efficient.

Chapter "Personal Assistants in a Virtual Education Space" is written by Jordan Todorov, Vladimir Valkanov, Stanimir Stoyanov, Borislav Daskalov, Ivan Popchev, and Daniela Orozova (Bulgaria). The Virtual Education Space is implemented as an IoT ecosystem. The personal assistant will offer more effective and efficient support to its users. Besides, the new version will be able to effectively adapt to a group of users needing different additional help in their everyday activities as, e.g., for people with different physical disabilities. To attain this goal, the IoT LISSA system will be actively supported by the guards in the space.

Chapter "Practical Guidelines for Design of Human-in-the-Loop Systems: Lessons Learned" written by Vassily Moshnyaga (Japan) considers not only the typical autonomous smart systems that keep humans out of the control loop by allowing the user to intervene, if necessary. The human involvement in the system control can be passive, active, and hybrid. The author's approach is one of the so-called "human-in-the-loop" in which the user's involvement may be much more active and effective. For this, the systems must be easy to deploy, maintain, and extend. A large variety of such systems have been developed, among them are user-aware computer display, viewer-conscious TV set, smart door, smart carpet, in-home patient monitoring system, medication adherence monitoring system, and many more.

The next Chapter "NEO-Fuzzy Neural Networks for Knowledge Based Modeling and Control of Complex Dynamical Systems" written by Yancho Todorov and Margarita Terziyska (Bulgaria) deals with the NEO-fuzzy neural networks for knowledge-based modeling and control. It shows up the arguments that the most successful contemporary control systems are the fuzzy set based ones.

Chapter "Sign-Based Representation and World Model of Actor" is written by Gennady Osipov (Russia). The sign-based synthesis of behavior is described. Formalisms are proposed to enable the description of basic cognitive functions, such as introspection, reflection, goal setting, etc., and enhancing the extent of understanding of cognitive processes. A description is provided of a cognitive function named a goal setting function. The development of the proposed formalism is related to handling multi-agent and robotic systems.

Chapter "Responsive Production in Manufacturing: A Modular Architecture" written by Maria Marques, Carlos Agostinho (Portugal), Gregory Zacharewicz (France), Raul Poler (Spain), and Ricardo Jardim-Goncalves (Portugal) is entitled "Responsive Production in Manufacturing: A Modular Architecture". This research is widely applied in large and medium industrial projects. In a Smart Factory, everything is connected as production machines, humans, products, transportation means, and IT tools communicate with each other and are organized with the objective of improving the overall production level. The presented chapter identifies opportunities, challenges, and main characteristics of Industry 4.0 followed by an analysis of main barriers to its implementation. The developed work answers questions on specific needs and challenges that must be addressed to solve problems related to dynamic market changes which are intimately connected with the design and reconfiguration of the manufacturing enterprise.

In Chapter "Multisensor Data Association by Using the Polar Hough Transform", Ivan Garvanov (Bulgaria) explores the data association problem and its usage in advanced security applications. An algorithm is proposed and applied in a multiple input multiple output radar system. Experiments have been conducted with the so-called decentralized Hough detector in which the Hough association is used. The considered application is also very up to date from the point of view of problems and challenges of data sciences.

In Chapter "Scientific Research Funding Criteria: An Empirical Study of Peer Review and Scientometrics", the authors D. Devyatkin, R. Suvorov, I. Tikhomirov, and O. Grigoriev (Russia) apply some original data science views and approaches. The most significant criteria are identified that affect decisions about research funding. They include the importance of expected results and quality of previous research performed by applicants. The described technique can be used by scientific foundations to evaluate new criteria before they are included into a review form (by simulation). The authors' goal is that all these additions together with traditional peer-review framework will make the funding decisions more transparent and difficult for cheating.

Chapter "Comparing Robots with Different Levels of Autonomy in Educational Setting" is written by Mirjam de Haas, Alex Mois Aroyo (The Netherlands), Pim Haselager (Italy), Iris Smeekens, and Emilia Barakova (The Netherlands). The authors discuss innovative applications of robots in education and assistive technologies. One of the key features of every intelligent system, the autonomous behavior, is compared to the robots' ability to learn and show intelligence. Recognition, playing strategies, and other applications have been discussed. The authors have decided to use a design solution consisting of two main conditions: the robot with a low autonomy that is completely remotely controlled by the experimenter, and the robot with a high autonomy that uses pattern recognition based on visual information to recognize objects. The authors explore the effects of an autonomous robot on the children's perception of the robot and on a described game. The interplay with children and robots paves the way to an interactive learning between cognitive and hardware agents; this evolutionary approach still is not well explored. A clear advantage is that a teacher/therapist can use a robot with increasing levels of autonomy instead of a remotely controlled robot.

Chapter "Clustering Non-Gaussian Data Using Mixture Estimation with Uniform Components" written by Ivan Nagy and Evgenia Suzdaleva (Czech Republic) considers the problem of clustering non-Gaussian data with fixed bounds via a recursive mixture estimation under the Bayesian methodology. An algorithm is proposed aiming at the enrichment of clustering and classification tools. Realistic simulations from the transportation microscopic simulator Aimsun (www.aimsun.com) are used for testing the proposed algorithm.

We would like to express our gratitude to all the authors for their interesting, novel, and inspiring contributions. Peer-reviewers also deserve a deep appreciation, because their insightful and constructive remarks and suggestions have considerably improved many contributions.

And last but not least, we wish to thank Dr. Tom Ditzinger, Dr. Leontina di Cecco, and Mr. Holger Schaepe for their dedication and help to implement and finish this large publication project on time maintaining the highest publication standards.

Sofia, Bulgaria   Vassil Sgurev
Sofia, Bulgaria   Vladimir Jotsov
Warsaw, Poland   Janusz Kacprzyk
Summer 2017

Contents

Non-conventional Control Design by Sigmoid Generated Fixed Point Transformation Using Fuzzy Approximation

Adrienn Dineva, József K. Tar, Annamária Várkonyi-Kóczy, János T. Tóth and Vincenzo Piuri

Abstract Lyapunov's 2nd or Direct method is recognized as being the primary tool of adaptive control of nonlinear dynamic systems. The great majority of the adaptive nonlinear control design rest on Lyapunov's stability theorem. Recent findings have revealed that the Robust Fixed Point Transformation-based method can succesfully replace the Lyapunov technique. Later the *"Sigmoid Generated Fixed Point Transformation (SGFPT)"* has been introduced. This systematic method has been proposed for the generation of whole families of Fixed Point Transformations. Its extension from Single Input Single Output (SISO) to Multiple Input Multiple Output (MIMO) systems has also been given. In recent times, the great majority of model building issues are replaced by *"Soft Computing"* techniques. In contrast to the classical mathematical methods the intelligent methodologies are able to cope with ill-defined systems, disturbances and missing information by an efficient and robust way. Especially fuzzy logic has become to be used to model complex systems. This

A. Dineva (✉)
Kandó Kálmán Faculty of Electrical Engineering, Óbuda University, Budapest, Hungary
e-mail: dineva.adrienn@kvk.uni-obuda.hu

A. Dineva
Department of Information Technologies, Doctoral School
of Computer Science, Università degli Studi di Milano, Crema, Italy

J. K. Tar
Antal Bejczy Center for Intelligent Robotics (ABC iRob), Óbuda University,
Budapest, Hungary
e-mail: tar.jozsef@nik.uni-obuda.hu

A. Várkonyi-Kóczy · J. T. Tóth
Department of Mathematics and Informatics, J. Selye University,
Komárno, Slovakia
e-mail: varkonyi-koczy@uni-obuda.hu

J. T. Tóth
e-mail: tothj@mail.ujs.sk

V. Piuri
Department of Information Technologies, Università degli Studi di Milano,
Crema, Italy
e-mail: vincenzo.piuri@unimi.it

© Springer International Publishing AG, part of Springer Nature 2018
V. Sgurev et al. (eds.), *Practical Issues of Intelligent Innovations*,
Studies in Systems, Decision and Control 140,
https://doi.org/10.1007/978-3-319-78437-3_1

contribution makes an attempt to utilize the advantages of fuzzy approximation in the SGFPT control design. The theoretical investigations are validated by the adaptive control of the inverted pendulum. Comparative analysis have been carried out between the "*affine*" and the "*soft computing-based*" models. Results of numerical simulations confirm the applicability and efficiency of the proposed method.

Keywords Sigmoid Generated Fixed Point Transformation · Fuzzy modeling Nonlinear control · Iterative learning · Fixed point transformation

1 Introduction

Lyapunov's *direct* method (or commonly referred to as the *second method* of Lyapunov) is a widely used technique in the analysis of the stability of the motion of the non-autonomous dynamic systems of equation of motion as $= f(x, t)$. Its main advantage that it does not require to analytically solve the equations of motion [1, 2]. Instead of the analytical solution, in most of the practical applications the uniformly continuous nature and non-positive time-derivative of a positive definite Lyapunov-function V is constructed of the tracking errors and modeling errors of the systems parameters that are assumed in the $t \in [0, \infty)$ domain. From this point the convergence $\dot{V} \to 0$ can be concluded according to Barbalat's lemma. However, in spite of its advantages, the application of Lyapunov's second method is far difficult and requires great practice for finding the appropriate Lyapunov function candidate. Additionally, the automization of such algorithm is challenging and also the computational need of realizing the method is significant.

Though, most of the control design techniques are based on Lyapunov's theorem. For instance, the *Model Reference Adaptive Controller (MRAC)* applies it for tuning the feedback parameters [3–6]. The classical methods, such as the *Slotine-Li Adaptive Robot Controller (SLARC)* and also the *Adaptive Inverse Dynamics Robot Controller (AIDRC)* control strategies apply Lyapunov's second method for tuning the parameters of the analytical system model [7]. An exhaustive study on the application of Lyapunov's direct method discussing the relating practical issues can be found in [8]. Originally, the Robust Fixed Point Transformation (RFPT) based control method has been introduced for replacing Lyapunov's second method [9–11]. The base of the theory of the Fixed Point Transformation-based control approach relies on the Expected-Realized Response Scheme. Some control tasks can be formulated by using the concept of the appropriate excitation of the controlled system to which it is expected to respond with some desired response r^d. The appropriate excitation can be computed by using the inverse dynamic model of the system. Since normally this inverse model is neither complete nor exact, the actual response determined by the systems dynamics, results in a realized response r^r that differs from the desired one. The idea of the RFP-based iterative adaptive learning approach is that the controller normally can manipulate or deform the input value from r^d to $r_\star^d$, so that $r^d(r_\star^d)$. This situation can be maintained by using some local deformation that has been realized by fixed point transformation [8]. The other specific feature of the RFPT-based control method, that it strictly separates the "kinematic" and "dynamic" aspects of the control task and predefines the "desired system response" by the use of purely kinematic terms.

The necessary dynamic terms are estimated using the approximate dynamic model of the system under control. The realization of the kinematic specification then based on these dynamic terms. The undesired effects of modeling imprecisions and unknown external disturbances are compensated by the above mentioned adaptive deformation of the *desired response* r^{Des} as the input of the approximate model. By this deformation, the desired response can be obtained as the realized one. The adaptive deformation is carried out by some kinematically expressed trajectory error reduction, i.e. *desired response* r^{Des} and the *actually observed response* r^{Act} are compared. Then, the *actually observed response* is deformed due to the exact system under control, formally $r^{Act} = f(r^{Des}, \dots)$ where the symbol "..." stands for the unknown state variables, for e.g. coupled subsystems, disturbances, etc. Due to the additive noises, etc., $r^{Act} \neq r^{Des}$. Hence, the input of the response function from r^{Des} to $r_{\star}$ is deformed using *Banach's Fixed Point Theorem* [12] in order to obtain $r^{Act} \equiv f(r_{\star}) = r^{Des}$.

Following, the *Sigmoid Generated Fixed Point Transformation (SGFPT)* has been introduced for adaptive control of nonlinear *Single Input–Single Output (SISO)* and *Multiple Input–Multiple Output (MIMO)* systems [13], which allows generating a whole family of fixed point transformations. Besides, a new function has been presented for the adaptive deformation, that gives significant enhancement of the performance [14].

The great popularity of soft computing methods, such as fuzzy logic, neural networks, genetic algorithms, is due to their simplicity and versatility; these methods have the ability of modeling and analyzing ill-defined problems by a cost-effective way [15]. In the past it has been proved that most of the deterministic model buildings can be effectively replaced by soft-computing methods [16], especially with fuzzy models [17]. Fuzzy modeling is of primary importance in fuzzy theory and has many interpretations [18–20]. For instance, the widely known Takagi-Sugeno models are great universal approximators for some nonlinear behaviour [21]. By the use of Lyapunov stability theorem, a large number of papers have been published on the synthesis of fuzzy modeling and control [22]. In comparison with the classical hard-computing methods the intelligent methodologies are able to deal with imprecisions and uncertainties.

In this short contribution we demonstrate the possible cooperation of fuzzy modeling and the SGFPT control design. Investigations focus on both the "*affine*", the "*soft computing-based*" and the "*fully soft computing-based*" models of the inverted pendulum. Finally, numerical simulations have been shown for validating the usability of the proposed method.

Fig. 1 The fixed points of *F*(*x*)

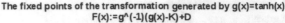
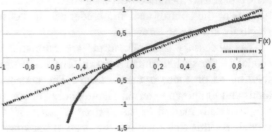

The fixed points of the transformation generated by g(x)=tanh(x)
F(x):=g^(-1)(g(x)-K)+D

2 Combination of Fuzzy Modeling and Sigmoid Generated Fixed Point Transformation

2.1 The Concept of Sigmoid Generated Fixed Point Transformation

This section gives a brief summary of the mathematical background of the "*Sigmoid Generated Fixed Point Transformation*" control method. At first, let us consider a monotonic increasing, bounded and smooth $g(x) : \mathbb{R} \mapsto \mathbb{R}$ "sigmoid" function. For some $K > 0$ and $D > 0$ consider a function $F(x) \stackrel{def}{=} g^{-1}(g(x) - K) + D$, in which the inverse function of $g()$ is denoted by $g^{-1}()$. The solution of $F(x) = x$ is the *fixed point* of $F(x)$. For the case of $g(x) \stackrel{def}{=} \tanh(x)$, $K = 0.5$, and $D = 0.6$ an example is displayed in Fig. 1

It is clear from Fig. 1, that this function has two fixed points. The first one at $\left|\frac{dF}{dx}\right| > 1$ and the second one at $\left|\frac{dF}{dx}\right| < 1$. A function is called *contractive* in the case of real, differentiable $\varphi : \mathbb{R} \mapsto \mathbb{R}$ functions if $\exists 0 \leq K < 1$, hence $\forall a, b \in \mathbb{R}$ $|g(a) - g(b)| < K|a - b|$. It is evident, if $\exists 0 \leq K < 1$ so that $\left|\frac{d\varphi}{dx}\right| < K$, therefore the following integral estimation can be applied:

$$|\varphi(b) - \varphi(a)| = \left|\int_a^b \frac{d\varphi}{dx} dx\right| \leq \int_a^b \left|\frac{d\varphi}{dx}\right| dx \leq K|b - a| \ . \tag{1}$$

Clearly, $F(x)$ is contractive around the fixed point at about $x = 0.7$, hence the sequence $\{x_0, x_1 \stackrel{def}{=} F(x_0), \ldots, x_{n+1} \stackrel{def}{=} F(x_n), \ldots\}$ is a Cauchy sequence since $\forall L \in \mathbb{N}$

$$\begin{aligned} |x_{n+L} - x_n| &= |F(x_{n-1+L}) - F(x_{n-1})| \leq \\ &\leq K|x_{n-1+L} - x_{n-1}| \leq \ldots K^n |x_L - x_0| \to 0 \\ &\text{as } n \to \infty \ . \end{aligned} \tag{2}$$

Since the real numbers forms a complete space there $\exists x_* \in \mathbb{R}$ so that $x_n \to x_*$. Based on these considerations, the following fixed point transformation is proposed using $F(x)$ for adaptive control in the case of slowly varying desired response r^{Des}:

$$r_{n+1} = G(r_n) \overset{def}{=} F\left(A\left[f(r_n) - r^{Des}\right] + x_*\right) + r_n - x_*, \tag{3}$$

where $A \in \mathbb{R}$ is a parameter. When $r_\star$ is the solution of the control task, i.e. $f(r_\star) - r^{Des} = 0$ then $G(r_\star) = r_\star$. Since $F(x_*) = x_*$, this fixed point of the function G is this solution. In order to ensure the convergence of the series $\{r_n\}$, function G must be contractive, i.e. the relation $\left|\frac{dG}{dr}\right| < 1$ must be guaranteed. It has been shown, that this contractivity can be achieved by properly setting the value of parameter A:

$$\frac{dG}{dr} = A\frac{dF\left(A\left[f(r) - r^{Des}\right] + x_*\right)}{dx}\frac{df}{dr} + 1 . \tag{4}$$

A simple possibility for applying the same idea of adaptivity systems for Multiple Input–Multiple output systems has been presented in [13], when $f, r \in \mathbb{R}^n$, $n \in \mathbb{N}$ (in which n denotes the degree of freedom (DoF) of the controlled system). In this case the iteration generates sequences as $\{r(i) \in \mathbb{R}^n | i = 0, 1, ...\}$ by applying a sigmoid function projected to the direction of the *response error* in the ith control cycle as:

$$h(i) \overset{def}{=} f(r(i)) - r^{Des}, e(i) \overset{def}{=} \frac{\mathfrak{A}h(i)}{\|\mathfrak{A}h(i)\|}, \tag{5}$$

in which the $\|h\|$ is the norm (in Frobenius sense):

$$r(i+1) = \tilde{G}(r(i)) \overset{def}{=}$$
$$\overset{def}{=} \left[F(A\|\mathfrak{A}h(i)\| + x_*) - x_*\right] e(i) + r(i) . \tag{6}$$

The solution of the control task is the fixed point of $\tilde{G}(r)$ when $f(r_\star) - r^{Des} \equiv h(i) = 0$ then $r(i+1) = r(i) = r_\star$.

The convergence properties of the controller can be enhanced by appropriately tuning the diagonal matrix with positive main diagonals $\mathfrak{A}$ in Eqs. 5 and 6. In this soft-computing based method it has been modified. In the original SGFPT it was the unit matrix [13]. Its matrix elements can be tuned by observing little fluctuations in the convergence of the adaptive signal when these main diagonals are too big. These fluctuations can be recognized as a negative content in a forgetting buffer as it was done in [23]. Furthermore, the stability conditions for MIMO systems were discussed in an earlier paper [13]. Because the former technique worked in bounded region of attraction around $r_\star$ that formally could not guarantee global stability. In order to address this question, the so-called Strectched Sigmoid Function has been investigated [14]. It was assumed that for the first element of the iteration x_0 there exists x_1 for which $g(x_0) - K = g(x - D)$. It was found to be valid for most of the x_0

values but not for each of them. This limitation could be evaded by introducing the following stretched function in [24]

$$F(x) = B\tanh(a(x+b)) + K \tag{7}$$

with $a, b > 0$. For allowing further improvements and a more precise positioning of the function in the vicinity of the solution of the control task, the next, new type of function has been introduced in [24]

$$F(x) = a\tanh(\tanh(x+D)/2). \tag{8}$$

2.2 Application Example: Adaptive Control of the Inverted Pendulum

For the numerical simulation the inverted pendulum system is under consideration. The inverted pendulum is a well-defined benchmark for nonlinear control and many variants of this problem is documented. The generalized coordinates are q_1 [rad], q_2 [rad] and the generalized forces are torque signals: Q_1 [N · m], Q_2 [N · m]. The equations of motion are given in (9).

$$\begin{pmatrix} mL^2 & 0 \\ 0 & mL^2 \sin^2 q_1 \end{pmatrix} \begin{pmatrix} \ddot{q}_1 \\ \ddot{q}_2 \end{pmatrix} + \begin{pmatrix} -mL^2 \sin q_1 \cos q_1 \dot{q}_2^2 + mgL \sin q_1 \\ 2mL^2 \sin q_1 \cos q_1 \dot{q}_1 \dot{q}_2 \end{pmatrix} = \begin{pmatrix} Q_1 \\ Q_2 \end{pmatrix} \tag{9a}$$

The system parameters are given in Table 1.

2.3 Realization of the Proposed Method

The simulations have been carried out by the use of the package "*Julia*". This dynamic language allows a very fast evaluation. For the calculations the Euler integration method have been used with a fixed step length 10^{-4} s. The constant control parameters are collected in Table 2.

Table 1 The system model and its parameters

Parameter	Exact value	Approximate value
Mass m (kg)	0.5	1.2
Length L (m)	1.6	1.7
Gravitational acceleration g ($\frac{m}{s^2}$)	9.81	10

Table 2 The constant
control parameters

Parameter	Value
Λ	$4\,\mathrm{s}^{-1}$
D	0.3
δt time delay in learning	$10^{-3}\,\mathrm{s}$

The PID-type relaxation can be defined for the tracking error of q. Let $\mathbb{R} \ni \Lambda > 0$, and let

$$e(t) \overset{def}{=} q^N(t) - q(t) \ , \tag{10a}$$

$$e_{int}(t) = \int_{t_0}^{t} \left(q^N(\xi) - q(\xi)\right) d\xi \ , \tag{10b}$$

$$\left(\Lambda + \frac{\mathrm{d}}{\mathrm{d}t}\right)^2 e(t) = 0 \Rightarrow \tag{10c}$$

$$\ddot{q}^{Des} = \ddot{q}^N + \Lambda^3 e_{int}(t) + 3\Lambda^2 e(t) + 3\Lambda \dot{e}(t) \ . \tag{10d}$$

Firstly, the adaptive controller deforms $\ddot{q}^{Des}$ into $\ddot{q}^{Def}$. After, it applies the approximate dynamic model for the calculation of the control forces for this deformed value. According to (9), for this excitation the controlled system responses with the "*Realized*" response $\ddot{q}$. The function $\ddot{q} = f\left(\ddot{q}^{Def}\right)$ is called as the "*response function*" of the system under control. The adaptive deformation relies on the measurements of these responses. For the adaptive deformation the $F(x) = \mathrm{atanh}\left(\frac{\mathrm{tanh}(x+D)}{2}\right)$ function have been used.

3   Simulation Investigations

3.1   *Performance Using the Affine Model*

At first, we investigate the affine model. The trajectory tracking and trajectory tracking error are depicted in Figs. 2, 3 and 4. The adaptive strategy has been shown in comparison with a non-adaptive one. It can be seen, that the imprecisions of trajectory tracking are due to the errors of the computed torque signals that caused by the modeling errors (Fig. 5).

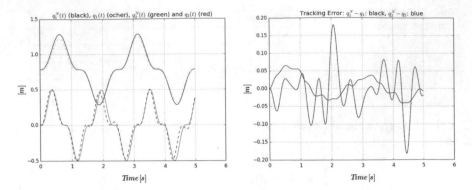

Fig. 2 Trajectory tracking of the non-adaptive controller for the "*affine model*"

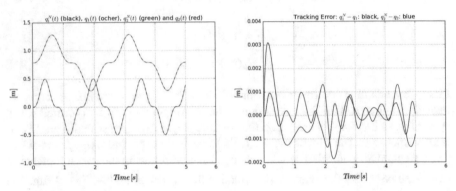

Fig. 3 Trajectory tracking of the adaptive controller for the "*affine model*"

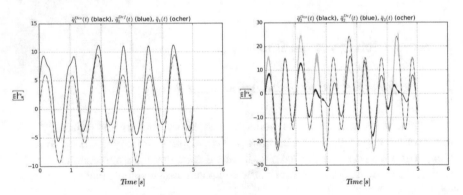

Fig. 4 The $\ddot{q}$ values of the adaptive controller for the "*affine model*"

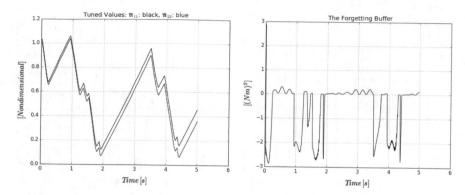

Fig. 5 The tuned parameters of the adaptive controller and the content of the forgetting buffer for the "*affine model*"

Fig. 6 Functions $\sin(x)$ (black), $\mu_s(x)$ (blue), $\cos(x)$ (green), and $\mu_c(x)$ (red)

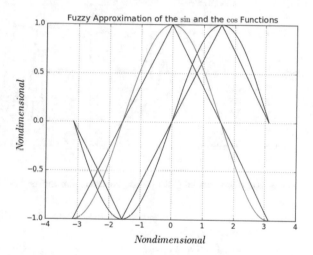

3.2 Performance Using the Soft Computing-Based Model

In order to avoid the above mentioned errors the $\sin(x)$ and the $\cos(x)$ functions were approximated by fuzzy rules as $\mu_s(x)$ and $\mu_c(x)$ over a bounded region according to Fig. 6.

In the *approximate dynamic model* the additional terms remained constants just as in the case of the "affine model". In the inertia matrix the function $\sin(x)$ was replaced by fuzzy modeling as follows (Figs. 7 and 8):

```
function ApprMod(q,q_p,q_ppDes)
    global ma
    global La
    global ga
```

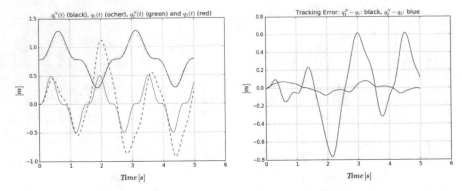

Fig. 7 Trajectory tracking of the non-adaptive controller for the *"soft computing-based model"*

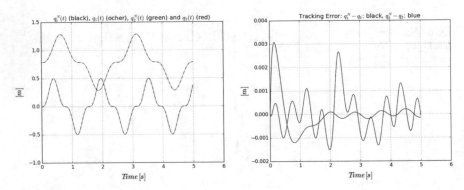

Fig. 8 Trajectory tracking of the adaptive controller for the *"soft computing-based model"*

```
      H=zeros(2,2)
      sq1=mus(q[1,1])
      H[1,1]=ma*La^2;
      H[2,2]=ma*La^2*sq1^2;
      h=[1.0;1.0];
      return H*q_ppDes+h;
end
```

From Figs. 9 and 10 it can be seen, that the fuzzy modeling improved the precision.

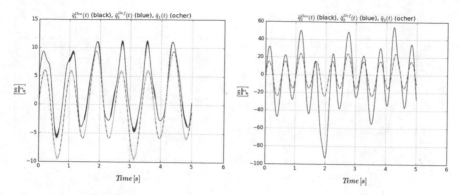

Fig. 9 The $\ddot{q}$ values of the adaptive controller for the "*soft computing-based model*"

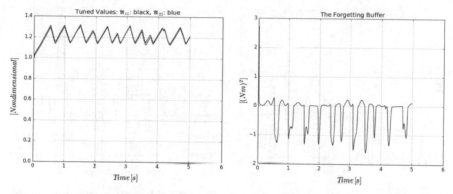

Fig. 10 The tuned parameters of the adaptive controller and the content of the forgetting buffer for the "*soft computing-based model*"

3.3 Performance Using the Fully Soft Computing-Based Model

In the case of the *fully soft computing-based model* in the *approximate dynamic model* each term of the functions sin(x) and cos(x) were approximated by the fuzzy $\mu_s(x)$ and $\mu_c(x)$ functions according to the code below:

```
function ApprMod(q,q_p,q_ppDes)
    global ma
    global La
    global ga
    H=zeros(2,2)
    sq1=mus(q[1,1])
    cq1=muc(q[1,1])
    H[1,1]=ma*La^2;
```

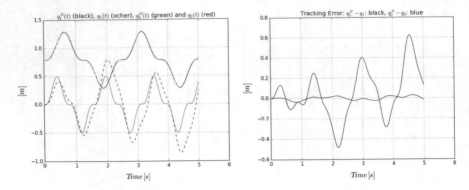

Fig. 11 Trajectory tracking of the non-adaptive controller for the "*soft computing-based model*"

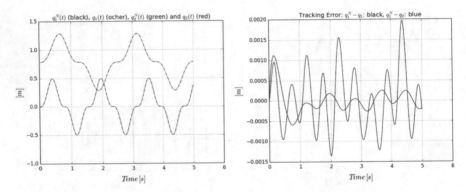

Fig. 12 Trajectory tracking of the adaptive controller for the "*fully soft computing-based model*"

```
H[2,2]=ma*La^2*sq1^2;
h=[1.0;1.0];
h[1,1]=-ma*La^2*sq1*cq1*q_p[2,1]^2+
+ma*ga*La*sq1;
h[2,1]=2*ma*La^2*sq1*cq1*q_p[1,1]*q_p[2,1];
return H*q_ppDes+h;
end
```

The results revealed by Figs. 11, 12, 13 and 14 validate, that by the combination of fuzzy modeling and the *Sigmoid Generated Fixed Point Transformation* design can ensure high performance and robustness to modeling imprecisions.

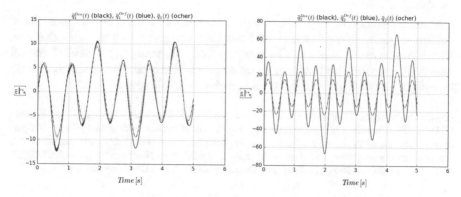

Fig. 13 The $\ddot{q}$ values of the adaptive controller for the *"fully soft computing-based model"*

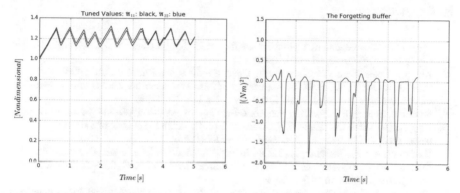

Fig. 14 The tuned parameters of the adaptive controller and the content of the forgetting buffer for the *"fully soft computing-based model"*

4 Conclusions

We have presented a new adaptive control method by the combination of fuzzy modeling and the *Sigmoid Generated Fixed Point Transformation*. Recent years, the *"Sigmoid Generated Fixed Point Transformation (SGFPT)"* has been introduced that can serve as an efficient alternative of the Lyapunov technique in nonlinear control design. This systematic method has been proposed for the generation of whole families of Fixed Point Transformations. In addition it has been extended from Single Input Single Output (SISO) to Multiple Input Multiple Output (MIMO) systems. Many studies regarding modeling issues highlight the efficiency of soft computing methods. In order to handle errors occurring due to modeling imprecisions we have introduced fuzzy approximation in this technique. For the numerical simulations the inverted pendulum system served as a benchmark problem. The *"affine"* model has been investigated in comparison with the *"soft computing-based"* model. In the soft computing-based model the trigonometric terms of the approximate dynamic model were replaced by fuzzy rules. In the case of the *"fully soft computing-based"* model each term the functions $\sin(x)$ and $\cos(x)$ were approximated by the fuzzy rules. Additionally, the SGFPT type control has been compared to a non-adaptive one. The results have confirmed that by applying fuzzy modeling in the SGFPT control design the performance level can be significantly increased. Taken together, these investigations confirm the usability and efficiency of this new method.

Acknowledgements This work has been sponsored by the Hungarian National Scientific Fund (OTKA 105846). This publication is also the partial result of the Research and Development Operational Programme for the project "Modernisation and Improvement of Technical Infrastructure for Research and Development of J. Selye University in the Fields of Nanotechnology and Intelligent Space", ITMS 26210120042, co-funded by the European Regional Development Fund.

References

1. Lyapunov, A.: A general task about the stability of motion (in Russian). Ph.D. Thesis, University of Kazan, Tatarstan (Russia) (1892)
2. Khalil, H.: Nonlinear Systems, 2nd edn. Upper Saddle River, Prentice Hall (1996)
3. Kamnik, R., Matko, D., Bajd, T.: Application of model reference adaptive control to industrial robot impedance control. J. Intell. Robot. Syst. **22**, 153–163 (1998)
4. Somló, J., Lantos, B., Cát, P.: Advanced Robot Control. Akadémiai Kiadó, Budapest (2002)
5. Nguyen, C., Antrazi, S., Zhou, Z.-L., Campbell Jr., C.: Adaptive control of a stewart platform-based manipulator. J. Robot. Syst. **10**(5), 657–687 (1993)
6. Hosseini-Suny, K., Momeni, H., Janabi-Sharifi, F.: Model reference adaptive control design for a teleoperation system with output prediction. J. Intell. Robot. Syst. 1–21 (2010). https://doi.org/10.1007/s10846-010-9400-4
7. Slotine, J.-J.E., Li, W.: Applied Nonlinear Control. Prentice Hall International, Inc., Englewood Cliffs, New Jersey (1991)
8. Tar, J.: Adaptive Control of Smooth Nonlinear Systems Based on Lucid Geometric Interpretation (DSc Dissertation). Hungarian Academy of Sciences, Budapest, Hungary (2012)
9. Tar, J.K.: Replacement of Lyapunov function by locally convergent robust fixed point transformations in model based control, a brief summary. J. Adv. Comput. Intell. Intell. Inf. **14**(2), 224–236 (2010)
10. Tar, J., Bitó, J., Nádai, L., Machado, J.T.: Robust fixed point transformations in adaptive control using local basin of attraction. Acta Polytechnica Hungarica **6**(1), 21–37 (2009)
11. Tar, J.K., Bitó, J.F., Rudas, I.J.: Replacement of Lyapunov's direct method in model reference adaptive control with robust fixed point transformations. In: Proceedings of the 14th IEEE International Conference on Intelligent Engineering Systems, Las Palmas of Gran Canaria, Spain, pp. 231–235 (2010)
12. Banach, S.: Sur les opérations dans les ensembles abstraits et leur application aux équations intégrales (About the operations in the abstract sets and their application to integral equations). Fund. Math. **3**, 133–181 (1922)
13. Dineva, A., Tar, J., Varkonyi-Koczy, A., Piuri, V.: Generalization of a sigmoid generated fixed point transformation from siso to mimo systems. In: IEEE 19th International Conference on Intelligent Engineering Systems (INES2015), 3–5 Sept 2015, Bratislava, Slovakia, pp. 135–140 (2015)
14. Dineva, A., Tar, J.K., Várkonyi-Kóczy, A., Piuri, V.: Adaptive control of underactuated mechanical systems using improved sigmoid generated fixed point transformation and scheduling strategy. In: IEEE 14th International Symposium on Applied Machine Intelligence and Informatics, 21–23 Jan 2016, Herlány, Slovakia, pp. 193–197 (2016)

15. Wong, P., Aminzadeh, F., Nikravesh, M. (eds.): Soft Computing for Reservoir Characterization and Modeling. Springer, Berlin, Heidelberg (2002)
16. Seki, H.: Nonlinear function approximation using fuzzy functional SIRMs inference model. In: AIKED'12 Proceedings of the 11th WSEAS International Conference on Artificial Intelligence, Knowledge Engineering and Data Bases, pp. 201–206 (2012)
17. Zadeh, L.A.: The concept of a linguistic variable and its application to approximate reasoning. Inf. Sci. 8(3), 199–249 (1975)
18. Cpalka, K., Lapa, K., Przybyl, A., Zalasinki, M.: A new method for designing neuro-fuzzy systems for nonlinear modeling with interpretability aspects. Neurocomputing 135, 203–217 (2014)
19. de Soto, A.R.: A hierarchical model of linguisti variable. Inf. Sci. Spec. Issue Interpret. Fuzzy Syst. 181, 4394–4408 (2011)
20. Matia, F., Al-Hadithi, B., Jimenez, A., Segundo, P.S.: An affine fuzzy model with local and global interpretations. Appl. Soft Comput. 11(6), 4226–4235 (2011)
21. Takagi, T., Sugeno, M.: Fuzzy identification of systems and its application to modeling and control. IEEE Trans. Syst. Man Cybern. (SMC) 15(1), 116–132 (1985)
22. Tanaka, T., Wang, H.: Fuzzy Control Systems Design and Analysis: A Linear Matrix Ineqality Approach. Wiley, New York, NY, USA (2001)
23. Kósi, K., Tar, J., Rudas, I.: Improvement of the stability of RFPT-based adaptive controllers by observing "precursor oscillations". In: Proceedings of the 9th IEEE International Conference on Computational Cybernetics, Tihany, Hungary, pp. 267–272 (2013)
24. Dineva, A., Varkonyi-Koczy, A.R., Tar, J.K., Piuri, V.: Sigmoid generated fixed point transformation control scheme for stabilization of Kapitza's pendulum system. In: IEEE 20th Jubilee International Conference on Intelligent Engineering Systems: INES 2016, Budapest, Hungary, 30 June–02 July 2016

From von Neumann Architecture and Atanasoff's ABC to Neuromorphic Computation and Kasabov's NeuCube. Part II: Applications

Maryam Gholami Doborjeh, Zohreh Gholami Doborjeh,
Akshay Raj Gollahalli, Kaushalya Kumarasinghe, Vivienne Breen,
Neelava Sengupta, Josafath Israel Espinosa Ramos, Reggio Hartono,
Elisa Capecci, Hideaki Kawano, Muhaini Othman, Lei Zhou,
Jie Yang, Pritam Bose and Chenjie Ge

Abstract Spatio/Spector-Temporal Data (SSTD) analyzing is a challenging task, as temporal features may manifest complex interactions that may also change over time. Making use of suitable models that can capture the "hidden" interactions and interrelationship among multivariate data, is vital in SSTD investigation. This chapter describes a number of prominent applications built using the Kasabov's NeuCube-based Spiking Neural Network (SNN) architecture for mapping, learning, visualization, classification/regression and better understanding and interpretation of SSTD.

M. Gholami Doborjeh (✉) · Z. Gholami Doborjeh · A. R. Gollahalli · K. Kumarasinghe
V. Breen · N. Sengupta · J. I. E. Ramos · R. Hartono · E. Capecci
Knowledge Engineering and Discovery Research Institute,
Auckland University of Technology, Auckland, New Zealand
e-mail: maryam.gholami.doborjeh@aut.ac.nz; mgholami@aut.ac.nz

H. Kawano
Kyushu Institute of Technology, Kitakyushu, Japan

M. Othman
Universiti Tun Hussein Onn Malaysia, Parit Raja, Malaysia

L. Zhou · J. Yang
Institute of Image Processing and Pattern Recognition,
Shanghai Jiao Tong University, Shanghai, China

P. Bose
University of Trento, Trento, Italy

C. Ge
Shanghai Jiao Tong University, Shanghai, China

© Springer International Publishing AG, part of Springer Nature 2018
V. Sgurev et al. (eds.), *Practical Issues of Intelligent Innovations*,
Studies in Systems, Decision and Control 140,
https://doi.org/10.1007/978-3-319-78437-3_2

1 Introduction

NeuCube [1, 2] is the first machine learning system to analyze integrated *space* and *time* aspects of big data to deliver deeper insights. Inspired by the human brain, the most evolved learning system there is, NeuCube does the same advanced pattern recognition of complex data streams in just seconds. Much like the brain, NeuCube uses a network of virtual neurons connecting to each other or disconnecting depending on the timing of signals encoded in incoming data streams. Continuous streaming data can be fed into NeuCube which learns as it goes by constantly evolving this network of neurons. Learning is represented as chains of connected neurons that 'fire' in sequence by transmitting the incoming signal via their interconnections. Once patters in the data are represented in NeuCube as chains of 'firing' neurons, these are learned and recognized. Then new incoming data is constantly compared to the learned patterns and in this way NeuCube can predict future events as they unfold.

NeuCube consists of a set of independent mandatory and optional modules [2], some of them are:

- Module M1: Generic prototyping and testing;
- Module M2 and M3: PyNN simulator for implementation on neuromorphic hardware;
- Module M4: 3D visualization and mining;
- M5 module (I/O and information exchange) for interaction between modules.

The full configuration of the NeuCube is explained in chapter: "*From von Neumann architecture and Atanasoffs ABC to Neuromorphic Computation and Kasabov's NeuCube: Principles and Implementations*" in Springer book and it is graphically illustrated in Fig. 1.

This system is the first of its kind that can:

1. Learn and predict patterns from analyzing space and time aspects of data.
2. Use principles of the human nervous system to increase computational efficiency and reduce resource usage.
3. Facilitates understanding and rule extraction through virtual reality visualization of the model.

NeuCube has been successfully used in a number of application areas including:

- Application of NeuCube in brain data modelling;
- NeuCube and brain computer interfaces (BCI) with neurofeedback for neurorehabilitation;
- NeuCube personalized modelling in neuroinformatics and bioinformatics;
- Risk of stroke prediction;
- Predicting and understanding response to treatment in biomedical environments;
- Seismic data modelling for earthquake prediction;
- NeuCube spatiotemporal pattern recognition from satellite images in remote sensing

The above applications are briefly described in the following sections.

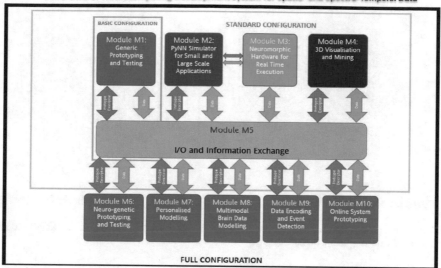

Fig. 1 The NeuCube software development architecture for SNN applications on spatio and spectrotemporal data

2 Application of NeuCube in Brain Data Modelling

NeuCube has been successfully applied to various case studies of Spatio-Temporal Brain data (STBD), the most prominent of which includes Electroencephalography (EEG) and Functional Magnitude Resonance Imaging (fMRI) data. Due to the complex spatiotemporal nature of STBD, it is often abstruse to explore the predictive potential factors using standard machine learning techniques, which are often used to examine EEG and fMRI data These techniques lack the ability to: classify neurological dynamics that occur over the time, identify the involved brain areas through meaningful brain-like visualization, and also quantify the information involved. However, NeuCube based SNN architecture is shown to be capable of such tasks and leads to better understanding of the human behavior through brain data modelling, exemplified as follows:

Progression of Alzheimer's Disease (AD) [3]: Motivated by the dramatic rise of neurological disorders, we proposed an SNN architecture to model EEG data collected from people affected by Alzheimer's disease (AD) and people diagnosed with mild cognitive impairment (MCI). The model developed allows for studying the AD progression and predicting whether the MCI patients are likely to be developed to AD over time. Figure 2 shows the spatiotemporal connections created in the SNN models, one is trained with the initial measurement of EEG data (time t_0) and the other model corresponds to the EEG data recorded after three months (time t1) from the same person. Referring to [3], the model enabled us to precisely visualize the

M. Gholami Doborjeh et al.

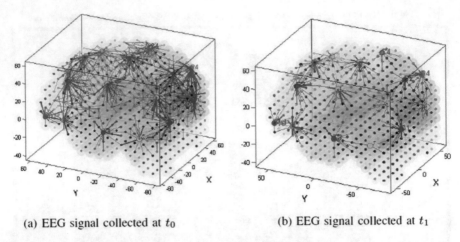

(a) EEG signal collected at t_0 (b) EEG signal collected at t_1

Fig. 2 The progression of Alzheimer disorder is captured in decrement of SNN model connectivity from t_0 to t_1. Figure from [3]

alternation of EEG band-frequencies (Alpha, Beta, theta and Delta) influenced by physiological brain ageing in AD patients.

Recognition of Attentional Bias using EEG Data [4]: Inspired by importance of the attentional bias principle in human choice behavior, we formed a NeuCube based SNN model for efficient recognition of attentional bias as influential factor in consumers' preferences. The model was tested on a case study of EEG data collected from a group of moderate drinkers when they were presented by different drink product features. Our case study findings suggest that a product brand name may not significantly impress consumers by itself. However, when the name of a brand comes along with an additional context, such as design, color, alcoholic or non-alcoholic features, etc. it may direct the consumers attention to certain features and lead the consumers to choose a product. In this particular case study, we found that attentional bias towards alcoholic-related features had more outstanding effects on the brain activity than the non-alcoholic features, as shown in the SNN connectivity in Fig. 3.

Analysis of Perception and Production of Facial Expressions [5]: This is a feasibility study of using the NeuCube SNN architecture for modelling EEG data related to a facial expression-related task. Making use of the NeuCube model allowed for the first time to discover the association between perceiving a particular facial expression and mimicking the same expression. Our finding confirms the biological principle of the Mirror Neurons System (MNS) [6] in human brain. As illustrated in Fig. 4, we identified the role of mirror neurons can be dominant in sadness emotion when compared with other emotions. Very similar areas of the brain will be activated when someone perceiving sadness emotion *versus* mimicking the same. Figure 4d shows the biggest differences between SNN models of perceiving and mimicking the sadness emotion.

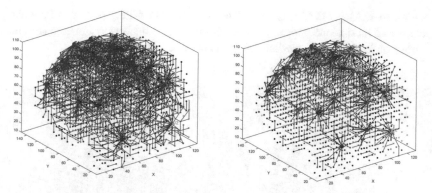

(a) SNNcube connectivity based on the alcoholic feature   (b) SNNcube connectivity based on the non-alcoholic feature

Fig. 3 NeuCube based SNN models trained on EEG data of alcoholic-related features in (**a**) *versus* non-alcoholic-related features in (**b**). Figure form [4]

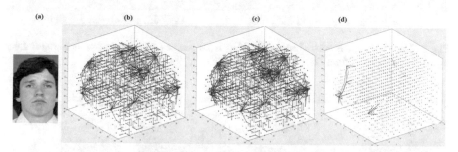

Fig. 4 a Exposing emotional facial expressions on a screen (sadness in this example); **b** connectivity of a SNN model trained on EEG data related to perceiving the facial expression images by a group of subjects; **c** connectivity of a SNN model trained on EEG data related to mimicking the facial expressions by a group of subjects; **d** subtraction of the SNN models from **a** and **b** to visualize, study and understand the differences between perceiving and mimicking an emotion. Figure form [5]

Predicting the Outcome of Methadone Treatment in Addict Patients [7]: We applied the NeuCube based SNN architecture to a case study of EEG data collected during a cognitive task performed by three groups of subjects: (a) untreated opiate addicts; (b) those undergoing methadone maintenance treatment (MMT) for opiate dependence; and (c) a healthy control group. The experimental results proved the following phenomena: (1) the NeuCube-based models obtained superior classification accuracy when compared with traditional machine learning methods. (2) The brain activity patterns of healthy volunteers were significantly different from people with history of opiate dependence. The differences appeared less pronounced in people undertaking MMT compared to those current opiate users. (3) The brain functional pathways of the healthy volunteers were greater and broader than either people undertaking MMT or those opiate users. (4) The STBD patterns of people on low dose of methadone appeared more comparable to healthy volunteers compared to those on high dose of methadone (as shown in Fig. 5).

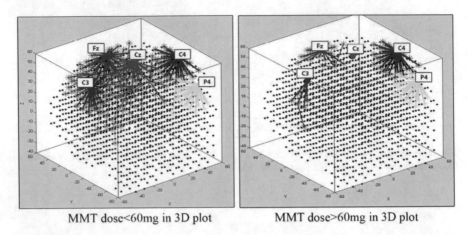

MMT dose<60mg in 3D plot   MMT dose>60mg in 3D plot

Fig. 5 The SNN models are trained on EEG data from people on low (left) and high (right) dose of methadone. Figure from [7]

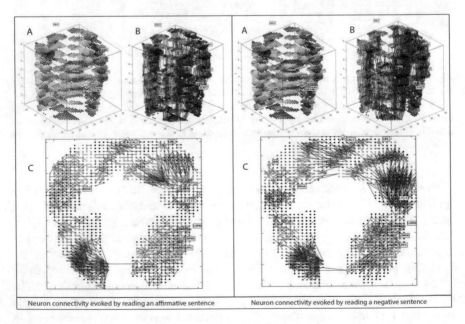

Neuron connectivity evoked by reading an affirmative sentence   Neuron connectivity evoked by reading a negative sentence

Fig. 6 The initial (A) and final (B) connectivity of a SNN model after training with two different data sets, related correspondingly to: affirmative sentence *versus* negative sentence. The final connectivity is also shown as a 2D projection (C). Positive connections are shown in blue while the negative connections are in red. Figure form [8]

Modelling and Classification of Cognitive FMRI Data [8]: We utilized the NeuCube SNN architecture for modelling the benchmark STAR/PLUS fMRI dataset [9] collected from subjects when reading affirmative versus negative sentences.

The trained connections in the SNN model (shown in Fig. 6) represent dynamic spatiotemporal interactions derived by the fMRI voxels variables over time. In this study, tracing the 3-D SNN model connectivity enabled us for the first time to capture prominent brain functional pathways evoked in language comprehension.

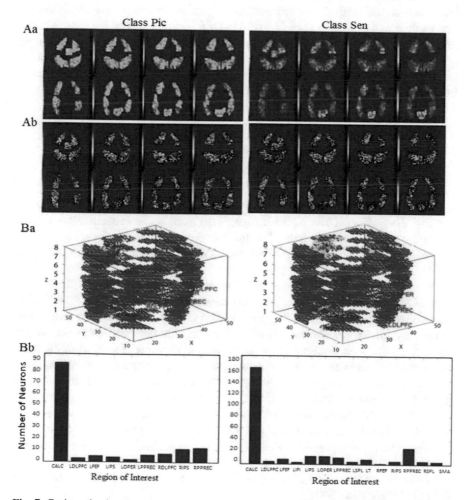

Fig. 7 Brain activation detection and brain regions mapping in the SNN model trained by fMRI data; (Aa) the 2-D SNN model activation maps for each class: watching a picture (Class Pic) or reading a sentence (Class Sen); (Ab) Probability map estimated by t-test for Class Pic (left) and Class Sen (right); (Ba) Locations of activation neurons in the averaged SNN model; (Bb) Histogram of activated neurons with respect to different regions of interest (ROIs) for each class. Figure from [10]

We found stronger spatiotemporal connections between Left Dorsolateral Prefrontal Cortex (LDLPFC) and Left Temporal (LT) while reading negated sentences than affirmative sentences. The NeuCube SNN model resulted also in a superior classification accuracy of 90% when compared with traditional AI and statistical methods.

In another research [10], we proposed a novel method based on the NeuCube SNN architecture for which the following new algorithms were introduced: fMRI data encoding into spike sequences; deep unsupervised learning of fMRI data in a 3-D SNN reservoir; classification of cognitive states; connectivity visualization and analysis for the purpose of understanding cognitive dynamics. The method was applied to the STAR/PLUS fMRI dataset of seeing a picture *versus* reading a sentence. The results are partially presented in Fig. 7 and fully explained in [10]. The evolution of neurons' activation degrees and the deep learning architecture formed in the SNN model is visualized at https://kedri.aut.ac.nz/neucube/fmri.

3   NeuCube and Brain Computer Interfaces (BCI) with Neurofeedback for Neurorehabilitation

In every 6 s, someone in the world becomes physically disabled due to a stroke. To improve the quality of life of these stroke survivors, Neurorehabilitation aims at rebuilding the affected brain motor functions through regular exercises. This intends to strengthen the remaining neural connections by utilizing the brain's ability to build new neural pathways.

Decoding movements of the same limb is an important problem in BCI for neurorehabilitation. Due to the non-invasiveness and high temporal resolution, EEG has been widely used for decoding movements in BCI. However, less spatial resolution caused by the limited number of electrodes is a challenge for pattern recognition. Previous studies on neural activities in motor-related areas of the brain during physical movements provide evidences that approximately the same areas of brain are activated during the movements of the same limb. Thus, classification of movements of same limb from EEG results in less accuracy and limits applicability of BCI for Neurorehabilitation. The state-based online classification module of the NeuCube addresses this limitation and facilitates a BCI platform for Neurorehabilitation. Using this approach we aim to detect the patient's intention to move his or her hand and pass the command to the rehabilitation robot. Figure 8 depicts a basic overview of this approach which facilitates a brain state-based classification of EEG signals using SNN.

The module encloses a Finite State Machine which acts as a finite memory to the model and a biologically plausible NeuCube SNN architecture to decode the state transitions over the time. The module follows the cue based (synchronous) BCI paradigm. While the subject is performing the task, EEG signals are recorded and

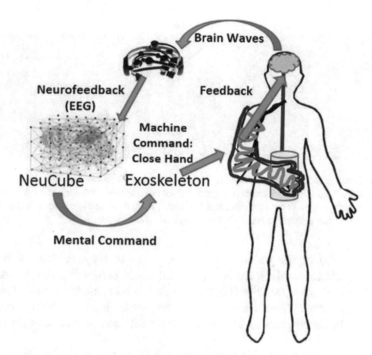

Fig. 8 Basic functional flow of BCI based neurorehabilitation through NeuCube SNN architecture

classified. This classification output is used to control the rehabilitation robots through human thoughts or intentions and also provides neurofeedback to help them to improve their brain functions.

In line with development of the NeuCube based Neurorchabilitation, two cognitive games (Grasp and NeuroRehab [11]) and one portable BCI have been developed. The concept of cognitive game does not only give a "fun" factor to the patients, but also trains them with the functionality of the product. These applications were developed for patients who have no voluntary muscular movements. The patients are trained with an imaginary task, which involves them to imagine moving a part of their body or a series of relatively complex muscle movements. A patient is equipped with EEG cap on the scalp followed by the instruction on what to imagine, so that the instructor can record the neural activity of the brain. Based on the recorded data, a NeuCube model is trained, which can be used to control objects. Once the training process is completed, the instructor performs an online classification with a new EEG data. The classified output is converted into a control signal, which controls the movements in the game.

Figure 9a is the Grasp game virtual environment, where a user is trained on how to hold a glass through EEG data using the NeuCube.

Figure 9b shows the NeuroRehab game virtual environment [11] as a two class problem, the aim of which is to move a ball either left or right depending on the

(a)   **(b)**

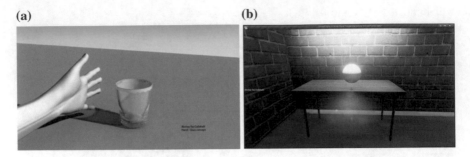

Fig. 9 **a** Grasp game virtual environment, where a user is trained on how to hold a glass using NeuCube with EEG data; **b** NeuroRehab game virtual environment, where a subject is trained to move a ball left or right. If a wrong direction is chosen, a negative mark is given. These exercises are used to help the patients to improve their cognitive abilities

thought patterns of the patient. The patient can get the overview of how the NeuCube SNN connections are being formed while he/she is trying to move an object. Our preliminary studies [12] showed that in compared to standard machine learning algorithms, NeuCube enabled us to obtain higher pattern recognition accuracy, a better adaptability to new incoming data and a better interpretation of the models.

For the purpose of making our software and hardware inter-compatible, but keeping in mind of the cost and better power consumption, we use Portable BCI's.

Portable BCI devices can be used for different application areas such as Neurorehabilitation, cognitive gaming or to control a prosthetic limb. Currently we are developing a portable BCI using the NeuCube SNN architecture to dynamically extract knowledge from brain data in real time. NeuCube being a multiplatform software, it can be easily integrated with the Raspberry Pi, which is cost effective and is widely used for prototyping software to hardware interactions.

4 NeuCube Personalized Modelling in Neuroinformatics and Bio-informatics

NeuCube advanced data analytics offers improved personal outcome prediction, personalization of treatment and understanding through identifying the most predictive factors for a person.

In Neuroinformatics, the NeuCube personalized spiking neural network (PSNN) model presents for the first time the integration of static data and dynamic STBD using SNN architecture and the approach from [13] and [14]. We hypothesize that personalized modelling with SNN could be successfully used, if the models learn from the most informative STBD samples, which are selected based on clustering of integrated static-dynamic data. In this approach, instead of building a global model and training it with STBD of the whole subject population, for every person, we will build a PSNN model to train it only on STBD of those subjects who have similar integrated factors:

- For an individual, a neighborhood of samples is collected based on similarity in integrated static-dynamic data variables.
- A model is built using NeuCube on streaming data from the neighboring samples to predict an outcome for the individual.

NeuCube based PSNN user interface is graphically shown in Fig. 10. In [15], the proposed personalized modelling approach was applied to a case study of "response to treatment" using EEG data for predicting the outcome of Methadone treatment in addicts. The PSNN models trained on a subset of informative EEG data resulted in a higher classification accuracy when compared with global SNN models. In addition, they can be used to reveal individual characteristics on brain activities that can be used to find the best patient- oriented treatment.

In bioinformatics, personalized modelling within NeuCube was successfully applied to the determination of functional dysrhythmias of the stomach, whilst preserving the spatial/temporal relationships present. The contraction of muscles that facilitate the movement required in the stomach are generated by pacemaker cells and propagated via electrical slow waves. The disruption to the normal rhythms of these waves results in various digestive disorders which include gastroparesis, unexplained nausea and vomiting, and functional dyspepsia [16], which do not have biological or bacterial causes.

Gastric slow waves are recorded using Electrogastrography (EGG) on the skin surface. This study sampled the slow waves at 100 Hz with the patient at rest, utilizing a sensor mesh of 851 nodes covering the entire stomach. In stage 1 only 4 different types of dysfunction were tested, and then expanded to 6 types in stage 2 with the inclusion of irregular irregularities. Two aspects of personalized modelling set it apart as the application of choice are the ability to model successfully with low

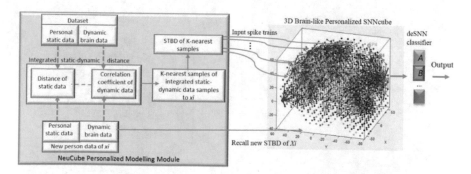

Fig. 10 A block diagram of the NeuCube based personalised modelling approach. Vector-based static data is available, each vector represents personal static clinical features. For every new input person *xi*, K-nearest samples are selected based on similarity in integrated static and dynamic STBD to sample *xi*. Then the STBD of neighbouring subjects are used to train the personalised SNN model

sample numbers, and the prediction of single samples. Couple this with the ability to specify node locations within a functional network in NeuCube, a greater degree of inherent complexity and interaction is retained by the model. In this study only 7 samples were available for each dysfunction. In stage 1 a determination accuracy of 100% was achieved for each dysfunction, but only after the introduction of a specific flexible structure within the NeuCube network. The 851 input nodes were located according to their physical locations, and a "computational" cube added to help differentiate each dysfunction. In Stage 2 various structural dimensions of the network, number of training cycles, and parameter optimization is included. On first inspection the results were surprising in that a smaller computational cube (Fig. 11a) was better along with a single training cycle. The pattern of input node activation was recorded and can be used to assist in the understanding if wave propagation throughout the system. The overall pattern of node activation was seen to be different for each dysfunction. Figure 11b shows one such pattern, for coverage of the results of stage 1 as explained in [17].

Spike encoding using the moving window method was used throughout with a threshold set to capture small changes in input signal. This allowed the distinction amongst dysfunctions especially as some dysrhythmias can occur at the same frequency as normal activity [18]. All but one dysfunction in stage 2 were predicted accurately. Reentry dysfunctions, both anterior and posterior, are known to be the most dynamic and therefore difficult to determine in conjunction with their often very close resemblance to Ectopic Pacemaker signals. This was evident in our results, along with the successful prediction of non-dysfunctional time segments.

To the best of our knowledge this study is the first to apply this type of modelling to EGG slow wave signals. It also demonstrates the diversity of the NeuCube architecture, and that irregular irregularities in signals are detectible where previously they have been notoriously difficult.

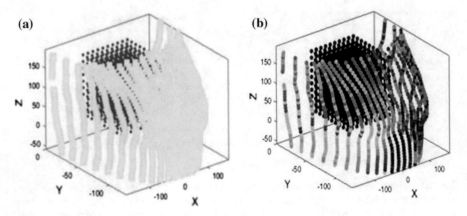

Fig. 11 a NeuCube node layout. Yellow nodes are input, blue are computational nodes; **b** example of average input node activation

5 Risk of Stroke Prediction

Stroke is a silent killer and a major cause of disability. About 80% of strokes can be prevented through control of modifiable risk [19, 20]. Many studies [21–23] discovered associations between environmental variables toward increment of stroke risk.

Moving toward personalized preventive measures, we applied an individualized approach during two seasons (winter and spring) based on individual's risk factors (hypertension, smoking, alcohol, diabetes, obesity and high cholesterol) through various environmental variables (weather characteristics, solar activity, air pollution) measured daily over 60 days before the stroke onset. Daily environmental data were collected through the following 12 variables: wind speed; wind chill; dry bulb temperature; wet bulb temperature; temperature max; temperature min; humidity; atmospheric pressure; sulphur dioxide (SO_2); nitrogen dioxide (NO_2); ozone (O_3); and solar radiation. Using the NeuCube-based model, we created personalized models of 46 randomly selected individuals to validate whether the combinations of inclement environment condition increase the risk of stroke occurrence in an individual with modifiable risk factors. This model also assisted to understand the relationship and interactivity exist in the combined environmental factors on individual level of risk. Finally we determined the earliest time point to best predict the risk of stroke incident for individual as preventive measures. Based on biological plausibility of association between stroke and weather/environmental characteristics, the time window between days 60 and 40 before the stroke event was used as 'low risk' days and the days in the interval between 2nd and 20th day before the event—as 'high risk' days.

Figure 12 shows the low risk and high risk deep patterns in two learned SNN models for one subject in the winter season (subject id: 9). These patterns assist us in interpreting the specific risk triggering environmental factors for individual. For

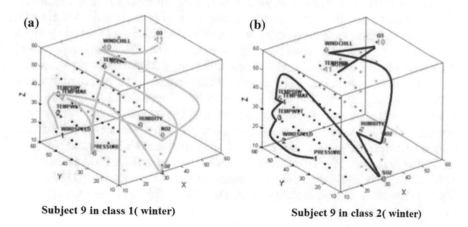

| Subject 9 in class 1(winter) | Subject 9 in class 2(winter) |

Fig. 12 Individual analysis of subject 9 for winter case study in low risk class in (a) and high risk class in (b)

example high risk can be predicted for this subject if atmospheric pressure changes first followed by wind speed, temperature wet, temperature max, temperature dry, sulfur dioxide, humidity, nitrogen dioxide, wind chill, ozone gas and temperature min sequentially.

Using the NeuCube based models for classification problem (class 1: low risk and class 2: high risk), we obtained excellent total accuracy of 95% in winter and 85% in spring for one day ahead stroke risk prediction.

6   Predicting and Understanding Response to Treatment in Biomedical Environment: A Case Study of Clozapine Monotherapy

This study was conducted as part of a large cross-sectional study investigating clozapine (CLZ) response in people with treatment-resistant schizophrenia (TRS) using EEG, MRI and genetic information. CLZ is uniquely effective for treatment-resistant schizophrenia. However, many people still suffer from residual symptoms or do not respond at all (ultra-treatment resistant schizophrenia; UTRS) to CLZ. In this study, our aim was to build a predictive model for discriminating CLZ monotherapy respondent and non-respondent individuals using multimodal brain data.

For the purpose of our investigation, we used a subset of data (resting state fMRI and DTI data with the intention of classifying subjects into groups with either TRS or UTRS. Both fMRI and DTI data for each subject were registered to a subject specific structural image and normalized to the MNI-152 2 mm atlas [24, 25].

As the fMRI data was collected during resting-state, the mean activity and deviation of activity from the voxels over time is negligible compared to task-driven fMRI data. Since a major component of our model is time dependent, we hypothesize that the discriminatory information is hidden in the voxels with significant variation in the activity over time. We selected a set of voxels with an absolute mean standard deviation of greater than 105. The final preprocessed dataset consists of one fMRI trial and one DTI trial of 2318 voxels per subject.

To create a personalized SNN model of the NeuCube, we proposed the new aiSTDP learning algorithm to train a set of 1000 computational spiking neurons, randomly scattered around the input neurons. The experimental results were reported after a grid based hyper-parameter search using the leave-one-out validation protocol. The best model achieved an overall cross validated accuracy of 72%. The area under the ROC curve for this model was 0.72. Evaluation of the confusion matrix showed equally distributed true positive/negative (UTRS: 73%, TRS: 71%) and false positive/negative (UTRS: 27%, TRS: 29%) rates.

Table 1 Comparison of classification performance by different pattern recognition methods on the binary classification task

Method	Data	Accuracy (%)	TP rate (%)	TN rate (%)
Personalized SNN + aiSTDP	fMRI + DTI	72	73	71
Personalized SNN + STDP	fMRI	56	55	57
SVM [23]	fMRI	64	64	71
AutoMLP [24]	fMRI	60	60	64.2

We have further compared the classification performance of the model built on fMRI and DTI with models built using only fMRI through a number of pattern recognition algorithms (see Table 1). For modelling fMRI data, we have used three different algorithms. The personalized SNN + STDP method uses the canonical STDP to update the weights of the SNN model in the NeuCube architecture. The other two algorithms used are the standard machine learning algorithms like SVM and MLP. The proposed personalized SNN + aiSTDP outperformed the other algorithms, not only in the overall accuracy of the model but in the true positive and true negative metrics, which allows the model to be the most robust of all. Furthermore, we have individually scrutinized the connection weights of the SNN models trained on TRS and the UTRS groups, generated by the aiSTDP learning algorithm. Figure 13 shows a comparison of the strongest mean connection weights of the TRS and the UTRS groups. The majority of the strong connections are created in the lower cerebellum and thalamus. It has been shown that by connections via the thalamus, the cerebellum innervates with motor cortical, prefrontal and parietal lobes [26]. Following cerebellar damage, neurocognitive symptoms and a cognitive affective syndrome including blunted affect and inappropriate behavior have been shown [27]. Our findings confirm the recent fMRI and PET studies that have demonstrated the involvement of cerebellum and thalamus in sensory discrimination [28], attention [29], and complex problem solving. All these functional modules are impaired in people with schizophrenia. Also a large density of strong connections is observed in the cerebellum region in the UTRS group compared to the TRS group. Similarly, larger number of strong connections are present in the thalamus region of the TRS as opposed to UTRS.

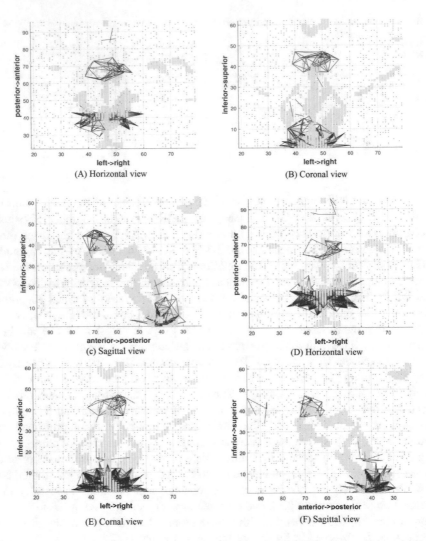

(A) Horizontal view

(B) Coronal view

(c) Sagittal view

(D) Horizontal view

(E) Cornal view

(F) Sagittal view

Fig. 13 Visual comparison of the strongest connections (mean weight across subjects within a group) formed in the SNN model of the TRS (the top) and the UTRS group (bottom row). The yellow colored cluster represents the input neurons and the green neurons are the computational spiking neurons

7 Seismic Data Modelling for Earthquake Prediction

Several computational intelligence approaches have extracted features from earthquake records of a particular region to predict aftershocks (smaller earthquakes happening hours to weeks after a major event), using empirical relations from geophysics such as the b-value (Gutenberg-Richter Law), Båth's Law, and Omori's Law.

(a) **(b)**

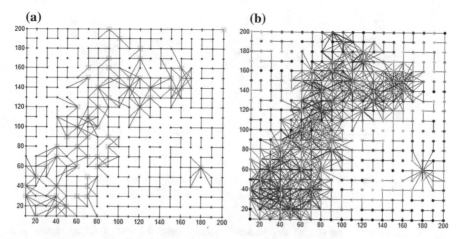

Fig. 14 a The SNN model is trained on a single earthquake in Christchurch area, from 5 days before and up to 1 h before the actual event; **b** the SNN model is trained on seismicity data from 52 sites across New Zealand, from 5 days before and up to 1 h before historical large earthquakes in Canterbury region The NeuCube SNN models were trained by seismicity data from 52 sites across New Zealand, from 5 days before and up to 1 h before historical large earthquakes in Canterbury (illustrated in Fig. 14). The dynamics of the SNN model learning process is visualized at https://kedri.aut.ac.nz/neucube/seismic

Recently we used multiple time-series readings of seismic activity prior to the earthquake, applying the NeuCube based SNN architecture towards earthquake prediction in New Zealand. Seismometer readings from the GeoNet web services (by GNS science, New Zealand) have been used for earlier prediction of an earthquake. We have used the NeuCube architecture to build an early prediction model and tested the prediction performance on the retrospective events in the Christchurch region of New Zealand. This region experienced major earthquake from 2010 to 2015. The NeuCube models predict severe earthquakes with remarkable accuracy, ranging from 75% at 24 h before the event, to 85% at 6 h before, and 91.36% at 1 h before.

8 NeuCube Spatio-Temporal Pattern Recognition from Satellite Images Remote Sensing

Spatio-temporal pattern recognition in remote sensing is a complex problem and the most commonly used models for dealing with temporal information. However, based on Hidden Markov Models (HMM) and traditional artificial neural networks (ANN), they have limited capacity to achieve the integration of complex and long temporal spatial/spectral components because they usually either ignore the temporal dimension or over simplify its representation.

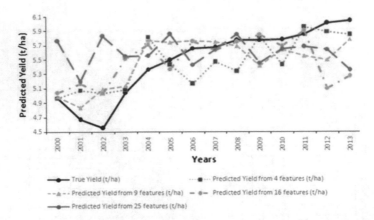

Fig. 15 Comparative analysis of the predicted yield versus the true yield for every year (2000–2013) using different numbers of features. Figure form [30]

SNN explicitly encodes temporal information by transforming input data into trains of spikes that represent time sensitive events. Our work introduced the very first SNN computational model for crop yield estimation from normalized difference vegetation index image time series. It presented the development and testing of a methodological framework which utilized the spatial accumulation of time series of Moderate Resolution Imaging Spectroradiometer 250 m resolution data and historical crop yield data to train an SNN to make timely prediction of crop yield. The research also included an analysis on the optimum number of features needed to optimize the results from our experimental data set. The proposed approach was applied to estimate the winter wheat (*Triticum aestivum L.*) yield in Shandong province, one of the main winter-wheat-growing regions of China. Our method was able to predict the yield around six weeks before harvest with a very high accuracy. Our methodology provided an average accuracy of 95.64%, with an average error of prediction of 0.236 t/ha and correlation coefficient of 0.801 based on a nine-feature model [30] (Fig. 15).

9 Conclusions

This chapter describes the feasibility study of the Kasabov's NeuCube based SNN architecture for different prominent applications of spatio/spectro temporal data. NeuCube SNN development system along with a benchmark EEG data are available at http://www.kedri.aut.ac.nz/neucube.

Acknowledgements The research is supported by the Knowledge Engineering and Discovery Research Institute of the Auckland University of Technology (www.kedri.aut.ac.nz), New Zealand.

References

1. Kasabov, N.: NeuCube: a spiking neural network architecture for mapping, learning and understanding of spatio-temporal brain data. Neural Netw. **52**, 62–76 (2014)
2. Kasabov, N., Scott, N.M., Tu, E., Marks, S., Sengupta, N., Capecci, E., Othman, M., Doborjeh, M.: Evolving spatio-temporal data machines based on the NeuCube neuromorphic framework: design methodology and selected applications. Neural Netw. **78**, 1–14 (2016)
3. Cappeci, E., Gholami Doborjeh, Z., Mammone, N., Foresta, F., Morabito, F., Kasabov, N.: Longitudinal study of Alzheimer's disease degeneration through EEG data analysis with a NeuCube spiking neural network model. In: Proceedings of IJCNN, Vancouver (2016)
4. Gholami, Z., Doborjeh, M., Kasabov, N.: Efficient recognition of attentional bias using EEG data and the NeuCube evolving spatio-temporal data machine. In: Proceedings of ICONIP, Kyoto (2016)
5. Kawano, H., Seo, A., Gholami Doborjeh, Z., Kasabov N., Doborjeh, M.: Analysis of similarity and differences in brain activities between perception and production of facial expressions using EEG data and the NeuCube spiking neural network architecture. In: Proceedings of ICONIP, Kyoto (2016)
6. Gallese, V., Fadiga, L., Fogassi, L., Rizzolatti, G.: Action recognition in the premotor cortex. Brain **119**(2), 593–609 (1996)
7. Doborjeh, M.G., Wang, G., Kasabov, N., Kydd, R., Russell, B.R.: A spiking neural network methodology and system for learning and comparative analysis of EEG data from healthy versus addiction treated versus addiction not treated subjects. IEEE Trans. Biomed. Eng. **63**(9), 1830–1841 (2016)
8. Kasabov, N., Doborjeh, M., Gholami, Z.: Mapping, learning, visualisation, classification and understanding of fMRI data in the NeuCube evolving spatio temporal data machine of spiking neural networks. IEEE Trans. Neural Netw. Learn. Syst. (2016)
9. Mitchel, T., Wang, W.: StarPlus fMRI data (2016). Accessed 21 Oct 2016
10. Kasabov, N., Zhou, L., Doborjeh, M.G., Gholami, Z., Yang, J.: New algorithms for encoding, learning and classification of fMRI data in a spiking neural network architecture: a case on modelling and understanding of dynamic cognitive processes. IEEE Trans. Cogn. Dev. Syst. (2016)
11. Gollahalli, A.: NeuroRehab (2016). https://github.com/akshaybabloo/NeuroRehab. Accessed 29 Nov 2016
12. Gollahalli, A.: Brain-computer interfaces for virtual Quadcopters based on a spiking-neural network architecture —NeuCube. AUT University, Auckland (2015)
13. Kasabov, N., Hou, Z., Feigin, V., Chen, Y.: Improved method and system for predicting outcomes based on spatio/spectro-temporal data. Patent PCT patent, WO 2015030606 A2 (2015)
14. Kasabov, N., Valery, F., Hou, Z.-G., Yixiong, h, Linda, L., Rita, K., Muhaini, O., Priya, P.: Evolving spiking neural networks for personalised modelling, classification and prediction of spatio-temporal patterns with a case study on stroke. Neurocomputing **134**, 269–279 (2014)
15. Doborjeh, M., Kasabov, N.: Ersonalised modelling on integrated clinical and EEG spatio-temporal brain data in the NeuCube spiking neural network system. In Proceedings of IJCNN, Vancouver (2016)
16. O'grady, G., Wang, T.H.H., Du, P., Angeli, T., Lammers, W.J., Cheng, L.K.: Recent progress in gastric arrhythmia: pathophysiology, clinical significance and future horizons. Clin. Exp. Pharmacol. Physiol. **41**(10), 854–862 (2014)
17. Vivienne, B., Kasabov, N., Peng, D., Stefan, C.: A spiking neural network for personalised modelling of electrogastrography (EGG). In: IAPR Workshop on Artificial Neural Networks in Pattern Recognition (2016)
18. Du, P., O'grady, G., Paskaranandavadivel, N., Tang, S.J., Abell, T., Cheng, L.K.: Simultaneous anterior and posterior serosal mapping of gastric slow wave dysrhythmias induced by vasopressin. Exp. Physiol. **101**(9), 1206–1217 (2016)

19. Kate, M., Jesse, D., Matt, W.: What is it with the weather and stroke? Expert Rev. Neurother. **10**(2), 243–249 (2010)
20. Tu, E., Kasabov, N., Othman, M.: Improved predictive personalized modelling with the use of spiking neural network system and a case study on stroke occurrences data. In: International Joint Conference on Neural Networks (IJCNN) (2014)
21. Gill, R.S., Hambridg, H.L., Schneide, E.B., Hanff, T., Tamargo, R.J., Nyquist, P.: Falling temperature and colder weather are associated with an increased risk of aneurysmal subarachnoid hemorrhage. World neurosurgery **79**(1), 136–142 (2013)
22. Cevik, Y., Dougan, N., Dacs, M., Ahmedali, A.: The association between weather conditions and stroke admissions in Turkey. Int. J. Biometeorol. **59**(7), 899–905 (2015)
23. Feigin, V.L., Parmar, P.G., Barker-Collo, S., Bennett, D., Anderson, C., Thrift, A., Stegmayr, B., Rothwell, P.M., Giroud, M., Bejot, Y.: Geomagnetic storms can trigger stroke evidence from 6 large population-based studies in Europe and Australasia. Stroke **45**(6), 1639–1645 (2014)
24. Fonov, V., Evans, A.C., Botteron, K., Almli, R., McKinstry, R., Collins, L.: Unbiased average age-appropriate atlases for pediatric studies. NeuroImage **54**(1), 313–327 (2011)
25. Fonov, V.S., Evans, A.C., McKinstry, R., Almli, C., Collins, D.: Unbiased nonlinear average age-appropriate brain templates from birth to adulthood. NeuroImage **47**, S102 (2009)
26. Ou, W., Cameron, P., Thomas, D.: Natomical evidence for cerebellar and basal ganglia involvement in higher cognitive function. Biol **2**, 227 (1992)
27. Baillieux, H., Verslegers, W., Paquier, h, De Deyn, P.P., Marien, P.: Cerebellar cognitive affective syndrome associated with topiramate. Clin. Neurol. Neurosurg. **110**(5), 496–499 (2008)
28. Gao, J.-H., Parsons, L.M., Bower, J.M., Xiong, J.: Cerebellum implicated in sensory acquisition and discrimination rather than motor control. Science **272**(5261), 545 (1996)
29. Courchesne, E., Akshoomoff, N., ownsend, J., Saitoh, O.: A model system for the study of attention and the cerebellum: infantile autism. Suppl. Electroencephalogr. Clin. Neurophysiol. **44**, 315–325 (1994)
30. Bose, P., Kasabov, N., Bruzzone, L., Hartono, R.: Spiking neural networks for crop yield estimation based on spatiotemporal analysis of image time series. IEEE Trans. Geosci. Remote Sens. **54**(11), 6563–6573 (2016)

Data-Driven Interval Type-2 Fuzzy Modelling for the Classification of Imbalanced Data

Adrian Rubio-Solis, Ali Baraka, George Panoutsos and Steve Thornton

Abstract The design and implementation of Data-Driven Fuzzy Models (DDFMs) to learn balanced industrial/manufacturing data has demonstrated to be a popular machine learning methodology. However, DDFMs have also proven to perform poorly when it comes to learn from heavily imbalanced data, particularly in manufacturing systems. In order to tackle real-world imbalanced problems, we propose a DDFM for rail manufacturing classification. This framework includes Feature Selection, iterative information granulation, and a Fuzzy Decision Engine (FDE) that is based on an Interval Type-2 Radial Basis Function Neural Network (IT2-RBF-NN). The proposed modelling framework is then tested against a real manufacturing case study provided by TATA Steel, UK. Simulation results showed the proposed framework outperformed the generalisation properties of various well known methodologies including a DDFM that employs the RBF-NN of type-1.

Keywords Granulation · RBF Neural Networks of Type-1 (RBF-NN) and Interval Type-2 (IT2-RBF-NN) · Feature Selection (FS) · Imbalanced data · Fuzzy Decision Engine (FDE)

A. Rubio-Solis · G. Panoutsos
Department of Automatic Control and Systems Engineering,
University of Sheffield, Sheffield S1 3JD, UK
e-mail: a.rubiosolis@sheffield.ac.uk

G. Panoutsos
e-mail: g.panoutsos@sheffield.ac.uk

A. Baraka (✉)
British Steel, Redcar, Middlesbrough TS6 7RP, UK
e-mail: ali.baraka@britishsteel.co.uk

S. Thornton
TATA STEEL, Brigg Road, Scunthorpe, North Lincolnshire DN16 1BP, UK
e-mail: steve.thornton@tatasteel.com

© Springer International Publishing AG, part of Springer Nature 2018
V. Sgurev et al. (eds.), *Practical Issues of Intelligent Innovations*,
Studies in Systems, Decision and Control 140,
https://doi.org/10.1007/978-3-319-78437-3_3

1 Introduction

The popularity of data-driven models usually lies on the exploitation of their learning capabilities, for example via the integration of several machine learning methodologies. Particularly, in pattern recognition Bayesian networks, Neural fuzzy systems, decision trees, nearest neighbour and support vector machines have played an important role in the classification of balanced data [1]. Nevertheless, to most classifier learning algorithms an imbalanced class distribution is not assumed [1]. The problem of learning from imbalanced data is characterised as having more instances of a number of classes over others. This leads to a significant compromise to the performance of most machine learning methodologies. Notably in systems such as security networks, manufacturing processes, internet and finance, the understanding of knowledge discovery and analysis from raw imbalanced data is crucial [2]. The nature of imbalanced data is exhibited when the distribution of classes is unequal. As mentioned in [2], the common proportion of class imbalances are on the order of $100 : 1, 1000 : 1$ and $10000 : 1$. In this paper we focus on a two-class imbalanced learning problem for Rail data classification. In particular, a DDFM offers a number of advantages that include the mechanism for explicit knowledge representation in the form of transparent IF-THEN rules, a mechanism of human-like reasoning expressed through linguistic terms, and the ability to approximate non-linear functions [3, 4]. The proposed DDFM framework takes advantage of the state-of-the-art in granular computing [5, 6], feature selection and bootstrapping [7] in order to create a system that is tolerant to imbalanced data. The proposed approach is tailored to the manufacturing case study of steel making and rail manufacturing. However, the framework is generic and could be used in other similar cases.

The remaining of this article is structured as follows: Sect. 2 provides an overview of the rail manufacturing case study and associated challenges. Section 3, describes the framework of the proposed Data Driven Fuzzy Model (DDFM). In order to verify the performance and applicability of the DDFM, a number of simulation results are presented in Sect. 4. Finally, Sect. 5 conclusions and recommendations for future work are presented.

2 Overview of the Rail Production Data

The case study in this article, and the associated data set are relevant to the manufacturing of rails, which includes: (1) steel making, (2) continuous casting, and (3) rolling and finishing sub-processes as illustrated in Fig. 1. Different grades of steel can be obtained through a chemical reduction process of iron ore and scrap by using an integrated steel manufacturing process that removes impurities and adds alloying elements. Primary in **steel making**, a basic oxygen process consists of a top-blown oxygen furnace in which molten blast-furnace iron and scrap are refined into liquid iron usually called hot metal. Chemical and physical reactions, including oxidation

Fig. 1 Railway
Manufacturing Route (taken
from [7])

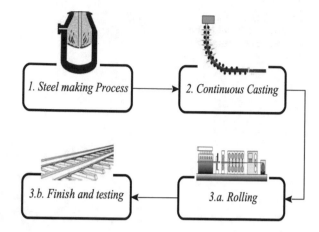

that results from this process reduces impurities such as carbon, phosporus and manganese. Thus, the molten iron is then converted into steel where its chemical structure is adjusted via a secondary steel making process that consists of desulphurisation and alloy addition. In the process of **continuous casting**, the molten metal is transported to multi-strand continuous casting machines where is solidified into 8-tonne steel blooms and then re-heated for subsequent rolling in the finishing mills which produce rails of over 100 m in length. Continuous casting is also used to improve the quality, cost efficiency and to increase productivity in rail manufacturing. Non-destructive testing (NDT) is carried out to verify whether each rail follows strict standards for internal and surface quality.

2.1 Rail Manufacturing Data

The original data provided by TATA Steel, UK consists of more than 218,000 data records which includes 137 process attributes. Preliminary analysis reveals a heavily imbalanced two-class distribution—where the majority class represents approximately 96% out of the total number of rails whose quality is considered good or no-defect. Therefore, the minority class in this particular data sample represents less than 4% of rails that do not meet the standards of quality ('defect'). A pre-processing stage was carried out in order to normalise the data between [0, 1] and a sub-set consisting of 9,000 complete records for the purpose of modelling was extracted. As it is illustrated in the flow diagram in Fig. 2, a Feature Selection (FS) step that is detailed in Sect. 3 is then performed to select the most relevant and less redundant features that better represent the nature of the rail data set. The major aim of the FS step is to reduce the complexity and computational load of of the extracted data set to be processed.

Fig. 2 Strategy for
constructing the proposed
DDFM for imbalanced rail
data

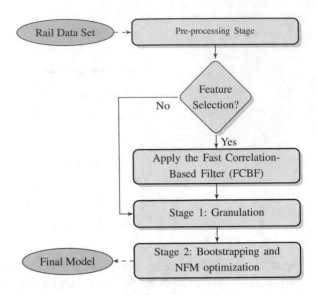

3   Data-Driven Fuzzy Modelling (DDFM)

In this section, the framework of the proposed DDFM is described in terms of two
main stages as illustrated in Fig. 2 [8, 9], i.e.: (1) a process of iterative data granulation for knowledge discovery and (2) the creation of a Neural Fuzzy Model (NFM)
whose main FDE can be viewed either as an IT2-RBF-NN or a RBF-NN for rail
quality prediction. Granulation aims to produce a set of compact granules used as
labels for an ensemble of fuzzy sets [9] as semantic conditions that any fuzzy model
should meet to achieve a low-level interpretability. Therefore, each granule is dis-
tinctive enough from each other to represent a set of input vectors via the creation of
a multidimensional linguistic term with a clear semantic meaning along the universe
of discourse [9].

We exploit granulation in order to achieve a specific level of interpretability and
to create a set of initial fuzzy rules for the NFM. Under some mild conditions [10], in
the second stage an IT2-RBF-NN is used as a NFM which is functionally equivalent
to Fuzzy Logic Systems of type-2 (T2-FLSs). In order to construct the NFM, a two-
part procedure must be performed. In the first part, an oversampling bootstrapping
approach is used to balance the pre-processed input data. In the second part, the
resulting balanced data is fed into the IT2-RBF-NN in order to optimise the fuzzy
linguistic rules provided by the granulation [5]. The framework for the creation of
the DDFM is illustrated in Fig. 2.

Algorithm 1 Fast Correlation-Based Filter Strategy.

Input:
$S(F_1, F_2, \ldots, F_N, C)$ ▷ Training data set
δ ▷ Pre-defined threshold
Output: S_{best}
1: **procedure :**
2:   **for** $i = 1$ *to* N **do**
3:    Calculate $SU_{i,c}$ for F_i;
4:    **if** $(SU_{i,c} \geq \delta)$ **then**
5:     append F_i to S'_{list};
6:   order S'_{list} in descending $SU_{i,c}$ value;
7:   $F_p = getFE(S'_{list})$
8:   **do**
9:    $F_q = getNE(S'_{list}, F_p)^1$;
10:   **if** $(F_q <> NULL)$ **then**
11:    **do**
12:     $F'_q = F_q$;
13:     **if** $(SU_{p,q} \geq SU_{q,c})$ **then**
14:      remove F_q from (S'_{list});
15:      $F_q = getNE(S'_{list}, F'_q)$;
16:     **else** $F_q = getNE(S'_{list}, F_q)$;
17:    **while** $(F_q <> NULL)$
18:    $F_p = getNE(S'_{list}, F_p)$;
19:   **while** $(F_p <> NULL)$
20:   $S_{best} = S'_{list}$

3.1 Fast Correlation-Based Filter (FCBF) for Feature Selection

In this research work, a fast correlation-based filter strategy is introduced [3] in order to select a subset of the most representative features of the process dynamics. The algorithm yields the most relevant and less redundant features out of the pre-processed initial data set. The steps involved during the FCBF are illustrated in Algorithm 1 [3]. The FCBF estimates the associated uncertainty of a random variable based on the information-theoretical concept of entropy [11]. Thus, the entropy of the attribute X_I in relation to the attribute X_J can be obtained as:

$$H(X_I|X_J) = -\sum_{j=1}^{n_1} P(x_j) \sum_{i=1}^{n_1} P(x_i|x_j) log_2(P(x_i|x_j)) \tag{1}$$

where $X_I = \{x_1, \ldots, x_{n_1}\}$, $n = \{n_1, n_2, \ldots, n_m\}$ is the total number of elements in the attribute I, $P(x_i)$ and $P(x_i|x_j)$ is the prior and posterior probability of X_I, respectively. The "*Information Gain*" defined in Eq. (2) can be interpreted as the decrease of entropy of the attribute X_I given X_J.

$$IG(X_I|X_J) = H(X_I) - H(X_I|X_J) \tag{2}$$

If $IG(X_I|X_J) > IG(X_K|X_J)$, then X_J is more correlated to X_I than to X_K. In other words, the resulting information gain 'IG' is biased to those features that contain more values. The FCBF uses a Symmetrical Uncertainty measure SU that is normalized to the range [0, 1] in order to compensate for information gain's bias toward features with more values [3]. Where 1 represents the highest correlation value for any two features and 0 indicates that are independent.

---

[1] *getFE* and *getNE* denote the operation for getting the First and Next Element of the list.

$$SU(X_I, X_J) = 2 * \frac{IG(X_I|X_J)}{H(X_I) + H(X_J)} \tag{3}$$

The SU is used as a goodness index for feature selection (including the class C) and it involves two main steps:

1. Evaluate the relevance of each feature.
2. Discriminate features that are redundant.

Here, the majority class is employed to evaluate the associated relevance and redundancy of each attribute/feature. More specifically a data set S contains n features and a class C. $SU_{i,C}$ denotes the correlation between the feature F_i and the class C. A subset S' of relevant attributes can be extracted from S based on a predefined threshold δ, such that $\forall F_i \in S'$; $1 \leq i \leq N$, $SU_{i,C} \geq \delta$ [12]. In relation to the redundancy that may be caused by each feature, in [3] the authors proposed the concept of predominant correlation to remove those features that do not contribute with information *iff* there exist a $SU_{i,C} \geq \delta$, and $\forall F_j \in S'(j \neq i)$ such that there is no F_j where $SU_{j,i} \geq SU_{i,C}$.

3.2 Iterative Information Granulation

Iterative information granulation is a clustering technique that uses a compatibility measure that defines how good is the merging operation of any two granules A and B [6, 13, 14], that is:

$$compat(A, B) = D_{MAX} - d_{A,B} \times e^{\left(-\alpha \times \frac{card_{A,B}/Cardilanity_{MAX}}{L_{A,B}/Length_{MAX}}\right)} \tag{4}$$

where

$$d_{A,B} = \frac{\sum_{k=1}^{n} w_k(max(u_{Ak}, u_{Bk}) - min(l_{Ak}, l_{Bk}))}{n} \tag{5}$$

Such as D_{MAX}, $Length_{MAX}$ and the term $Cardinality_{MAX}$ is the maximum possible distance and length of a granule and the total number of granules in the data set respectively. $d_{A,B}$ is the weighted multidimensional average distance of the resulting granule with w_k playing the importance weight for the dimension k, $k = 1, \dots, n$. In Eq. 4, α weights the requirements between distance and cardinality/length. In Eq. 5, l_{Ak} and u_{Ak} are the lower and upper limits (corners) of the granule A respectively and L_{AB} is the multidimensional length of the resulting granule. Hence, an iterative process of granulation can be stated as:

- Find the two most 'compatible' information granules A and B and merge them together as a new information granule containing both original granules.
- Repeat the process of finding the two most compatible granules until a satisfactory data abstraction level is achieved. Thus, the final set of the derived information granules are used as the initial rule-base of the DDFM.

3.3 Bootstrapping

In this article, we focus on a two-class imbalanced real data set for Rail quality classification. Particularly in this data set, a high proportion of rails are considered to fulfil the quality standards (majority class) and only a small number of rails are rejected (minority class). Thus, the quality classification may not be reached with the current data set structure. In a like manner to [15, 16], here an oversampling bootstrapping technique with the same mean is suggested since [17]:

- Oversampling changes the class distribution of the Rail data set.
- The oversampling is straightforward to implement and avoids any loss in information.

3.4 Parameter Identification of the NFM via the Application of an Oversampling Bootstrapping

The basic idea behind the implementation of the oversampling bootstrapping for the parameter identification of the NFM is described in the flow diagram presented in Fig. 3. According to Fig. 3, the application of the bootstrapping involves two main loops. The outer loop initialises an index R_M to control the level of imbalanced data and its associated limits $[R_{min}, R_{max}]$. R_M represents the ratio of the number

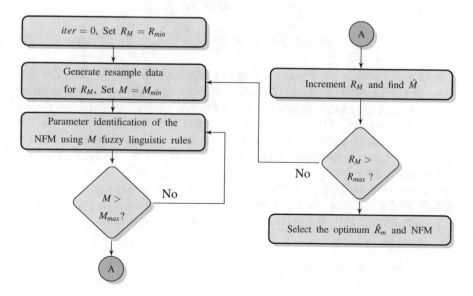

Fig. 3 Parameter Identification of the NFM by using the Bootstrapping strategy

of samples (nodefect rails) that belong to the majority class ($R_{majority}$) to those samples (defect rails) that belong to the minority class ($R_{minority}$). For our case study, $R_M = R_{majority}/R_{minority} = 19$. Initially, $R_M = R_{min}$ and then increased with a predefined step Δ_{iter} until $R_M = R_{max}$. The inner loop consists of identifying the parameters of a NFM via a two-step fuzzy process that is repeated $M_{max} - M_{min}$ times. The first step involves the granulation of the pre-processed input data which may contain the most relevant and less redundant features. The output of the granulation process is a number of fuzzy linguistic constraints M that are used as the initial Fuzzy Decision Engine (FDE) in the NFM. In the second step, the IT2-RBF-NN [13, 18] optimises the FDE by using an adaptive gradient descent approach. At each iteration of the outer loop the parameters that produce the highest classification performance are recorded, thus the value of M ranges in the interval $[M_{min} - M_{max}]$. Finally, to evaluate the performance of each binary classification experiment we use the metrics: (a) specificity, (b) sensitivity and (c) accuracy [16, 19]. For binary classification of the rail quality we use **0** and **1** to denote the majority and minority class respectively. While *specificity* measures the proportion of *negative samples* (majority class, *TN*) that are correctly classified by the NFM, *sensitivity* measures the proportion of *positive samples* (minority class, *TP*) that are identified correctly, and accuracy is the overall percentage of both categories.

Thus, such metrics can be calculated as follows:

$$Specificity = \frac{TN}{TN + FP} \tag{6}$$

$$Sensitivity = \frac{TP}{TP + FN} \tag{7}$$

$$Accuracy = \frac{TN + TP}{TP + TN + FP + FN} \tag{8}$$

where *FP* and *FN* are good rails predicted as rejected rails and rejected rails predicted as good rails respectively. Therefore, the output of the bootstrapping strategy is the most appropriate number of fuzzy linguistic rules '$\hat{M}$' and the optimum $\hat{R}_m$. The power of the RBF-NN viewed as a NFM [10] lies on its ability to add expert knowledge and approximate non-linear functions by using kernel functions. In Fig. 4a, a typical taxonomy of the RBF-NN of type-1 is illustrated. The balanced input vector $\vec{x}_p = \{x_1, \ldots, x_n\}$ includes data obtained by the application of the bootstrapping methodology. The NFM is viewed as a multi-input-single-output (MISO) FLS $f : U \subset R^n \to R$ is considered having n inputs $x_k \in [x_1, \ldots, x_n]^T \in U_1 \times U_2 \times .. \times U_k.. \times U_n \triangleq U$ where the *ith* fuzzy rule has the form:

$$R^i : IF\ x_1\ is\ F_1^i\ and\ \ldots\ x_k\ is\ F_k^i\ and\ \ldots$$

$$and\ x_n\ is\ F_n^i\ THEN\ y\ is\ G^i;\ i = 1, \ldots, M \tag{9}$$

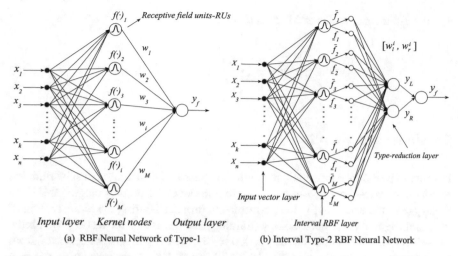

(a) RBF Neural Network of Type-1   **(b) Interval Type-2 RBF Neural Network**

Fig. 4 Radial Basic Function Neural Network of **a** Type-1 and **b** Interval Type-2 used as FDE

And $F_1^i \times ... \times F_n^i = A^i$, hence Eq. (9) can be expressed as:

$$R^+ : F_1^i \times ... \times F_n^i \rightarrow G^i = A^i \rightarrow G^i; i = 1, ..., M \tag{10}$$

A rule R^i is described by $\mu_{R^i}(\vec{x}_p, y) = \mu_{R^i}[x_1, ..., x_n, y]$, where $\vec{x}_p = [x_1, ..., x_n] \in X_1, ..., X_p = R^p$. The following implication (Mamdani) can be used:

$$\mu_{R^i}(\vec{x}_p, y) = \mu_{A^i \rightarrow G^i}(\vec{x}_p, y) = \left[T_{k=1}^n \mu_{F_k^i}(x_k) \star \mu_{G^i}(y) \right] \tag{11}$$

where each firing strength f_i is defined as $\mu_{R^i}(\vec{x}_p, y) = \mu_{A^i \rightarrow G^i}(\vec{x}_p, y) = f_i$. Compared to the RBF-NN of type-1, the IT2-RBF-NN directly handle rule uncertainties. For that reason, an IT2-RBF-NN with variable standard deviation and a Karnik-Mendel type-reducer is used (See Fig. 4b). The membership function (MF) used for the simulation results is a n-dimensional Gaussian primary MF having a fixed mean m_k^i and an uncertain standard deviation σ_i defined as

$$f_i(\vec{x}_p) = exp\left[-\frac{\|\vec{x}_p - m_k^i\|^2}{(\sigma_i)^2} \right], \ \sigma_i \in [\sigma_i^1, \sigma_i^2] \tag{12}$$

Correspondingly, the n-dimensional upper MF $\overline{f}_i$ is

$$exp\left[-\frac{\sum_{k=1}^n \left(x_k - m_k^i \right)^2}{(\sigma_i^1)^2} \right] \triangleq \overline{f}_i(m_k^i, \sigma_i^1; \vec{x}_p) \tag{13}$$

and the n-dimensional lower MF $\underline{f}_i$ is computed as:

$$exp\left[-\frac{\sum_{k=1}^{n}\left(x_k - m_k^i\right)^2}{(\sigma_i^2)^2}\right] \triangleq \underline{f}_i(m_k^i, \sigma_i^2; \bar{x}_p) \tag{14}$$

4   Simulation Results

In order to investigate the efficiency of the proposed DDFM a number of simulation results were carried out by using a different number of fuzzy rules ranging from 3 to 6 and presented according to the performance produced during the identification of the parameters of each methodology (training) and the associated testing. We used a two-class imbalanced data set of $218,000 \times 137$ vectors which were collected from a rail production line and provided by TATA Steel, UK. A representative rail data set that consists of 9000×137 dimensional complete feature vectors was extracted. Consequently, the pre-processed rail data set was divided into two sets, namely: 60% of the data set was used for training and 40% for testing the performance of the proposed DDFM. As described in the previous section, feature selection was applied via the FCBF with a correlation threshold $\delta = 0.2$ to the training data. In Table 1, the most relevant and less redundant features found by applying the FCBF are presented. As can be seen from Table 1, those parameters that are associated to temperature, flow, position and speed appear to be the reduced set of variables that better describe the rail manufacturing process.

In Table 2, we compare the simulation results between the proposed DDFM and a number of other popular methods including the RBF-NN of type-1. Results show that the highest accuracy was produced by the IT2-RBF-NN, while the ANFIS exhibited the poorest performance [16]. A technique based on fuzzy support vector machines (F-SVMs) proposed in [15] to address the rail quality classification that is highly imbalanced is also presented in Table 2. From our results, it was found that the optimal number of fuzzy rules is 3, and the appropriate ratio $R_M = 3$. In Fig. 5a and b, the accuracy performance of the integrated bootstrapping methodology during the parameter identification of the NFM and the associated testing stage that uses the IT2-RBF-NN are illustrated.

We observed from Fig. 5 that the associated accuracy is a compromise between specificity and sensitivity. In other words, the accuracy tends to decrease when the number of oversamples is reduced. Thus, the classification performance that corresponds to the testing stage is presented in Fig. 6. We use a *classification threshold* of 0.5 for the correct identification of the majority (left section to the red line, 95.22%) and minority (right section, 92.45%) classes. As can be seen from Fig. 5 and 6, the accuracy decreased sharply when $R_M > 4$. Fig. 7 is an example of the model's simulation for the features that correspond to the degasser 1 and the tundish 2. It is clear in Fig. 7 the positive correlation of the tundish 2 temperature with poor real quality, however the input of the degasser 1 is less pronounced. Finally, the rule-base fuzzy

Table 1 Input ranking using the FCBF for feature selection

List of the most relevant and less redundant features

No	Feature	No	Feature	No	Feature
1	Grade	16	Tundish W Ar Press Mean	31	Casting Speed Max
2	MSM Bloom	17	Tundish W Ar Flow Mean	32	Cast Change Bloom
3	Strand Cut	18	Arrive Caster Time Stamp	33	Arrived degas time
4	Time 1000 to Discharge	19	Steel Grades	34	Gross Fill Rate Max
5	Last Time Ladle used	20	Tundish 1	35	Area
6	Pre-heat Time	21	Vacuum Mean	36	Tundish 2
7	Arrive STIR Time Stamp	22	Time Stir 1 Time Stamp	37	Time Deg 1 Time Stamp
8	Time Tundish 5 Time Stamp	23	Tundish E Ar Press Mean	38	Degasser
9	Time Tundish 4 Time Stamp	24	Prev freecutter	39	Mould Level SDev
10	Tundish Stopper SDev	25	Time Tundish 3 Time Stamp	40	Degasser 1
11	Section Code Description	26	Tundish 4	41	Discharge Time
12	Tundish on Gas Time	27	Time out of furnace	42	Ladle Wt Min
13	Gas Generation	28	Tundish 3	43	Bloom Length
14	Time Tundish 1 Time Stamp	29	Tundish 6	44	
15	Tundish E Ar Flow Mean	30	Time LMF 1 Time Stamp	45	

Table 2 Performance of the proposed DDFM with a $R_M = 3$

NFM Model	R_M	Specificity %	Sensitivity %	Accuracy
Training				
RBF-NN	3	94.25	93.01	93.75
IT2-RBF-NN	3	92.12	96.89	94.50
ANFIS [16]	2.5	67.00	62.00	64.50
F-SVM [15]	–	80.00	72.00	76.00
Testing				
RBF-NN	3	92.10	93.67	93.14
IT2-RBF-NN	3	92.45	95.22	95.82
ANFIS	2.5	65.00	63.00	64.00
F-SVM	–	74.00	64.00	69.00

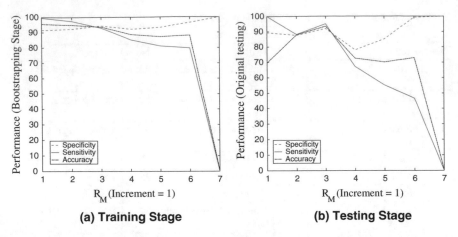

Fig. 5 Accuracy performance for the sensitivity and specificity of the DDFM

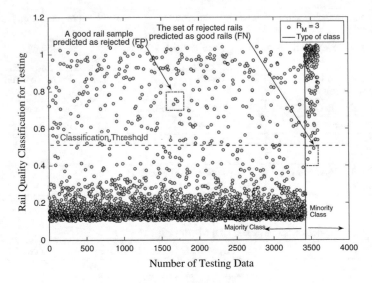

Fig. 6 Testing Performance by using a DFE based on the IT2-RBF-NN

model that is created during the optimisation of the IT2-RBF-NN is presented in Fig. 8.

It can be seen in Fig. 8 that the three generated information granules/rules are associated to a unique rail quality outcome (poor, medium, good), and each outcome is associated to a specific set of features/inputs. The proposed DDFM exploits approximate similarity measures based on interval Gaussian functions to generate an interpretable and simple system.

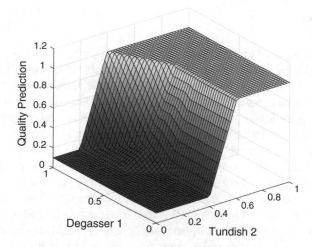

Fig. 7 Fuzzy output surfaces and data distribution for the two most relevant features

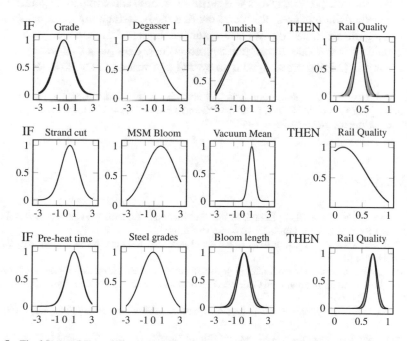

Fig. 8 Final Interval Type-2 Fuzzy Decision Engine using a representative set

5   Conclusion

In this article, a Data-Driven Fuzzy Model (DDFM), whose main Fuzzy Decision Engine (FDE) is based on the Interval Type-2 RBF Neural Network (IT2-RBF-NN), is used as the core facet of a modelling framework for imbalanced data classification. The proposed classification is carried out through an iterative procedure that includes a stage for identifying the most relevant and not redundant features, a stage for capturing knowledge/information and creating the initial structure for the model via an iterative information granulation approach, and finally a a bootstrapping-guided methodology used to tackle data imbalance and identify the optimal parameters for the Neural Fuzzy Model (NFM). The overall framework is designed to specifically address the issue of imbalanced data in data-driven modelling for manufacturing/industrial processes. The case study under investigation is steel-making and rail manufacturing, towards the data-driven classification of manufacturing defects. Simulation results are created to demonstrate the performance of the proposed framework on data gathered from an active industrial process at TATA Steel, UK. The performance of the proposed work was further explored and compared against other popular methods in this field. Results show that the bootstrapping-guided approach can have significant impact on the performance of the data-driven model, particularly in the ability of the model to recognise (predict) new data (data that were not used in the learning/training process) with an overall performance of $> 90\%$ accuracy.

Acknowledgements The authors would like to acknowledge **Innovate UK** for the financial support, under grant agreement 101947, SPEEAK-PC and **TATA STEEL, UK** for providing the manufacturing case study and associated data.

References

1. Yanmin, S., Wong, A.-K.C., Kamel, M.S.: Classification of imbalanced data: a review. Int. J. Pattern Recognit. Artif. Intell. **23**(4), 687–719 (2009)
2. He, H., Garcia, E.: Learning from imbalanced data. IEEE Trans. Knowl. Data Eng. **21**(9), 1263–1284 (2009)
3. Yu, L., Liu, H.: Feature selection for high-dimensional data: a fast correlation-based filter solution. In: Twentieth International Conference in Machine Learning, Washington DC, vol. 3, pp. 856–863 (2003)
4. Zadeh, L.A.: Fuzzy logic = computing with words. IEEE Trans. Fuzzy Syst. **4**(2), 103–111 (1996)
5. Rubio-Solis, A., Panoutsos, G.: Fuzzy uncertainty assessment in RBF Neural Networks using neutrosophic sets for multiclass classification. In: 2014 IEEE International Conference on Fuzzy Systems (FUZZ-IEEE). IEEE (2014)
6. Rubio-Solis, A., Panoutsos, G.: Iterative information granulation for novelty detection in complex datasets. In: 2016 IEEE International Conference on Fuzzy Systems (FUZZ-IEEE), pp. 953–960 (2016)
7. Rubio-Solis, A., Panoutsos, G., Thornton, S.: A Data-driven fuzzy modelling framework for the classification of imbalanced data. In: 2016 IEEE 8th International Conference on Intelligent Systems (IS), pp. 302–307 (2016)

8. Yang, Y.Y., Mahfouf, M., Panoutsos, G.: Development of a parsimonious GA-NN ensemble model with a case study for Charpy impact energy prediction. Adv. Eng. Softw. **42**(7), 435–443 (2011)
9. Shang Ming, Z., Gan, J.Q.: Low-level interpretability and high-level interpretability: a unified view of data-driven interpretable fuzzy system modelling. Fuzzy Sets Syst. **159**(23), 3091–3131 (2008)
10. Jang, JSR., Sun, C.T.: Functional equivalence between radial basis function networks and fuzzy inference systems. IEEE Trans. Neural Netw. **4**(1), 156–159 (1993)
11. Martinez-Hernandez, U., Dodd, T., Prescott, T.J., Lepora, N.: Active Bayesian perception for angle and position discrimination with a biomimetic fingertip. In: International Conference on Intelligent Robots and Systems (IROS), pp. 5968–5973. IEEE (2013)
12. Martinez-Hernandez, U., et al.: Bayesian perception of touch for control of robot emotion. In: 2016 International Joint Conference on Neural Networks (IJCNN), pp. 4927–4933. IEEE (2016)
13. Rubio-Solis, A., Panoutsos, G.: Granular computing neural-fuzzy modelling: a neutrosophic approach. Appl. Soft Comput. **13**(9), 4010–4021 (2013)
14. Pedrycz, W., Andrzej, B.: Granular clustering: a granular signature of data. IEEE Trans. Syst. Man Cybern. Part B: Cybern. **32**(2), 212–224 (2002)
15. Zughrat, A., Mahfouf, M., Yang, Y.Y.: Support vector machines for class imbalance rail data classification with bootstrapping-based over-sampling and under-sampling. In: 19th World Congress of the International Federation of Automatic Control, Cape Town (2014)
16. Yang, Y.Y., Panoutsos, G., Mahfouf, M., Zang, Q., Thorton, S.: Adaptive neural-fuzzy inference system for classification of rail quality data with bootstrapping-based over-sampling. In: 2011 IEEE International Conference on Fuzzy Systems (FUZZ), pp. 2205–2212 (2011)
17. Moore, D.S., McCabe, G.P., Duckworth, W.M., Sclove, S.L.: Bootstrap methods and permutation tests. In: The Practice of Business Statistics, Freeman, Ch. 18, pp. 1–73 (2003)
18. Rubio-Solis, A., Panoutsos, G.: Interval type-2 radial basis function neural network: a modelling framework. IEEE Trans. Fuzzy Syst. **23**(2), 457–473 (2015)
19. Saah, A.J., Hoover, D.R.: "Sensitivity" and "specificity" reconsidered: the meaning of these terms in analytical and diagnostic settings. Ann. Intern. Med. **126**(1), 91–94 (1997)

Network Flows and Risks

Vassil Sgurev and Stanislav Drangajov

Abstract Several network flow models are considered in the present work for transportation of resources—material, financial, communication, commodities, services etc. and the risks related to them at transportation on the network the goal being to achieve maximal flow of minimal risk and minimal payment for the risk, e.g. minimal insurance cost. Five models are proposed. In some of them the flow is maximized and risk is not considered as a network flow in others it is otherwise—the risk is minimized and the resource flow does not observe the flow conservation requirements in the separate vertices. In another model the different probability of an adverse event on the separate sections of the network is not taken into account and it is assumed to be the same on all arcs. Model M4 is the most general one. In it the different probability of an adverse event on each arc is taken is taken into account as well as the maximization of the resource flow on the network. For this purpose a two commodity network flow model of resource and risk is proposed, the risk being considered as network flow also, which keeps all conservation equations in the network vertices. Model M3 supposes the same probability of an adverse event on each arc of the network which decreases to a significant extent the computational complexity of the problem. In model M5 a two stage approach is considered based on both models M3 and M4. It is of lesser computational complexity than M4 and renders fully satisfactory results for practical applications. In it the different values of the probability of an adverse event on each section of the network are taken into account.

Keywords Network flow programming · Risks on network flows
Measures of risk

V. Sgurev · S. Drangajov (✉)
Institute of Information and Communication Technologies, Bulgarian Academy
of Science, Acad. G. Bonchev str., Bl. 2, 1113 Sofia, Bulgaria
e-mail: sdrangajov@gmail.com

V. Sgurev
e-mail: vsgurev@gmail.com

© Springer International Publishing AG, part of Springer Nature 2018   53
V. Sgurev et al. (eds.), *Practical Issues of Intelligent Innovations*,
Studies in Systems, Decision and Control 140,
https://doi.org/10.1007/978-3-319-78437-3_4

1 Preliminary Notes and Denotations

Although not always explicitly defined the network flow models [7, 11] are widely used in many cases in control of resources and risks. We will not consider particular transportation schemes and logistics in which the usage of such models is evident, neither will we consider pipe networks, financial, information or work power etc. At transportation of any resource some risk exists of an adverse event on any section of the route of the network which may result in heavily bad sequels. Such an approach is related to project management also—any activity carried out in due time in a given node reflects favorably on the execution of the activities in the successive nodes. As it is not possible exhaustive examples to be given for all possible applications we restrict in their number and we consider an abstract network flow no matter of what kind the resource flow is.

An exemplary simplified directed graph will be used for illustration further in this work shown below. It consists of 6 vertices (nodes) and 9 arcs (edges).

The following denotations will be used:

- $G(X, U)$ directed graph with a set of vertices (nodes) $X = \{x_i/i \in I\}$ and a set of arcs (edges) $U = \{x_{ij}/(i, j) \in G\}$, where I is the set of indices of all vertices from $X - I = \{i/x_i \in X\}$, and G is the set of pair of indices of all arcs U of the graph— $G = \{(i, j)/x_{i,j} \in U\}$. For each arc $x_{i,j}$ vertex x_i is start, and vertex x_j—end.

The whole bulk of resources, man power, goods, money, raw materials etc. is assumed to be generated in a single start vertex s, called source of the network and it is collected in another vertex of the network t, called consumer (sink). The network flow conservation equations are observed in all other vertices of the network $\{X\backslash\{s, t\}\}$, i.e. the amount of resource entering into the node is equal to the amount outgoing from it. Some amount of resource passes along each arc and some cost must be paid for insurance of the risk for passing along the arc, i.e. there are two flow functions on the network [4]

- f_{ij}—function of the resource arc flow on the arc x_{ij};
- r_{ij}—arc risk flow function of the section x_{ij}.

As shown in [1, 2] a relation exists between f_{ij} and r_{ij}:

$$r_{ij} = p_{ij} f_{ij}, \tag{1}$$

$$0 \leq p_{ij} \leq 1; \ (i, j) \in G. \tag{2}$$

where p_{ij} is the probability for an adverse event to occur at transportation of the resource f_{ij} along the arc x_{ij}. Risk r_{ij} in (1) is considered as a product of two measures—the amount being transported and the probability of an adverse event during this operation [9, 10].

The following denotations are introduced concerning the directed graph $G(X, U)$:

$X = \{x_i/i \in I\}$ \qquad set of all vertices of the network with indices from I;

S	initial node of the network (source);
t	terminal node of the network;
$\Gamma_i^1 = \{j/(i,j) \in G\}$	direct mapping, that is the set of all indices of the end vertices of arcs outgoing from vertex x_i;
$\Gamma_i^{-1} = \{j/(j,i) \in G\}$	reverse mapping, the set of all indices of the start vertices of arcs whose end vertex is x_i [3];
c_{ij}'	upper bound of capacity of the resource admissible on arc x_{ij};
c_{ij}''	upper bound of capacity of the admissible risk on arc x_{ij};
a_{ij}'	valuation (cost) for passing of an unit of resource along arc x_{ij};
a_{ij}''	valuation (e.g. insurance) of the risk for passing an unit of resource along arc x_{ij};
v and v_{max}	flow and max flow of resources;
v' and v'_{max}	flow and max flow of risks.

The requirement for observing the risk constraints $\{c_{ij}''\}$ and for the resource flow $\{c_{ij}'\}$ leads to the following generalized constraint for resources coordinated with the constraint for risk:

$$c_{ij}^s = \min\left[c_{ij}', \frac{c_{ij}''}{p_{ij}}\right]; \tag{3}$$

and keeping the constraints for resources $\{c_{ij}'\}$ in the constraints for risk $\{c_{ij}''\}$—to:

$$c_{ij}^r = \min\left[p_{ij}c_{ij}', c_{ij}''\right]. \tag{4}$$

It follows from the two relations above that if both sides of (3) are multiplied by p_{ij} the following two relations will be received:

$$c_{ij}^r = p_{ij}c_{ij}^s; \quad c_{ij}^s = \frac{c_{ij}^r}{p_{ij}}. \tag{5}$$

Both coordinated values c_{ij}^s and c_{ij}^r will be considered as arc capacities of resources and risks on the arc x_{ij}. Both network flows, resource—$\{f_{ij}\}$ and risk—$\{r_{ij}\}$, are upper bounded on the separate arcs in the following way:

$$f_{ij} \leq c_{ij}^s \text{ for each } (i,j) \in G; \tag{6}$$

$$r_{ij} \leq c_{ij}^r \text{ for each } (i,j) \in G. \tag{7}$$

In Sect. 2 models M1 and M2 are considered in which one of the flows—risk or resource does not keep the requirements for "flowability", i.e. the incoming flow in each vertex is not exactly equal to the outgoing one from the same vertex, in other words there are some losses in the network. Then from Sect. 3 on a detailed exposition with numerical examples begins for the more general models M3, M4, and M5 and their specific features and computational complexity are exposed.

2 Models M1 and M2

Optimal distribution of resources—man-power, commodities, money etc., considering the risk, related to that is not possible without creating models in which the following two peculiarities are considered:

(a) The necessity of network-flow interpretation of the resource distribution (man-power, commodities, money etc.)
(b) Effective consideration of the risk and its network distribution.

   We assume that the whole bulk of resources—man-power, commodities, money etc. is generated in a single initial node s, called source of the network and it is collected in another node of the network t called consumer (sink). In any one of the other nodes of the network $\{X \backslash \{s, t\}\}$ the conservation equations are observed, i.e. the amount of resource entering this node is equal to the amount leaving it.

Each section of the network $x_{ij} \in U$ is of limited capacity c_{ij}, i.e.

$$f_{ij} \leq c_{ij};\ (i,j) \in G. \tag{8}$$

At distributing the resources physical constraints exist

$$f_{ij} \leq c'_{ij};\ (i,j) \in G, \tag{9}$$

and at distributing the network risk the following is valid:

$$r_{ij} \leq c''_{ij};\ (i,j) \in G. \tag{10}$$

Besides for the realization of a unit of resource along the arc $x_{ij} \in U$ costs a'_{ij} units of value and for a unit of risk r_{ij}—a''_{ij} units of value.

The following network-flow model will be used for optimal distribution of the resources and the risks, related to them, in which the objective function L and the respective constraints are defined in the following way:

$$L = \sum_{(i,j) \in G} a_{ij} f_{ij} \to \max (\min) \tag{11}$$

with constraints: for each $i \in I$ and $(i, j) \in G$

$$\sum_{j\in\Gamma_i^1} f_{ij} - \sum_{j\in\Gamma_i^{-1}} f_{ji} = \begin{cases} v, & \text{if } x_i = s; \\ 0, & \text{if } x_i \neq s, t; \\ -v, & \text{if } x_i \in t; \end{cases} \tag{12}$$

$$f_{ij} \leq c_{ij} \text{ for each} (i,j) \in G; \tag{13}$$

$$f_{ij} \geq 0 \text{ for each} (i,j) \in G; \tag{14}$$

where v is a flow from the source s to the sink t.

The proportion between the resource f_{ij} on the section $x_{ij} \in U$, the probability p_{ij}; $(i, j) \in G$ of an adverse event, and the risk r_{ij}; $(i, j) \in G$ is defined by the following relation: for each $(i, j) \in G$

$$r_{ij} = p_{ij} f_{ij}, \tag{15}$$

where

$$r_{ij} \geq 0; p_{ij} \geq 0. \tag{16}$$

It follows from (15) that the risk r_{ij} is expressed by the same measure like f_{ij}. Methods for finding optimal resource or risk flows were proposed in [1, 2] for which all requirements concerning the capacities $\left\{ c_{ij}' \right\}$ and $\left\{ c_{ij}'' \right\}$ sub (upper) script s will be denoted the parameters related to the resource flows, and by a script r—the parameters of the risk flows.

$$f_{ij} a_{ij}^s = a_{ij}' f_{ij} + a_{ij}'' r_{ij} = f_{ij} \left(a_{ij}' + p_{ij} a_{ij}'' \right); \tag{17}$$

$$a_{ij}^s = p_{ij}' + p_{ij} a_{ij}''. \tag{18}$$

If we put in (5) and (7): for each $(i, j) \in G$

$$c_{ij} = c_{ij}^s \text{ and } a_{ij} = a_{ij}^s, \tag{19}$$

then the optimal resource flow may be defined through relations (5)–(8) under the requirements (9)–(14). Further on this will be called model M1.

In this case the distribution of the resources $\{f_{ij}\}$ on the network sections corresponds to the requirements of a network flow and as so observes the conservations Eq. (12). That in the general case is not observed for the results received for $\{f_{ij}\}$ through relation (15) for the risk $\{r_{ij}\}$. If all f_{ij}; $(i, j) \in G$ in (12) are substituted by $p_{ij} r_{ij}$; $(i, j) \in G$ from (15), then a generalized flow of risks with gains and losses will be obtained for which the flow v_s in the source s is not equal, in the general case, to the analogic flow v_t in the sink t, i.e. $v_s \neq v_t$. This results in an undesirable distribution of the risk on the separate cuts of the network.

If in relations from (11) to (14) all variables $\{f_{ij}\}$ according to (15) are substituted by $\left\{\frac{1}{p_{ij}} r_{ij}\right\}$ respectively, then a generalized flow with gains and losses will be achieved for which the flowability is not kept for $\{r_{ij}\}$, i.e. $\sum_{j \in \Gamma_i^1} r_{ij} \neq \sum_{j \in \Gamma_i^{-1}} r_{ji}$. For a number of applications this is inadmissible.

To avoid this fault a risk network flow is used with the following relations:

$$L_r = r_{ij} \sum_{(i,j) \in G} a_{ij}^r \to \max(\min) \tag{20}$$

with constraints: for each $i \in I$ and $(i, j) \in G$

$$\sum_{j \in \Gamma_i^1} r_{ij} - \sum_{j \in \Gamma_i^{-1}} r_{ji} = \begin{cases} v_r, & \text{if } x_i = s; \\ 0, & \text{if } x_i \neq s, t; \\ -v_r, & \text{if } x_i \in t; \end{cases} \tag{21}$$

$$r_{ij} \leq c_{ij}^r; \ (i,j) \in G; \tag{22}$$

$$r_{ij} \geq 0; \ (i,j) \in G. \tag{23}$$

To render the resource constraints $\left\{c_{ij}'\right\}$ and the resource arc costs $\left\{a_{ij}'\right\}$ in the risk flow $\{r_{ij}\}$ it is necessary the following requirements to be observed: for each $(i, j) \in G$:

$$c_{ij}^r = \min\left[p_{ij} c_{ij}', c_{ij}''\right]; \tag{24}$$

$$r_{ij} a_{ij}^r = \left(a_{ij}'' r_{ij} + a_{ij}' \frac{r_{ij}}{p_{ij}}\right); \\ a_{ij}^r = a_{ij}'' + \frac{a_{ij}'}{p_{ij}}. \tag{25}$$

If we put in (20) and (22) the values $\left\{c_{ij}^r\right\}$ and $\left\{a_{ij}^r\right\}$ from (24) and (25) respectively, then the optimal network risk flow, considering the resource constraints and arc evaluations could be defined through relations (20)–(25). It will be further on called model M2.

Then the distribution of the risk $\{r_{ij}\}$ on the network arcs forms a network flow in which the conservations Eq. (16) are observed. But if in this model the values of the risk $\{r_{ij}\}$ were substituted in (15) by $\{p_{ij} f_{ij}\}$ respectively, then we reach a generalized network flow with gains and losses in which the resource flow in the source s is not, in general, equal to the same flow in the sink t. In this case the resource $\{f_{ij}\}$ values do not observe in general the conservation equations in each of the nodes $x_i \in X$. This means that strengthening or weakening of the resources in the separate nodes $x_i \in X$ should be carried out which is in practice a hard to realize and disadvantageous operation.

3 Intelligent Combined Resource and Network Risk Flow—Model M3

An intelligent generalized risk network flow is proposed in which the disadvantages of models M1 and M2 are avoided, namely the requirements for the network flow to be observed for one of the flows only and the in the second no zero balance to be achieved in the corresponding nodes of the network. Removing this disadvantage provides a possibility the resource and the related to it risk flow to have almost permanent values in the separate cuts of the network keeping zero balance in the separate nodes $\{X\backslash\{s, t\}\}$ of this network. This is an important advantage in the various applications of such networks.

The following relations will be proved:

Proposition 3.1 For each $(i, j) \in G$

$$c_{ij}^r = p_{ij} c_{ij}^s. \tag{26}$$

A chain of truths follows from (15), (3), and (24), namely:

$$p_{ij} c_{ij}^s = \min \left[p_{ij} c_{ij}', p_{ij} \frac{c_{ij}''}{p_{ij}} \right] = \min \left[p_{ij} c_{ij}', c_{ij}'' \right] = c_{ij}^r,$$

which results in (26).

Proposition 3.2 For each $(i, j) \in G$

$$a_{ij}^s = p_{ij} a_{ij}^r. \tag{27}$$

It follows from (15), (18), (19), and (25) that

$$\frac{a_{ij}^s}{p_{ij}} = a_{ij}'' + \frac{a_{ij}'}{p_{ij}} = a_{ij}^r,$$

which leads to (27).

Proposition 3.3 It follows from (12), (15), and (21) that

$$v_r = p_{ij} v. \tag{28}$$

If we substitute the values of risk $\{r_{ij}\}$ in (21) by $\{p_{ij}f_{ij}\}$ from (15) respectively and then divide both sides of (21) by $\{p_{ij}\}$ then we achieve a new notation of (12) from which (28) follows.

Proposition 3.4 A relation exists;

$$\sum_{(i,j)\in G} a^s_{ij}f_{ij} = \sum_{(i,j)\in G} a^r_{ij}r_{ij} \tag{29}$$

If in the right hand side of (29) all values of $\{r_{ij}\}$ are substituted by $\{p_{ij}\,f_{ij}\}$ according to (15) and according to (19) and (22) the valuations $\left\{a^r_{ij}\right\}$ − by $\frac{a^s_{ij}}{p_{ij}}$ then

$$\sum_{(i,j)\in G} a^r_{ij}r_{ij} = \sum_{(i,j)\in G} \frac{a^s_{ij}}{p_{ij}}p_{ij}f_{ij} = \sum_{(i,j)\in G} a^s_{ij}f_{ij}$$

from which relation (29) follows.

Proposition 3.5 Condition

$$p_{ij} = p_0 \text{ for each } (i,j) \in G, \tag{30}$$

is sufficient for one-to-one mapping of relations (11) to (14) and (19) for model M1, and from (20) to (25)—for model M2.

(a) If the following substitutions are made in Eq. (21), according to (12), (15), (19) and (28):

$$r_{ij} = p_{ij}f_{ij} \text{ and } v_r = p_{ij}v,$$

and after dividing both sides by p_{ij} Eq. (12) will be received. In the same way if in (12) we put respectively the substitutions:

$$f_{ij} = \frac{r_{ij}}{p_{ij}} \text{ and } v = \frac{v_r}{p_{ij}}$$

and then multiply by p_{ij} Eq. (21) will be received.

(b) If the respective substitutions are made in both sides of (17) according to (9) and (16): $r_{ij} = fij\,p_{ij}$ and $c^r_{ij} = p_{ij}c^s_{ij}$, then inequality (7) will be received. In the same way if substitutions are made in (7): $f_{ij} = \frac{r_{ij}}{p_{ij}}$ and $c^s_{ij} = \frac{c^r_{ij}}{p_{ij}}$, then inequality (17) will be received.

(c) If, according to (15) a substitution is made in inequality (23) by $r_{ij} = p_{ij}f_{ij}$, and then divided by p_{ij} then inequality (14) will be received. Analogically if a substitution is carried out in (14) $f_{ij} = \frac{r_{ij}}{p_{ij}}$, and then both sides of (14) are multiplied by p_{ij} (23) will be received.

(d) According to (29), observing the requirements of the two flows—$\{f_{ij}\}$ and $\{r_{ij}\}$ the same values of the objective function are received at equivalent realization of those flows corresponding to conditions (15), (26), (27), and (28).

The results of the four items above—from (a) to (d) prove the sufficiency for one-to-one mapping of the flows $\{f_{ij}\}$ and $\{r_{ij}\}$. This allows both—resources and risks of their control to be interpreted as network flows of zero balance in nodes $\{X \setminus \{s, t\}\}$.

When observing the requirements defined above these two flows are equivalent and mutually bound which allows when a network realization of one of them is known comparatively easily and quickly the network realization of the other one to be defined, and vice versa. In this case the relations (15) are of key importance, namely: $r_{ij} = p_{ij} f_{ij}$, and $f_{ij} = \frac{r_{ij}}{p_0}$. Hence the model proposed M3 contains a combined flow of intelligent properties allowing to simultaneously carry out control of resources and the risks related to their control. The intelligent network flow proposed M3 allows also the risks on the separate cuts of the network to be found, i.e. the current risk ensuing from (23):

$$v_r = p_0 v; \qquad (31)$$

where v is the current amount of commodity in the network.

If the maximum possible amount of commodity v^* in the network is found the maximum admissible current risk equals to

$$v_r^* = p_0 v. \qquad (32)$$

There may not be greater current risk than v_r^* the value of which is defined by the network parameters, i.e.

$$v_r \leq v_r^*.$$

3.1 Numerical Examples for Model 3

A network for distribution of resources and risks is shown in Fig. 1 consisting of 6 nodes and 9 arcs, including the source s and sink t.

All examples further were solved by the web based package Weboptim [6]. Developed in IICT-BAS (Fig. 2) [5].

On the base of the resource constraints $\left\{ c'_{ij}; (i,j) \in G \right\}$ and risk constraints $\left\{ c''_{ij}; (i,j) \in G \right\}$ in formulae (11) and (19) combined capacities are determined for

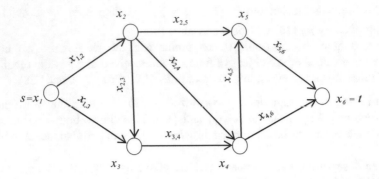

Fig. 1 .

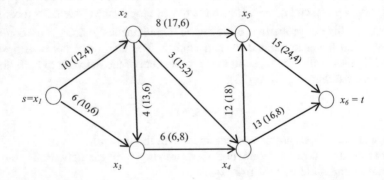

Fig. 2 .

the resources on arcs $\left\{c_{ij}^s; (i,j) \in G\right\}$ considering the risk constraints and for the risks on the arcs $\left\{c_{ij}^r; (i,j) \in G\right\}$ considering the resource constraints. Results are given in Table 1 for each arc of the network of Fig. 1.

Through the resource arc valuations $\left\{a_{ij}'; (i,j) \in G\right\}$ and the risk arc valuations $\left\{a_{ij}'; (i,j) \in G\right\}$ and formulae (13) and (20) the combined arc valuations for resource transportation may be found $\left\{a_{ij}^s; (i,j) \in G\right\}$ considering the risk and similar combined valuations of the risk $\left\{a_{ij}^r; (i,j) \in G\right\}$ considering the resource component.

The results of all calculations of the capacities and evaluations of the separate arcs of the network in Fig. 1 are shown in Table 1. They provide a possibility both flows—resource and risk to be defined. The following relations—from F_1 to F_{16} correspond to the network resource flow $\{f_{ij}\}$ from (12) to (14) with objective function L from (11).

Table 1 .

#	Arc param	(1,2)	(1,3)	(2,3)	(2,4)	(2,5)	(3,4)	(4,5)	(4,6)	(5,6)
1	c'_{ij}	10	8	4	4	8	7	12	14	15
2	c''_{ij}	2,2	1,2	1	0,6	1,8	1,2	2,5	2,6	3,2
3	c^s_{ij}	10	6	4	3	8	6	12	13	15
4	c^r_{ij}	2	1,2	0,8	0,6	1,6	1,2	2,4	2,6	3
5	a'_{ij}	10	9	12	14	16	6	16	14	20
6	a''_{ij}	12	8	8	6	8	4	10	14	22
7	a^s_{ij}	12,4	10,6	13,6	15,2	17,6	6,8	18	16,8	24,4
8	a^r_{ij}	62	53	68	76	88	34	90	84	122
9	p	0,2								
10	v_s									
11	v_r									

In the example the minimal cost for transportation and risk will be simultaneously found. For this purpose two problems—F and R will be solved. This provides a possibility the results for the example of Fig. 1 and Table 1 to be checked out for the proposed model M3.

Problem F

(F_1) $f_{1,2} + f_{1,3} = v_s$;

(F_2) $f_{2,3} + f_{2,4} + f_{2,5} - f_{1,2} = 0$;

(F_3) $f_{3,4} - f_{1,3} - f_{2,3} = 0$;

(F_4) $f_{4,5} + f_{4,6} - f_{2,4} - f_{3,4} = 0$;

(F_5) $f_{5,6} - f_{2,5} - f_{4,5} = 0$;

(F_6) $f_{5,6} + f_{4,6} = v_s$;

(F_7) $f_{1,2} \leq 10$;

(F_8) $f_{1,3} \leq 6$;

(F_9) $f_{2,3} \leq 4$;

(F_{10}) $f_{2,4} \leq 3$;

(F_{11}) $f_{2,5} \leq 8$;

(F_{12}) $f_{3,4} \leq 6$;

(F_{13}) $f_{4,5} \leq 12$;

(F_{14}) $f_{4,6} \leq 13$;

(F_{15}) $f_{5,6} \leq 15$;

(F_{16}) $L = \sum_{(i,j) \in G} a^s_{ij} f_{ij} \rightarrow \max (\min)$.

Problem R

Through the relations R_1 to R_{16} the risk network flow $\{r_{ij}\}$ may be defined that corresponds to requirements (16)–(18) with objective function L_r from (15)

$(R_1)\ r_{1,2} + r_{1,3} = v_r;$

$(R_2)\ r_{2,3} + r_{2,4} + r_{2,5} - r_{1,2} = 0;$

$(R_3)\ r_{3,4} - r_{1,3} - r_{2,3} = 0;$

$(R_4)\ r_{4,5} + r_{4,6} - r_{2,4} - r_{3,4} = 0;$

$(R_5)\ r_{5,6} - r_{2,5} - r_{4,5} = 0;$

$(R_6)\ r_{5,6} + r_{4,6} = v_r;$

$(R_7)\ r_{1,2} \leq 2;$

$(R_8)\ r_{1,3} \leq 1,2;$

$(R_9)\ r_{2,3} \leq 0,8;$

$(R_{10})\ r_{2,4} \leq 0,6;$

$(R_{11})\ r_{2,5} \leq 1,6;$

$(R_{12})\ r_{3,4} \leq 1,2;$

$(R_{13})\ r_{4,5} \leq 2,4;$

$(R_{14})\ r_{4,6} \leq 2,6;$

$(R_{15})\ r_{5,6} \leq 3;$

$(R_{16})\ L = \displaystyle\sum_{(i,j)\in G} a_{ij}^r r_{ij} \to \max\,(\min).$

Then the minimal cost for simultaneous transportation of resources and risk will be determined for these operations on the base of the above formulated problems F and R.

PROBLEMS F′ and F*: We assume that the resource flow v_s consequently accepts two different values $v_s = 15$ and $v_s = 16$. Then both network flow programming problems—F′ and F* from F_1 to F_{16} may be solved with the help of the program package WEBOPTIM [6] developed in the Institute of Information and Communication Technologies—BAS.

We show below a print out of the problem F computer solution for a flow value of 15. The results for both problems F′ and F* are shown in Table 2.

Problem F

$(F_1)\ f12 + f13 = 15;$

$(F_2)\ f23 + f24 + f25 - f12 = 0;$

Table 2 .

#	Problem	Arc val	(1,2)	(1,3)	(2,3)	(2,4)	(2,5)	(3,4)	(4,5)	(4,6)	(5,6)
1.	F′	$\{f_{ij}\}$	9	6	0	3	6	6	0	9	6
2.		v_s	15								
3.	F*	$\{f_{ij}\}$	10	6	0	3	7	6	0	9	7
4.		v_s	16								
5.	R′	$\{r_{ij}\}$	1,8	1,2	0	0,6	1,2	1,2	0	1,8	1,2
6.		v_r	3								
7.	R*	$\{r_{ij}\}$	2	1,2	0	0,6	1,4	1,2	0	1,8	1,4
8.		v_r	3,2								

(F_3) f34 − f13 − f23 = 0;
(F_4) f45 + f46 − f24 − f34 = 0;
(F_5) f56 − f25 − f45 = 0;
(F_6) f56 + f46 = 15;
(F_7) f12 ≤ 10;
(F_8) f13 ≤ 6;
(F_9) f23 ≤ 4;
(F_{10}) f24 ≤ 3;
(F_{11}) f25 ≤ 8;
(F_{12}) f34 ≤ 6;
(F_{13}) f45 ≤ 12;
(F_{14}) f46 ≤ 13;
(F_{15}) f56 ≤ 15;
(F_{16}) min: 12.4 f12 + 10.6 f13 + 13.6 f23 + 15.2 f24 + 17.6 f25 + 6.8 f34 + 18 f45 + 16.8 f46 + 24.4 f56;

Solution:

Objective value: 664.8

f12: 9
f13: 6
f23: 0
f24: 3
f25: 6
f34: 6
f45: 0
f46: 9
f56: 6

OPTIMAL

PROBLEMS R′ and R*: We assume that the risk flow v_r consequently accepts two different values $v_r = 3$ and $v_r = 3,2$. Then both network flow programming problems—R′ and R* from R_1 to R_{16} may be solved with the help of the program package WEBOPTIM developed in the Institute of Information and Communication Technologies—BAS.

We show below a print out of the problem R computer solution for a flow value of 3. The results for both problems R′ and R* are shown in Table 2.

Problem R

(R_1) r12 + r13 = 3;
(R_2) r23 + r24 + r25 − r12 = 0;
(R_3) r34 − r13 − r23 = 0;
(R_4) r45 + r46 − r24 − r34 = 0;
(R_5) r56 − r25 − r45 = 0;
(R_6) r56 + r46 = 3;
(R_7) r12 ≤ 2;
(R_8) r13 ≤ 1,2;
(R_9) r23 ≤ 0,8;
(R_{10}) r24 ≤ 0,6;
(R_{11}) r25 ≤ 1,6;
(R_{12}) r34 ≤ 1,2;
(R_{13}) r45 ≤ 2,4;
(R_{14}) r46 ≤ 2,6;
(R_{15}) r56 ≤ 3;
(R_{16}) min: 62 r12 + 53 r13 + 68 r23 + 76 r24 + 88 r25 + 34 r34 + 90 r45 + 74 r46 + 122 r56;

Solution:

Objective value: 664.8

r12: 1.8
r13: 1.2
r23: 0
r24: 0.6
r25: 1.2
r34: 1.2
r45: 0
r46: 1.8
r56: 1.2

OPTIMAL

The objective function L values of the four problems solved—F′, F*, R′, and R* are shown in Table 3.

Table 3 .

#	Problems	F′	F*	R′	R*
1.	Objective L value	664,8	719,2	664,8	719,2

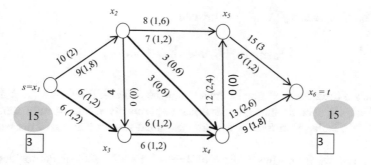

Fig. 3 .

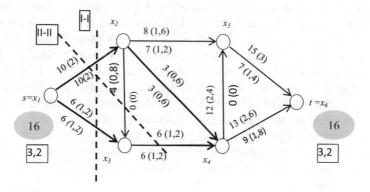

Fig. 4 .

In Fig. 3 the capacities' values of the resource flow for the respective arcs are given above the arc, and in brackets next to them—capacities of the risk flow. Below the arcs are given the values received for the resource flow functions $\{f_{ij}\}$ and next to them in brackets—the analogic values $\{r_{ij}\}$ for the risk flow. The results received correspond to the problems F' and R' solved.

Analogically, in Fig. 4 the capacities of the arc flow functions at max value of the flow and min value of the objective for both flows respectively—resource (problem F^*) and risk (problem R^*) are shown.

The network flow functions received after solving problems F^* and R^* are the maximum possible because, as seen in Fig. 4, two cuts with saturated arcs exist—I–I and II–II, such that:

(a)
$$f(X_0', \bar{X}_0') = v \text{ and } f(\bar{X}_0', X_0') = 0;$$

$$(X_0', \bar{X}_0') = \{x_{1,2}, x_{1,3}\}; \quad \left(\bar{X}_0', X_0' = \varnothing\right);$$

where Ø is an empty set.

(b) $f(X_0'', \bar{X}_0'') = v$ and $f(\bar{X}_0'', X_0'') = 0$; where

$$\left(X_0'', \bar{X}_0''\right) = \{x_{1,2}, x_{3,4}\}; \quad \left(\bar{X}_0'', X_0''\right) = \{x_{2,3}\}.$$

Hence a resource flow greater than $v*_s = 16$ and risk flow $v*_r = 3, 2$ does not exist. Data obtained through the numerical experiments in Sect. 4 fully confirm the theoretical results from Sect. 3 for the combined resource and risk network flow (Model 3).

As it follows from rows 3 and 4 of Table 1 equality (26) of Proposition 3.1 is observed. In the same way data from rows 7 and 8 of Table 1 confirm equality (27) from Proposition 3.2. Rows 9 to 11 of the same table confirm relation (28) of Proposition 3.3. Equality (29) in Proposition 3.4 is directly confirmed from the comparison of data in Tables 2 and 3. And these same data from the two latter tables explicitly confirm the sufficiency of condition (30) from Proposition 3.5 for one-to-one mapping of the respective resource and risk flow (Table 4).

In this way the intelligent combined flow thus defined provides a possibility to the flow properties to be preserved both for the resource and risk flow. It follows from this that in the Model M3 applied the disadvantages of models of M1 and M2 are avoided when always either risk or resource does not has network flow properties. The presence of such properties provides a possibility at one and the same time and for each cut on the network approximately equal values of the risk and the amount of resources to be kept which is an advantage of the Model M3 compared to the other models.

Table 4 .

#	Arc param	(1,2)	(1,3)	(2,3)	(2,4)	(2,5)	(3,4)	(4,5)	(4,6)	(5,6)
1	c_{ij}'	10	8	4	4	8	7	12	14	15
2	c_{ij}''	2,2	1,2	1	0,6	1,8	1,2	2,5	2,6	3,2
3	c_{ij}^s	10	4	4	2	6	3	10	13	15
4	c_{ij}^r	2	1,2	0,4	0,6	1,8	1,2	2,5	2,6	1,5
5	a_{ij}'	10	9	12	14	16	6	16	14	20
6	a_{ij}''	12	8	8	6	8	4	10	14	22
9	p_{ij}	0,2	0,3	0,1	0,3	0,3	0,4	0,25	0,2	0,1
10	v	$v = 9,4$								
11	v', v''	$v' = 7; v'' = 2,4$								

4 Two Commodity Network Flow of Resources and Risks—Model 4

A new class of problems arises for making decisions at which the expenses are considered for the transportation and allocation of the resource as well as when taking the corresponding risks. In previous works [1, 2] solutions were proposed for the problem pointed out through imposing some specific requirements which narrowed the class of problems being solved and the accuracy of the solutions obtained.

In the present Model M4 the most general precise solution is proposed for the specified problems through a two commodity network flow—from resources and corresponding risks. At that the requirement that distribution of resources and risks observe the conservation equations is essential, i.e. both resources and risks to have the property "flowability".

In model M3 [12] the case is considered when the probabilities for adverse events on all sections are one and the same, i.e. $p_{ij} = p_o$ for each $(i, j) \in G$. It is proved that in this case at observing certain additional conditions resources and the associated to them risks may be considered as one-to-one mapped network flows with common minimal total expenses.

In the present model M4 a more general solution is proposed of this problem when the observing the conditions of (30) is not required, i.e. probabilities $\{ p_{ij}/(i, j) \in G \}$ may be of arbitrary value in the interval $0 \le p_{ij} \le 1$; $(i, j) \in G$. At that as distinct from the single commodity models from M1 to M3 a two commodity model M4 is proposed in which the optimization procedures are simultaneously performed for both network flows—resource $\{f_{ij}\}$ and risk $\{r_{ij}\}$.

Besides the denotations listed in the introduction two more will be used:

v and v_{max} flow and max network flow of the resource;
v' and v'_{max} flow and max network flow of the risk

All other denotations are consistent with those formulated in the previous sections, mainly in Sect. 1. The relations from (3) to (7) are also valid for Model 4, being considered.

From the point of view of insurance bonds it is expedient to assume than the arc risk flow r_{ij} should not be of value less than the value of the real arc risk $p_{ij} f_{ij}$ received at defining the resource arc flow f_{ij}, i.e. it is necessary to observe the requirement:

$$p_{ij}f_{ij} \le r_{ij}; (i,j) \in G; \qquad (33)$$

which leads to the relation:

$$p_{ij}f_{ij} - r_{ij} \le 0; (i,j) \in G. \qquad (34)$$

Proposition 4.1 Inequalities (7) and (34) are sufficient conditions for keeping requirement (6).

It follows from (5) and (7) that

$$r_{ij} \leq p_{ij} c_{ij}^s. \tag{35}$$

The comparison of (33) and (35) leads to (6).

Corollary The chain of inequalities:

$$p_{ij} f_{ij} \leq r_{ij} \leq c_{ij}^r \leq c_{ij}^s. \tag{36}$$

immediately follows from relations (2), (5), (7), and (33).

It follows from (36) that the risk flow thus defined has bilateral constraints at which its lower, depending on the resource flow realization varies from 0 to $p_{ij} f_{ij}$. This provides a possibility a check-up for the risk flow existence to be avoided.

In the two commodity network flow the two flows—resource and risk are independently defined and with separate conservation equations, constraints' inequalities of the arc risk flows (7), and Eq. (34) binding the two flows $\{f_{ij}\}$ and $\{r_{ij}\}$. Then according to Proposition 4.1 it is not necessary to use inequalities (6).

The maximum flow v is considered as a sum of the analogical resource flow v' and risk flow v''

$$v = v' + v''. \tag{37}$$

In this case finding of the maximum two commodity flow is reduce to the following network flow programming problem:

$$L = v = v' + v'' \to \max \tag{38}$$

Under constraints: for each $i \in I$ and $(i, j) \in G$

$$\sum_{j \in \Gamma_i^1} f_{ij} - \sum_{j \in \Gamma_i^{-1}} f_{ji} = \begin{cases} v', & \text{if } x_i = s; \\ 0, & \text{if } x_i \neq s, t; \\ -v', & \text{if } x_i \in t; \end{cases} \tag{39}$$

$$f_{ij} \geq 0; (i, j) \in G; \tag{40}$$

$$p_{ij} f_{ij} - r_{ij} \leq 0; (i, j) \in G; \tag{41}$$

$$\sum_{j \in \Gamma_i^1} r_{ij} - \sum_{j \in \Gamma_i^{-1}} r_{ji} = \begin{cases} v'', & \text{if } x_i = s; \\ 0, & \text{if } x_i \neq s, t; \\ -v'', & \text{if } x_i \in t; \end{cases} \tag{42}$$

$$r_{ij} \leq c_{ij}^r; (i, j) \in G; \tag{43}$$

$$r_{ij} \geq 0; (i,j) \in G. \tag{44}$$

Analogically like in the classic mincut-maxflow theorem for a single commodity flow interpretation may be made of the minimal cut and the maximum flow after solving the problem from (38) to (44), i.e. keeping in mind the new constraints (41).

Definition 4.1 Flow (37) is maximal and the cut $(X_0, \overline{X_0})$ is minimal if after solving the problem from (38) to (44) the following is true:

$$v' = f(X_0, \overline{X_0}); v'' = r(X_0, \overline{X_0}); \tag{45}$$

$$f(\overline{X_0}, X_0) = 0; r(\overline{X_0}, X_0) = 0; \tag{46}$$

where

$$(X_0, \overline{X_0}) = \{x_{ij}/x_i \in X_0; \quad x_j \in \overline{X_0}; \quad x_{ij} \in U; \quad p_{ij}f_{ij} - r_{ij} = 0\}; \tag{47}$$

$$(\overline{X_0}, X_0) = \{x_{ji}/x_j \in \overline{X_0}; x_i \in X_0; x_{ji} \in U\}. \tag{48}$$

Other interpretations of the maximum flow and minimal cut are also possible with constraints (6), (7), and (33). It is expedient two more definitions to be introduced on the base of the definition of risk (1) and the two-commodity network flow interpretation from (39) to (44):

Definition 4.2 The quantity

$$p(f,r) = \frac{\sum_{(i,j) \in G} r_{ij}}{\sum_{(i,j) \in G} f_{ij}} \leq 1 \tag{49}$$

will be called average probability for adverse events on the network at some (f, r)— realization of two commodity flow of resources and risks.

Definition 4.3 The quantity

$$p'(f,r) = \frac{\sum_{(i,j) \in G} p_{ij}f_{ij}}{\sum_{(i,j) \in G} f_{ij}} \tag{50}$$

will be called average probability for adverse events on the network at some (f, r)— realization and not guaranteed flowability of the risk.

Definition 4.4 The quantity

$$Q(f,r) = \frac{\sum_{(i,j) \in G} a''_{ij}r_{ij}}{\sum_{(i,j) \in G} a'_{ij}f_{ij}} \tag{51}$$

will be called average ratio between the value of the payment for risk and the value paid for the transportation of the resource.

Definition 4.5 The arc $x_{ij} \in U$ will be called saturated if the condition

$$p_{ij}f_{ij} - r_{ij} = 0 \tag{52}$$

is observed for it.

After solving the problem for maximum two commodity flow—from (38) to (44) called **Problem A** on the base of the results received the following problem may be formulated.

Problem B Finding the maximum two commodity flow of minimal cost (mincost-maxflow).

The values received for v' and v'' after solving problem A are put as constants in Eqs. (39) and (42) and the objective function from (38) is substituted by the following one:

$$L = \left(\sum_{(i,j) \in G} a'_{ij} f_{ij} + \sum_{(i,j) \in G} a''_{ij} r_{ij} \right) \rightarrow \min \tag{53}$$

Then by solving Problem B—from (38) to (44) with objective function (53) a network flow realization of a maximum two commodity flow of minimal cost will be received.

4.1 Numerical Examples for Model 4

The exemplary small network from Fig. 1 like in the previous Model 3 with the same denotations of nodes and arcs will be used for the numerical examples here. On the same network a two commodity network flow of resources $\{f_{ij}\}$ and the risks $\{r_{ij}\}$ from their transportation will be defined. They are interconnected through the relations from (34).

The concrete values of the capacities $\left\{ c'_{ij}, c''_{ij} \right\}$ as well as of the calculated through (3) and (4) coordinated capacities $\left\{ c^s_{ij}, c^r_{ij} \right\}$ are given in Table 5. It also contains the arc evaluations $\left\{ a'_{ij}, a''_{ij} \right\}$ for the transportation of the resources and corresponding risks as well as the probabilities $\{p_{ij}\}$ an adverse event to occur on the separate sections of the network.

Problem A The maximum two commodity flow of resources and risks may be defined through the following network flow programming problem

$$L = v \rightarrow \max; \tag{54}$$

Table 5 .

#	Arc param	(1,2)	(1,3)	(2,3)	(2,4)	(2,5)	(3,4)	(4,5)	(4,6)	(5,6)
1	f_{ij}	4	3	0	2	2	3	0	5	2
2	p_{ij}	1,2	1,2	0	0,6	0,6	1,2	0	1,8	0,6
3	r_{ij}	0,2	0,3	0,1	0,3	0,3	0,4	0,25	0,2	0,1
4	$c_{ij}^s - f_{ij}$	6	1	4	0	4	0	10	8	13
5	$c_{ij}^r - r_{ij}$	0,8	0	0,4	0	1,2	0	2,5	0,8	0,9
6	$p_{ij}f_{ij} - r_{ij}$	−0,4	−0,3	0	0	0	0	0	-0,8	-0,4
7	$p'_{ij} = r_{ij}/f_{ij}$	0,3	0,4	0,1	0,3	0,3	0,4	0,25	0,36	0,3
8	$\Delta p_{ij} = p'_{ij} - p_{ij}$	0,1	0,1	0	0	0	0	0	0,16	0,2

observing relations from F_0 to F_{16} and from R_1 to R_{17}. For this purpose the web based software package Weboptim, developed in the Inst. of Information and Communication Technologies—BAS.

Resources flow

$(F_0)\ v' + v'' = v$;
$(F_1)\ f_{1,2} + f_{1,3} = v'$;
$(F_2)\ f_{2,3} + f_{2,4} + f_{2,5} - f_{1,2} = 0$;
$(F_3)\ f_{3,4} - f_{1,3} - f_{2,3} = 0$;
$(F_4)\ f_{4,5} + f_{4,6} - f_{2,4} - f_{3,4} = 0$;
$(F_5)\ f_{5,6} - f_{2,5} - f_{4,5} = 0$;
$(F_6)\ f_{5,6} + f_{4,6} = v'$;
$(F_7)\ 0,2\ f_{1,2} - r_{1,2} \leq 0$;
$(F_8)\ 0,3\ f_{1,3} - r_{1,3} \leq 0$;
$(F_9)\ 0,1\ f_{2,3} - r_{2,3} \leq 0$;
$(F_{10})\ 0,3\ f_{2,4} - r_{2,4} \leq 0$;
$(F_{11})\ 0,3\ f_{2,5} - r_{2,5} \leq 0$;
$(F_{12})\ 0,4\ f_{3,4} - r_{3,4} \leq 0$;
$(F_{13})\ 0,25\ f_{4,5} - r_{1,2,4,5} \leq 0$;
$(F_{14})\ 0,2\ f_{4,6} - r_{4,6} \leq 0$;
$(F_{15})\ 0,2\ f_{5,6} - r_{5,6} \leq 0$;
$(F_{16})\ f_{i,j} \geq 0$ for each $(i, j) \in G$

Risk flow

$(R_1)\ r_{1,2} + r_{1,3} = v''$;
$(R_2)\ r_{2,3} + r_{2,4} + r_{2,5} - r_{1,2} = 0$;
$(R_3)\ r_{3,4} - r_{1,3} - r_{2,3} = 0$;
$(R_4)\ r_{4,5} + r_{4,6} - r_{2,4} - r_{3,4} = 0$;
$(R_5)\ r_{5,6} - r_{2,5} - r_{4,5} = 0$;
$(R_6)\ r_{5,6} + r_{4,6} = v'$;
$(R_7)\ r_{1,2} \leq 2$;

(R$_8$) $r_{1,3} \leq 1,2$;
(R$_9$) $r_{2,3} \leq 0,4$;
(R$_{10}$)$\emptyset$ $r_{2,4} \leq 0,6$;
(R$_{11}$) $r_{2,5} \leq 1,8$;
(R$_{12}$) $r_{3,4} \leq 1,2$;
(R$_{13}$) $r_{4,5} \leq 2,5$;
(R$_{14}$) $r_{4,6} \leq 2,6$;
(R$_{15}$) $r_{5,6} \leq 1,5$;
(R$_{16}$) $r_{i,j} \geq 0$ for each $(i, j) \in G$
(R$_{17}$) $v \geq 0$; $v' \geq 0$; $v'' = \geq 0z$

The results received are reflected in Tables 4 and 5 for a two commodity flow and its resource, v' and risk, v'' components respectively. The bottom two lines of Table 4 show the maximum two commodity flow and its resource, v' and risk, v'' components respectively. The upper two rows in Table 5 show the values of the arc flow functions for the resource—$\{f_{ij}\}$ and for the risk—$\{r_{ij}\}$. The differences between the resource arc capacity c_{ij}^s and the resource flow f_{ij} are shown in row 4 of Table 5 and the analogical values—c_{ij}^r and r_{ij}—in row 5 of the same table. The differences $\{p_{ij} f_{ij} - r_{ij}\}$ showing the saturation with flow of arcs $\{x_{ij}/(i, j) \in G\}$ are in row 6. Ratios p'_{ij} between the risk and resource flows on each arc are shown in row 7 and the differences between the really received and initially assigned probabilities for adverse events on each arc are reflected in row 8.

The received value of the respective resource flow f_{ij} is shown above each arc and the arc capacity c_{ij}^s—to the right of it in brackets are shown in Fig. 5. The degree of the saturation of the arc is shown beneath the arc according to the difference $\{p_{ij} f_{ij} - r_{ij}\}$. In analogical way the risk parameters—r_{ij}, c_{ij}^r, and $\{p_{ij} f_{ij} - r_{ij}\}$ are reflected on the arcs in Fig. 6.

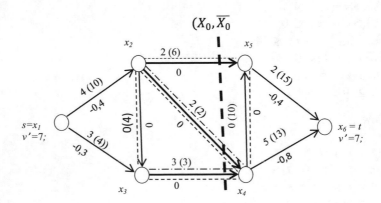

Fig. 5 .

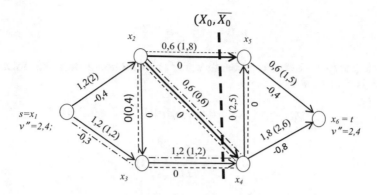

$(X_0, \overline{X_0}$

Fig. 6 .

The minimal cut $(X_0, \overline{X_0})$ is shown by a thick dash and line in Figs. 5 and 6 corresponding to the max flow $v = 9,4$ consisting of $v' = 7$ and $v'' = 2,4$—rows 8 and 9 of Table 1.

$$X_0 = \{x_1, x_2, x_3\}; \quad \overline{X_0} = \{x_4, x_5, x_6\}; \quad (X_0, \overline{X_0}) = \{x_{2,4}, x_{2,5}, x_{3,4}\};$$

$(\overline{X_0}, X_0) = \emptyset$ where $\emptyset$ is the empty set. Hence $f;(\overline{X_0}, X_0) = r(\overline{X_0}, X_0)$;

$$v' = f(x_{2,4}) + f(x_{2,5}) + f(x_{3,4}) = 2 + 2 + 3 = 7;$$
$$v'' = r(x_{2,4}) + r(x_{2,5}) + r(x_{3,4}) = 0,6 + 0,6 + 1,2 = 2,4;$$
$$v = v' + v'' = 7 + 2,4 = 9,4.$$

The saturated arcs according to inequality (34) are marked with dash lines in Figs. 5 and 6. These are $\{x_{2,3}, x_{2,4}, x_{2,5}, x_{3,4}, x_{4,5}\}$ and they are with zero difference $\{p_{ij} f_{ij} - r_{ij}\}$. Besides in both figures the arcs that are saturated according to (6)— for the resource flow and (7)—for the risk flow are marked with dash dot lines. For the first one these are $\{x_{2,4}, x_{3,4},\}$ from Fig. 5 and for the risk flow—$\{x_{1,3}, x_{2,4}, x_{2,5}\}$ from Fig. 6.

It may be defined from rows 1 and 2 of Table 5:

$$\sum_{(i,j) \in G} r_{ij} = (3 \times 1, 2) + (3 \times 0, 6) + 1, 8 = 7, 2$$

$$\sum_{(i,j) \in G} f_{ij} = 4 + (2 \times 3) + (3 \times 2) + 5 = 21. \tag{55}$$

Then it follows from (49) that the average probability for occurrence of adverse events on the whole network is equal to

$$p(f,r) = \frac{7,2}{21} = 0,34. \tag{56}$$

Analogically the average probability $p'(f, r)$ with non-guaranteed flowance of the risk is calculated by

$$\sum_{(i,j) \in G} p_{ij} f_{ij} = 2 + 0,9 + 0,6 + 0,6 + 1,2 + 1 + 0,2 = 6,5;$$

$$p'(f,r) = \frac{6,5}{21} = 0,31; \tag{57}$$

where the denominator value is calculated in (55).

Therefore at the given (f, r) realization the average probability for an adverse event $p(f, r)$ with guaranteed flowability of risk is 0,03 higher or about 10% than in the reverse case which in principle follows from the requirement (34).

For defining the quantities $Q(f, r)$ from (51) the following calculations are necessary:

$$\sum_{(i,j) \in G} a'' r_{ij} = 14,4 + 9,6 + 3,6 + 4,8 + 4,8 + 25,2 + 19,2 = 75,6; \tag{58}$$

$$\sum_{(i,j) \in G} a' f_{ij} = 40 + 27 + 28 + 32 + 18 + 70 + 40 = 255; \tag{59}$$

$$Q(f, r) = \frac{75,6}{255} = 0,296.$$

If we take for a base the expenses for payment for the transportation of the resource on the network and the risks of these operations then the average and calculated probability of adverse events is very near to 0,30, i.e. it is with 0,04 less than $p(f, r)$. In general at a given (f, r) realization the ratio between the quantities p (f, r), $p'(f,r)$, and $Q(f, r)$ depends on the particular source values of $\left\{ c_{ij}^s \right\}$, $\left\{ c_{ij}^r \right\}$, $\left\{ a_{ij}' \right\}$, and $\left\{ a_{ij}'' \right\}$. But according to (34) the inequality $p(f, r) \geq p'(f, r)$ is always true.

Using the data received from solving of Problem A solving may be carried out of the next:

Problem B Finding a maximum two commodity resource flow of minimal cost and risks of minimal cost (mincost-maxflow). This flow is defined through relations F_0 to F_{16} and from R_1 to R_{17}. In these relations the flows v' and v'' are considered as fixed constant values from rows 8 and 9 of Table 1, namely

$$v' = 7; \ v'' = 2, 4; \ v = 9, 4. \tag{60}$$

The objective function from Problem A is substituted by the following objective:

$$L' = 10f_{1,2} + 9f_{1,3} + 12f_{2,3} + 14f_{2,4} + 16f_{2,5} + 6f_{3,4} + 16f_{4,5} + 14f_{4,6} + 20f_{5,6} + 12r_{1,2}$$
$$+ 8r_{1,3} + 8r_{2,3} + 6r_{2,4} + 8r_{2,5} + 4r_{3,4} + 10r_{4,5} + 14r_{4,6} + 22r_{5,6} \text{min}.$$

$$\tag{61}$$

Solving the new network flow programming problems renders the same results for the realization of the resource and risk flows as shown in Table 5. The following value of the objective function is received:

$$L' = 330, 6. \tag{62}$$

This means that with the same v', v'', and v like in Problem A no other (f, r) flow realization exists with a lesser value than the objective function (61). This follows from the fact that the value of the objective function (62) is equal to the sum of the values from (58) and (59). Hence all research results considered in Problem A are thoroughly valid for Problem B also. The numerical example in this Sect. 4.1 proves in a persuasive way the theoretical research results received and described in the previous Sect. 4.

5 Two Stage Method for Network Flow Control of Resources and Risks—Model 5

Several network flow models for control of resources and risks are described in Sects. 2–4. Some of their advantages and disadvantages are exposed. Models 1 and 2 are of relatively not great computational complexity but they do not satisfy all requirements for the simultaneous control both of resources and risks. Model 3 allows greater generalization but imposes the restriction for equal probability of an adverse event when passing along each arc of the network. Model 4 is possibly most general taking into account the distinct of an adverse event on each arc and provides an exact optimal solution but that involves solving a problem for a multi commodity network flow which is of appreciable computational complexity. The two stage method proposed for defining the optimal resources network flow and the corresponding network risk flow is of less computational complexity than the most general model M4 at certain decrease of the precision of the solutions received. But this is quite admissible in a number of practical cases.

At creating the new model, called M5 the use of the theoretical results for model M3 is essential for the one-to-one mapping between the resource and the risk flows. The meaning of all denotations for the present Model 5 is the same, like in the previous sections.

One of the possible approaches in this case is finding the optimal solution of the resource network problem and searching with comparatively not complicated instrumental tools solution of the second problem for the network risks, based on data from the resource problem. A different approach is proposed in the present Model 5 in which the resource flow only is optimized and the results received are transferred in an appropriate manner to the risk flow, so that a good enough general solution is reached for both flows, observing the requirements for their network flow properties.

If the values $\left\{c'_{ij}\right\}$ and $\left\{c'_{ij}\right\}$ of the resource flow are appropriately coordinated with the respective indices $\left\{c''_{ij}\right\}$ and $\left\{a''_{ij}\right\}$ of the risk flow $\{r_{ij}\}$ then this provides a possibility one-to-one mapping between both flows—$\{f_{ij}\}$ and $\{r_{ij}\}$ to be achieved. It is necessary for this purpose, concerning (3) the following replacements to be made:

$$f_{ij}a^s_{ij} = a'_{ij}f_{ij} + a''_{ij}r_{ij},\tag{63}$$

from which it follows that

$$a^s_{ij} = a'_{ij} + p_{ij}a''_{ij}; \quad p_{ij} = \frac{r_{ij}}{f_{ij}};\tag{64}$$

where c^s_{ij} and a^s_{ij} are coordinated values of the capacities and the arc evaluations of the resource flow considering the analogical values of the risk flow.

Similarly, concerning (4) we may put for the risk flow:

$$r_{ij}a^r_{ij} = a''_{ij}r_{ij} + a'_{ij}\frac{r_{ij}}{p_{ij}},\tag{65}$$

from which it follows that

$$a^r_{ij} = a''_{ij} + \frac{a'_{ij}}{p_{ij}};\tag{66}$$

where c^r_{ij} and a^r_{ij} are coordinated values of the capacities and the arc evaluations of the risk flow considering the analogical values of the resource flow.

The optimal distribution of the resource flow being controlled, considering the coordinated indices of the risk flow, may be reduced to the following mathematical programming problem

$$L' = \sum_{(i,j)\in G} a^s_{ij}f_{ij} \to \max(min);\tag{67}$$

subject to constraints for each $i \in I$ and $(i, j) \in G$

$$\sum_{j \in \Gamma_i^1} f_{ij} - \sum_{j \in \Gamma_i^{-1}} f_{ji} = \begin{cases} v', \text{if } x_i = s; \\ 0, \text{if } x_i \neq s, t; \\ -v', \text{if } x_i \in t; \end{cases} \tag{68}$$

$$f_{ij} \leq c_{ij}^s; \quad (i,j) \in G; \tag{69}$$

$$f_{ij} \geq 0; \quad (i,j) \in G; \tag{70}$$

where v' is a resource flow from the source s to the sink t.

Similarly the optimal distribution of the risk flow may be defined, considering the coordinated indices of the resource flow. It is reduced to the following network flow programming problem:

$$L'' = \sum_{(i,j) \in G} a_{ij}^r r_{ij} \to \max(min); \tag{71}$$

subject to constraints for each $i \in I$ and $(i, j) \in G$

$$\sum_{j \in \Gamma_i^1} r_{ij} - \sum_{j \in \Gamma_i^{-1}} r_{ji} = \begin{cases} v', \text{ if } x_i = s; \\ 0, \text{ if } x_i \neq s, t; \\ -v', \text{ if } x_i \in t; \end{cases} \tag{72}$$

$$r_{ij} \leq c_{ij}^r; \quad (i,j) \in G; \tag{73}$$

$$r_{ij} \geq 0; \quad (i,j) \in G; \tag{74}$$

where v'' is a flow of risks from the source s to the sink t.

The separately solving of problems from (67) to (70) and from (71) to (74) in the general case gives solutions in which the respective requirements (1) to the relations between the resources and risks are not observed on each section of the network $\{x_{ij}\}$.

It is proved in Model 3 that (30) is a sufficient condition for one-to-one mapping between the corresponding values of $\{f_{ij}\}$ and $\{r_{ij}\}$. But following literally of requirement (30) means non-considering of the differences of $\{p_{ij}\}$ on the separate sections of the network and this is unwanted in some cases. These difficulties may be overcome through the following multistage scheme, proposed in this work:

1. Through relations from (67) to (70) the optimal resource flow $\{f_{ij}\}$ is defined whose indices $\{c_{ij}^s\}$ and $\{a_{ij}^s\}$ are coordinated with the analogical indices of the risk flow through relations from (3), (63), and (64). At that if a maximum resource flow is sought instead of (67) the following objective function is used:

$$L' = v \to \max. \tag{75}$$

In case that the flow v is known and it is of fixed value, and realization $\{f_{ij}\}$ of minimal value is sought then the objective L' of (67) is minimized.

2. The real risks $\left\{r'_{ij}\right\}$ are determined for each separate section $x_{ij} \in U$ of the network through the relations from (1) and by using the previously known probabilities of adverse events $\{p_{ij}\}$, i.e.:

$$r'_{ij} = p_{ij}f_{ij}; \quad (i,j) \in G. \tag{76}$$

In the most general case the values $\left\{r'_{ij}\right\}$ received do not observe the requirements for the flow property from (72) to (74)

3. An average value is defined for the probabilities of adverse events on the network on the base of the network flow values $\{f_{ij}\}$ and $\left\{r'_{ij}\right\}$:

$$p'_0 = \frac{\sum\limits_{(i,j) \in G} r'_{ij}}{\sum\limits_{(i,j) \in G} f_{ij}} \tag{77}$$

4. The requirement

$$p_{ij} = p'_0 \text{ for each } (i,j) \in G; \tag{78}$$

is supposed to be kept for the whole network. Then the following values of the risk on the separate sections $\{x_{ij}\}$ of the network may be defined:

$$r_{ij} = p'_0 f_{ij}; \quad (i,j) \in G. \tag{79}$$

According to the Model M3 they correspond to the requirements from (72) to (74) for the risk flow and they are in one-to-one mapping with relations from (68) to (70). Relations from (76) to (79) lead to

$$r_0 = p'_0 \sum\limits_{(i,j) \in G} f_{ij}; \quad f_0 = \sum\limits_{(i,j) \in G} f_{ij}; \tag{80}$$

$$r_0 = \sum\limits_{(i,j) \in G} r'_{ij} = \sum\limits_{(i,j) \in G} p_{ij}f_{ij}; \tag{81}$$

$$r_0 = p_0' f_0;$$ (82)

for each $(i, j) \in G$.

$$\Delta p_{ij} = p_0' - p_{ij}; \quad \Delta r_{ij} = \Delta p_{ij} f_{ij};$$ (83)

$$r_{ij} = r_{ij}' + \Delta r_{ij};$$ (84)

where f_0 is the total sum resource on the network;

r_0 total sum risk on the network;

Δp_{ij} the difference between the a priori arc probability p_{ij} and the average probability p_0' for an adverse event on the network;

Δr_{ij} the part of the arc risk which if added to $\left\{ r_{ij}' \right\}$ leads to risk values $\left\{ r_{ij} \right\}$ for which the requirements for flow properties from (72) to (74) are observed

It is to be taken in mind that the network parameters p_0', r_0, f_0, $\left\{ \Delta p_{ij} \right\}$, and $\left\{ \Delta r_{ij} \right\}$ in general have different values in different realizations of the resource flow $\{f_{ij}\}$. In the approach being proposed for building the risk flow—through the relations from (76) to (82) the values $\left\{ \Delta r_{ij} \right\}$ and $\left\{ \Delta p_{ij} \right\}$ show what deviations are admissible from the arc probabilities $\{p_{ij}\}$ and risks $\{r_{ij}\}$ compared to the real values of these quantities.

When defining the maximum resource flow through relations from (72) to (75) the maximum resource cut $(X_0, \overline{X_0})$ is also defined, which corresponds to the following requirements:

$$(X_0, \overline{X_0}) = \{x_{ij}/x_i \in X_0; \quad x_j \in \overline{X_0}; \quad (i,j) \in G\};$$
$$(\overline{X_0}, X_0) = \{x_{ji}/x_j \in \overline{X_0}; \quad x_i \in X_0; \quad (i,j) \in G\};$$
$$X_0 \cup \overline{X_0} = X; \quad X_0 \cap \overline{X_0} = \varnothing,$$

where $\varnothing$ is the symbol for the empty set.

Then according to Ford-Fulkerson's mincut-maxflow theorem [8] the maximum flow v_{max} and the minimal cut $(X_0, \overline{X_0})$ keep the requirements

$$v_{max} = f(X_0, \overline{X_0}) = c^s(X_0, \overline{X_0});$$ (85)

$$f(X_0, \overline{X_0}) = 0.$$ (86)

Parameters p_0', f_0, and r_0, as well as the network quantities $\left\{ \Delta p_{ij} \right\}$ and $\left\{ \Delta r_{ij} \right\}$ may be of different values in two not identical realizations of the resource flow $\{f_{ij}\}$. The solution proposed in the present Model 5 for the two commodity network flow of resources and risks reduces the problem to optimal solutions for the resource flow only, and depending on the results received—to comparatively good solutions for the risk flow. At that it is not necessary to simultaneously solve optimization problems for both flows, i.e. a kind of exchange is carried out of the accuracy of solutions against decreased number of variables and lesser computational complexity.

5.1 Numerical Examples for Model 5

The exemplary small network from Fig. 1 like in the previous models with the same denotations of nodes and arcs will be used for the numerical examples here.

This network together with the arc parameters is shown in Fig. 7. It will be explained in details further.

It is necessary the maximum resource flow v and the corresponding risk flow v_r, with or without minimal cost to be defined. At that relations through which both flows are connected between them should be taken in mind. The source data are given in Table 6 for the capacities $\left\{c'_{ij}, c''_{ij}\right\}$ and for the arc evaluations $\left\{a'_{ij}, a''_{ij}\right\}$ for the resource and the risk flows respectively. In rows #3 and #4 the respective coordinated capacities $\left\{c^s_{ij}, c^r_{ij}\right\}$ calculated through (3) and (4) are given for both flows respectively—arc and resource. Similarly, in rows #7 and #8 in the same Table 6 the coordinated evaluations $\left\{a^s_{ij}, a^r_{ij}\right\}$ calculated through (64) and (66) are shown for the resource and risk flows.

The values of the probabilities for adverse events on the separate arcs of the network $\{p_{ij}\}$ are given in row #9 of Table 6. In Fig. 7 on each arc of the network the value of the resource flow capacity $\left\{c^s_{ij}\right\}$ is shown and next to it in brackets— the arc evaluation of the same flow $\left\{a^s_{ij}\right\}$. The respective value of $\{p_{ij}\}$ indicated under each arc.

The following problems will be solved:

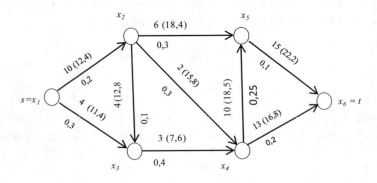

Fig. 7 .

Table 6 .

#	Arc param	(1,2)	(1,3)	(2,3)	(2,4)	(2,5)	(3,4)	(4,5)	(4,6)	(5,6)
1	c'_{ij}	10	8	4	4	8	7	12	14	15
2	c''_{ij}	2,2	1,2	1	0,6	1,8	1,2	2,5	2,6	3,2
3	c^s_{ij}	10	4	4	2	6	3	10	13	15
4	c^r_{ij}	2	1,2	0,4	0,6	1,8	1,2	2,5	2,6	1,5
5	a'_{ij}	10	9	12	14	16	6	16	14	20
6	a''_{ij}	12	8	8	6	8	4	10	14	22
7	a^s_{ij}	12,4	11,4	12,8	15,8	18,4	7,6	18,5	16,8	22,2
8	a^r_{ij}	62	38	128	52,66	61,33	19	74	84	222
9	p_{ij}	0,2	0,3	0,1	0,3	0,3	0,4	0,25	0,2	0,1
10	v	11								
11	v_r	2,563								

1. Problem of the maximum resource flow (min cut-max flow). The maximum flow may be found through the network flow programming methods with objective function (75) and observing the following constraints from F_1 to F_{15} below. Resources flow

   (F$_1$) $f_{1,2} + f_{1,3} = v$;
   (F$_2$) $f_{2,3} + f_{2,4} + f_{2,5} - f_{1,2} = 0$;
   (F$_3$) $f_{3,4} - f_{1,3} - f_{2,3} = 0$;
   (F$_4$) $f_{4,5} + f_{4,6} - f_{2,4} - f_{3,4} = 0$;
   (F$_5$) $f_{5,6} - f_{2,5} - f_{4,5} = 0$;
   (F$_6$) $f_{5,6} + f_{4,6} = v$;
   (F$_7$) $f_{1,2} \leq 10$;
   (F$_8$) $f_{1,3} \leq 4$;
   (F$_9$) $f_{2,3} \leq 4$;
   (F$_{10}$) $f_{2,4} \leq 3$;
   (F$_{11}$) $f_{2,5} \leq 6$;
   (F$_{12}$) $f_{3,4} \leq 6$;
   (F$_{13}$) $f_{4,5} \leq 10$;
   (F$_{14}$) $f_{4,6} \leq 13$;
   (F$_{15}$) $f_{5,6} \leq 15$;

   The value of the maximum resource flow is shown in row #10 of Table 6 and of the respective arc flow functions in row #1 of Table 6.

2. On the base of the solution received for the previous problem I a flow of minimal cost is sought, i.e. Min Cost—Max Flow problem. For this purpose the following objective function is defined with data from row #7 of Table 6:

$$L' = \sum_{(i,j) \in G} a_{ij}^s f_{ij} = 12, 4f_{1,2} + 11, 4f_{1,3} + 12, 8f_{2,3} + 15, 8f_{2,4} + 18, 4f_{2,5} + 7, 6f_{3,4}$$

$$+ 18, 5f_{4,5} + 16, 8f_{4,6} + 22, 2f_{5,6} \to \min.$$

The value received for the flow v received from the network flow programming problem in the previous Sect. 1 is put as a constant quantity in equalities F_1 and F_6. Then the maximum resource flow of minimal cost will be defined through the objective L' and the relations from F_1 to F_{15}.

The values of the arc resource functions $\{f_{ij}\}$ received are shown in row #2 of Table 6. They are identical to the respective numerical experiment of Sect. 1., i.e.— row #1 of Table 7. The objective value received is $L' = 515, 4$.

3. On the base of the results received from solving the previous problem II the risk flow and its parameters $\left\{r'_{ij}\right\}$, $\{r_{ij}\}$, $\{\Delta p_{ij}\}$ and $\{\Delta r_{ij}\}$ may be defined using relations from (76) to (84). Results achieved are shown in rows from #3 to #6 of Table 7. The arc risks $\left\{r'_{ij}\right\}$ from formula (76) which do not keep the flow property are shown in row #3 of the same table.

Quantities r_0 and f_0 from formula (80) are equal, respectively to

$$f_0 = \sum_{(i,j) \in G} f_{ij} = 33; \quad r_0 = \sum_{(i,j) \in G} r'_{ij} = 7, 7$$

and the respective parameter from (77)—$p'_0 = \frac{r_0}{f_0} = 0, 233$. It shows that the average probability for an adverse event on the whole network is equal to 0,233.

The arcs values of r_{ij} may be found through formula (79).They are put in row #4 of Table 7. They observe the requirements for flow property and represent the risk flow received through the resource flow $\{f_{ij}\}$.

Comparing p'_0 with the probabilities $\{p_{ij}\}$ from row #9 of Table 6 through formula (83) leads to defining the data for $\{\Delta p_{ij}\}$ and $\{\Delta r_{ij}\}$ which are shown in rows #5 and #6 of Table 7. It follows from these data that relation (84) is observed and that the differences of the risk $\{\Delta r_{ij}\}$ added with different signs to the initial arc

Table 7 .

#	Arc param	(1,2)	(1,3)	(2,3)	(2,4)	(2,5)	(3,4)	(4,5)	(4,6)	(5,6)
1	f_{ij} (v)	8	3	0	2	6	3	0	5	6
2	f_{ij}	8	3	0	2	6	3	0	5	6
3	r'_{ij}	1,6	0,9	0	0,6	1,8	1,2	0	1	0,6
4	r_{ij}	1,864	0,699	0	0,466	1,398	0,699	0	1,165	1,398
5	Δp_{ij}	0,033	−0,067	0,133	−0,067	−0,067	−0,167	−0,017	0,033	0,133
6	Δr_{ij}	0,264	−0,201	0	−0,134	−0,402	−0,501	0	0,165	0,798

risks $\left\{r'_{ij}\right\}$ result in receiving the risk flow $\{r_{ij}\}$ for which the requirements of the flow property are observed.

On the base of the data in Tables 6 and 7 both flows—resource and risk may be pictorially presented.

In Fig. 8 along each arc x_{ij} the value of the arc flow function f_{ij} is presented and next to it in brackets—the value of the arc risk flow r_{ij}. The respective value of the risk difference Δr_{ij} is given in brackets (due to the sign) beneath the arc x_{ij}. The values of the max resource flow v and the risk flow received v_r are noted at the source $s = x_1$ and the sink $t = x_6$. The arc values of both flows thoroughly correspond to the requirements for the flow properties. This follows from the flow balance in each node. It is illustrated in Fig. 8.

The ratio between v_r and v is equal to p'_0 and this is confirmed by

$$\frac{v_r}{v} = \frac{2,563}{11} = 0,233 = p'_0$$

which confirms the correctness of the formulation and the solving of the problem of two flows—resource and risk.

Comparison of the respective data from rows #4 of both tables shows that the risk arc flow function $\{r_{ij}\}$ has no saturated arcs, i.e.

$r_{ij} < c^r_{ij}$ for each $(i, j) \in G$ and the risk flow v_r is not maximal.

Comparison of the data from row #3 of Table 6 and row #1 of Table 7 shows that three only arcs are saturated with resource flow, namely $\{x_{2,4}; x_{2,5}; x_{3,4}\}$. As they form a cut then this cut, as follows from relations (85) and (86), is minimal and it is equal to the maximum flow $v = 11$.

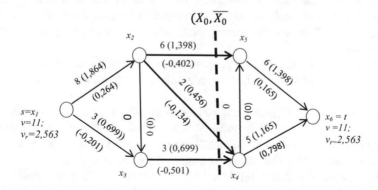

Fig. 8 .

$$X_0 = \{x_1, x_2, x_3\}; \quad \overline{X_0} = \{x_4, x_5, x_6\}; \quad X_0 \cup \overline{X_0} = X; \quad X_0 \cap \overline{X_0} = \emptyset;$$
$$(X_0, \overline{X_0}) = \{x_{2,4}; x_{2,5}; x_{3,4}\}; \quad (\overline{X_0}, X_0) = \emptyset;$$
$$f(X_0, \overline{X_0}) = f(x_{2.4}) + f(x_{2.5}) + f(x_{3.4}) = 2 + 6 + 3 = 11;$$
$$v = 11 = f(X_0, \overline{X_0}); f(\overline{X_0}, X_0) = 0.$$

It follows from the above that the cut $(X_0, \overline{X_0})$ is minimal and it is equal to the max resource flow. This cut is shown by a thick dashed line in Fig. 8.

6 Conclusion

The present work is dedicated to an investigation of the behavior of the flow of resources (people, goods, raw materials, money, information etc.) on a network keeping in concern the capacity and the transportation expenses, and the risks being taken on any section of the network. At that two inter related flows arise—of resources and risks. The purpose is those two flows to have a common minimal cost of expenses and to observe the conservation equations in the separate vertices of the network, i.e. to have the flowability feature, as well to satisfy the capacity requirements of the resources and the risk related to them. Thus a two commodity network flow of resources and risks emerges with specific requirements for which control is necessary—together and separately both of resources and risks. At that the risk on the separate sections of the network is considered as a product of two measures—the amount of resource and the probability of an adverse event to occur on the same section of the network. Such types of problems are known to have a good practical application.

A method is proposed for coordinating the values of the capacities and the arc evaluations of both flows respectively in which, in the examples given, the parameters of one of the flows are taken in mind in the parameters of the other one and vice versa. Relations are proved from which the expediency of the method proposed follows.

Two models—M1 and M2 are proposed for control of the two commodity flow; for defining a two commodity flow of min value, as well as its maximal admissible value.

In the first one—M1, the method proposed leads to a generalized risk flow with gains and losses, i.e. loss of flowability, while for the resource network flow this property is kept. In the second model M2 the situation is reverse—the resource flow loses flowability, while this feature is observed for the risk flow.

In the model M3 being developed it is assumed that the probability for an adverse event to occur is one and the same for all sections of the network. It is proved that in this case that solving the optimization problems for the resource flow leads to the corresponding optimal parameters of the risk flow, i.e. only single commodity problems are solved in model M3.

It is shown how in this model resource and risk flows of minimal value may be received, maximal admissible flows, as well such of minimal cost.

A most general model M4 is proposed that simultaneously provides optimal solutions for both network flows considering all constraints both on arc capacity and arc valuations. This model is of greatest polynomial computational complexity compared to all 5 models proposed in the present work.

A model M5 is being developed in which analogically to M3 from the optimal solutions for the resource flow the analogical solutions for the risk flow may be received. At that as a difference from model M3 the individual probability values for an adverse event to occur on the separate sections of the network are taken into account. In this case a two stage approach is used—first the resource flow optimal parameters are found, and then, on their base the average value of the risk for the whole network is determined and then the analogical optimal values of the risk flow are computed. At that all constraints for both network flows are observed. It is shown that this model gives more precise results than model M3.

The behavior of the resource and risk network flows is demonstrated on the base of a number of numerical examples when using the five models being proposed. Minimal values, maximal admissible flows, and min cost ones are received for both flows.

References

1. Sgurev, V., Drangajov, S.: Optimal control of the flow of risk on network. In: Proceedings of International Conference Automatics and Informatics'15, Bulgaria, Sofia, 4–7 Oct 2015, pp. 119–122. John Atanasoff Society of Automatics and Informatics. ISSN 1313-1850 (2015)
2. Sgurev, V., Drangajov, S.: Intelligent control of flows with risks on a network. In: Angelov, P. et al. (eds.) Proceedings of the 7th IEEE International Conference Intelligent Systems—IS'14, 24–26 Sept 2014, Warsaw, Poland. ISSN 2194-5357, ISSN 2194-5365 (electronic), ISBN 978-3-319-11309-8, ISBN 978-3-319-11310-4 (eBook). https://doi.org/10.1007/978-3-319-11310-4. Tools, Architectures, Systems, Applications, vol. 2. Advances in Intelligent Systems and Computing, vol. 323, pp. 27–35. Springer International Publishing, Switzerland (2014)
3. Christofides, N.: Graph Theory: An Algorithmic Approach. London, Academic Press (1986)
4. Sgurev, V.: Network Flows with General Constraints. Publishing House of the Bulgarian Academy of Sciences, Sofia (1991). (in Bulgarian)
5. Genova, K., Kirilov, L., Guliashki, V., Staykov, B., Vatov, D.: A prototype of a web-based decision support, system for building models and solving optimization and decision making problems. In: Rachev, B., Smrikarov, A. (eds.) Proceedings of XII International Conference on Computer Systems and Technologies—CompSysTech'11, Wien, Austria, 16–17 June 2011, ACM ICPS vol. 578, pp. 167–172. http://weboptim.iinf.bas.bg/. ACM ISBN 978-1-4503-0917-2
6. Sgurev, V., Drangajov, S.: Risk estimation and stochastic control of innovation processes. Cybern. Inf. Technol. (CIT) 14(1), 3–10. https://doi.org/10.2478/cait-2014-0012. Print ISSN 1311-9702; Online ISSN 1314-4081
7. Jensen, P.A., Barnes, J.W.: Network Flow Programming. Wiley, New York (1980)
8. Ford, L.R., Fulkerson, D.R.: Maximal flow through a network. Can. J. Math. 8, 399–404 (1956)

9. Hallikas, J., Virolainen, V.M., Tuominen, M.: Risk analysis and assessment in network environments: a dyadic case study. Int. J. Prod. Econ. **78**(1), 45–55. Elsevier (2002)
10. Sgurev, V., Drangajov, S.: Optimal control of technological innovations in postconflict regions by minimization of admissible network investments and risks, TECIS 2015. In: 16th IFAC Conference on Technology, Culture and International Stability, 24–27 Sept 2015, Sozopol, Bulgaria, p. 41
11. Phillips, D., Garcia-Diaz, A.: Fundamentals of Network Analysis, p. 474. Prentice Hall, Englewood Cliffs, NJ (1981). https://doi.org/10.1002/net.3230120210
12. Sgurev, V., Drangajov, S.: Two stage method for network flow control of resources and the risks related to them. In: Proceedings of the International Conference of Automatics and Informatics 2016, Bulgaria, Sofia, 4–5 Oct 2016. John Atanasoff Society of Automatics and Informatics, pp. 143–149. ISSN 1313-1850, CD: ISSN 1313_1869 © UAI

Decreasing Influence of the Error Due to Acquired Inhomogeneity of Sensors by the Means of Artificial Intelligence

Vladimir Jotsov, Orest Kochan and Su Jun

Abstract Sensors usually have the biggest error among all components in a measuring system. The paper considers the application of the methods of artificial intelligence, in particular, neural networks and data science applications for sensor data processing. The main attention is focused on improvement of measurement accuracy when using inaccurate sensors. The abovementioned methods illustrated on the example of improvement of measurement accuracy of the most widely used temperature sensor—the thermocouple. Neural networks and other methods of artificial intelligence ensure the improvement of accuracy of temperature measurements by an order of magnitude. However, they require considerable complication in both hardware and software.

1 Introduction

Scientific progress in the field of sensor production led to a considerable diversification of produced sensors. This is especially truly in the areas such as chemistry, biology, and security applications. However, the methods of sensor signal processing are rather complicated. Often even the obtaining of the final reading cannot be implemented as a functional transformation.

Even more complex problems occur in uncertainty estimation of sensor information as well as in error correction and compensation. In particular, during the estimation of errors of multisensors, the problem appears of required number of

V. Jotsov (✉)
University of Library Studies and Information Technologies,
69A, Ship. Prohod Blvd., 1574 Sofia, Bulgaria
e-mail: bgimcssmc@gmail.com

O. Kochan
Department of Information-Measuring Engineering,
Lviv Polytechnic National University, Bandera Str., 12, Lviv 79013, Ukraine

S. Jun
School of Computer Science, Hubei University of Technology, Wuhan, China

© Springer International Publishing AG, part of Springer Nature 2018
V. Sgurev et al. (eds.), *Practical Issues of Intelligent Innovations*,
Studies in Systems, Decision and Control 140,
https://doi.org/10.1007/978-3-319-78437-3_5

calibration points to identify their conversion characteristic (CC). In some cases, the value of one measured physical quantity influences the CC of a multisensor for the other physical quantity. For such multisensors this problem is very topical. In many cases, in particular, when using multisensors, it is reasonable to consider the pattern recognition of a sensor output [1–3]. In this case, it is reasonable to use data science.

In [4] 23 pattern recognition methods are considered based on the analysis from machine learning, statistical approaches, neural networks and carried out qualitative and/or quantitative comparisons. In particular, the considered methods are as follows: soft independent modeling of class analogy, linear discriminant analysis, nearest neighbors, artificial and probabilistic neural networks, learning vector quantization and different modifications. It should be noted there exist many suitable modifications of artificial neural networks (ANNs) [5] such as the multilayer perceptron with error backpropagation, recurrent ANNs, self-organizing Kohonen maps, Elman networks, time delay ANNs, adaptive resonance ANNs, and so on.

The results of the comparison show the method of ANNs as the slowest and the most complex part from the training point of view. However, it should be noted that in general the stages of training/recognition in neural networks can be separated [6]. The training stage can be carried out on a processing machine with higher performance. Therefore, the complex and long training of ANNs is not so significant drawback. For instance, in [6] a simple 8-bit microcontroller had been used for error correction caused by the sensor drift CC.

The other drawback of ANNs is the necessity of the choice of their optimal design. Usually the choice of the ANN deign is based on previous experiments and the study of recommendations how to choose a proper design [5]. On the other hand, ANNs almost always provide the required accuracy, which is important for sensor applications, since their error usually dominates among the component errors in measurement systems. In many cases different sensor errors determine the accumulated measurement error. Sometimes sensors are operated in special conditions such as nuclear power plants or gas turbines where the price of inaccurate measurements is very high [7, 8]. Therefore, the improvement of the sensor data processing is a topical task. The modern trend is to use data science aiming to improve sensor applications since the lack of accuracy can cause a catastrophe.

In particular, the proposal is to use an ANN for error correction in one parameter sensor from [9]. At the same time, the proposed approach considers such sensors as multisensors. Physical quantities that have the most influence on the sensor error are defined as input quantities for the multisensor. However, there is no need to measure them. The method considered in [9] ensures the correction of the temperature influence on photodiode used to measure the intensity of the ultraviolet radiation. In this case, the sensitivity of the photodiode considerably increases with temperature growth: up to 8% within the range of operating temperatures. In addition, it is reasonable to carry out the error correction caused by the temperature of the photodiode with respect to the temperature of its chip but not from its environment. Therefore, the proposal is to use the photodiode as a temperature sensor after measuring its short circuit current. It is proposed to carry out experiments just in 9 points out of 49 combinations of radiation flow/temperature to identify an individual reference function of the photodiode. The prognosis using additional ANNs is carried out in the rest 40 points. The proposed ANNs are trained using studies of 30 real reference functions of existing sensors. This approach decreases the error of the photodiode by approximately 35 times.

This chapter considers the usage of the simplest ANN such as the three-layer perceptron for decreasing the error of thermocouples (TCs), one of the simplest and widely used temperature sensors. Let's consider the TC errors to achieve this.

2 The TC Errors

Thermocouples are the most popular sensors of temperature used in measuring praxis for temperatures in the range 600–2500 °C [10], in spite of their drawbacks. The most important among them is their error [11], which is often too big for many cases in industry and science [12–14]. The total error of thermocouples is much greater than that of their measurement channels [15].

The main errors proper for thermocouples are as follows:

1. Considerable initial deviation of their reference function (RF) from the nominal one. The likely deviation in CC for the most popular type of thermocouples (type K) may reach 5.5 °C at 600 °C and 8 °C at 1100 °C [11].
2. Considerable drift of CC during operation at high temperatures, that is a change of CC in time. It may vary in the range of 0.5 °C to 10°C for the mentioned above type K of thermocouples during 1000 h at 600 °C or at 1100 °C respectively [16, 17].
3. Thermoelectric inhomogeneity of thermocouple legs acquired during operation at high temperatures [17–21]. Error due to inhomogeneity may reach 10 °C when measuring 1100 °C during 1000 h using type K thermocouples, or even more in some cases [19].

Correction of the first error can be carried out using the data obtained during the primary calibration in a few points of the measuring range [22]. Electromotive force (emf) developed by a TC is mostly determined by the chemical composition of its legs, therefore, it is a continuous function of temperature (without discontinuities and jumps). Praxis shows it is enough to carry out a calibration in 3 or 4 points for the measuring range from 100 to 600 °C or in 5 points for the measuring range from 100 to 1000 °C. It is reasonable to use the approximating ANN for error correction.

3 Correcting the Error Due to Drift of a TC CC

The correction of the second error is much more complicated. This is because of the considerable influence of the operating temperature and conditions on the rate of drift of the TC CC [16] as well as relatively big individual deviations of the error due to drift from the average trend. Therefore, the correction of the error due to drift based on the studies of the TCs of the same kind in similar conditions improves measurement accuracy very little. In addition, it has quite low metrological reliability. According to [23] error correction based on a calibration in a laboratory is inadmissible because of acquired during operation thermoelectric inhomogeneity (the third error).

There are known methods for TC calibration in situ.

In this case the third error, that is the error due to the acquired during operation inhomogeneity of TC legs [18–20], influences the error of calibration to much lesser extent.

The error of the thermocouple in situ can be determined as follows:

1. Using a reference thermocouple [16]. To use this method, it is necessary to have an additional channel into which the reference thermocouple is plugged. The drawback of the method is the usage of the reference thermocouple in operation conditions as well as the necessity to use an appropriate measuring channel.
2. Using a temperature fixed point cell [24, 25]. This is a hermetic capsule containing either a pure metal or an alloy with the accurately known temperature of phase transition (either melting or solidification). During steady heating (cooling) on the plot of developed Seebeck emf versus temperature appears a temperature plateau which makes possible to identify the moment of the phase transition. Therefore, the temperature of the measuring junction equals the phase transition temperature.
3. Without any reference instruments using the thermocouple with controlled profile of temperature field (TCPTF) [26]. In this case, the temperature field is purposefully stabilized during operation. It makes possible to eliminate the error due to acquired thermoelectric inhomogeneity of the legs of the main thermocouple (the thermocouple which measures the temperature of an object). The TCPTF can be designed in the way that it is possible to set such a temperature field, in which the area of the temperature gradient covers the sections of the legs, which usually operate at ambient temperature. The CC of these sections does not drift. Therefore, if the temperature of the measuring junction, while changing the temperature field remains constant, the difference between developed Seebeck emfs in two temperature fields equals the error due to drift of the conversion characteristic of the main thermocouple.

The abovementioned methods of the error determination make it possible to construct an individual mathematical model of CC drift of a thermocouple to make a forecast and correct the error. ANNs have very good prognostic abilities, therefore, it is reasonable to use them to make a forecast of error of thermocouples between determinations of their errors [27]. However, the traditional forecast using ANN requires a relatively big sample [5]—not less than 30 trials. It should be noted, that the sensitivity of thermocouples (10–70 $\mu V/°C$) [10] is not high and the speed of CC drift is low. That is why, because of a high level of random errors, it is not reasonable to carry out a calibration more often than once a week. Therefore, a reliable forecast of CC drift is possible after a half of a year of operation. It is possible to shorten this time using an approximating and a forecasting ANNs or the method of historical data integration [27].

The method of combining the approximating and the forecasting ANNs consists of the following steps:

1. Determining the function of the error of a TC CC due to drift in 5–6 points R1–Rn (see Fig. 1);
2. Training the approximating ANN using the data from point 1. The ANN can be, for instance, a three-layer perceptron consisting of one input distributive neuron, two or three neurons of the hidden layer, whose activation function is a sigmoid and one linear output neuron;
3. Generating 30–50 in-between points GI1-GIk, which correspond to the function of drift of a TC CC when the real determination of the TC error is not carried out, with the approximating ANN;
4. Training of the forecasting ANN using the data from the point 4;
5. Forecasting the error due to drift of the TC CC during operation.

The method makes it possible to reduce the number of required calibrations to construct individual mathematical models of a TC CC error due to drift in 5–6 times. Modeling showed that the error of the forecast of the error of a TC CC due to drift does not exceed 0.5–1.2 °C while the inter-calibration time had been increased in 4 times.

The method of historical data integration about the drift of a TC CC [27] is even more effective. It is necessary to have investigations of drift, that is, results of calibrations in certain moments, for 30–50 TCs in the same or similar operating conditions to apply the method. The application of the method can be split into a few stages. The first stage involves training of the integrating ANN. To achieve this, the following steps are required:

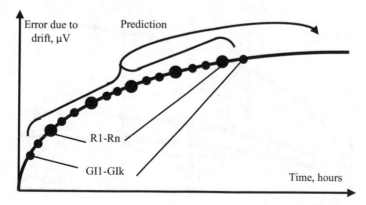

Fig. 1 The method of combining the approximating and the forecasting ANNs

1. to choose a TC among the like TCs whose drift of CC is temporarily assumed to be forecasted. Then the result of its first calibration is chosen;
2. to rank the calibration results of the other TCs in an ascending (descending order) with respect to the deviation of their errors from that of the chosen in point 1 TC;
3. to construct a vector of the training set for the integrating ANN. The error of the chosen in point 1 TC as well as the ranked errors of the like TCs from point 2 (the order itself is not important but it should always be the same for the further operations) are sent to the inputs of the ANN. The output of the ANN is compared to the second calibration of the TC chosen in point 1;
4. to repeat the points 1–3 for another TC to go over all TCs for the first calibration;
5. to train the integrating ANN using the set constructed in points 1–4.

A trained integrating ANN can make a forecast based on the results of the first calibration (at the time T1) of the thermocouple TCp, whose drift is forecasted, for the next result of a calibration at the time T2 (Fig. 2). To perform it, the first result of the TCp calibration should be taken into consideration in point 1, instead of the chosen TC that belongs to the group of the studied like TCs. In point 2 the result of the calibration of the thermocouple TCn, which deviates the most from the thermocouple whose error due to drift is forecasted, is eliminated.

The other integrating ANNs for another calibration times are trained in the same way. The results of calibrations of the thermocouple TCp, whose drift is forecasted, are the forecasted results of its calibrations at previous times for these ANNs. The study [27] showed that one neuron with the sigmoid activation function can be used as an integrating ANN.

Thus, the set of integrating ANN for all calibration times makes it possible to forecast the results of calibration of the thermocouple during a study of the like TCs. Then these results of calibrations can be used as input data for the approximating ANN, which, in turns, creates the training set for the forecasting ANN. The

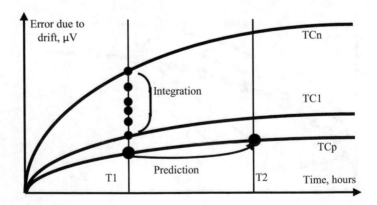

Fig. 2 The method of historical data integration

modeling shows that the error of a forecast of the error of TC CC due to drift using the proposed stage, which consists of the integrating, approximating and forecasting ANNs, in the worst case does not exceed 30% from the peak value of the TC CC error due to drift when the inter-calibration time is increased in 12 times.

4 The Method for Improving the Accuracy of Temperature Measurements Using Thermocouples with Inhomogeneous Legs

The error due to acquired inhomogeneity stems from changes in thermocouple legs at high temperatures in time. These changes are caused by the effect of chemical and physical processes (such as oxidation, diffusion, recrystallization etc.) in legs. If a thermocouple is split into imaginative sections, each section operates at its own temperature [28]. That is the reason why, during prolonged operation, CC of each section changes in time in accordance with its particular operating temperature. If the temperature field along the thermocouple legs changes, the temperatures of each section change correspondingly. Therefore, the error of each section is not constant, so the total error of a thermocouple varies even when the temperatures of the measuring and the reference junctions remain constant. This is an appearance of the error due to thermoelectric inhomogeneity when the developed emf depends on the distribution of temperature along thermocouple legs. Many researchers consider that it is the main reason of thermocouple error [18, 19, 21]. Sometimes the high thermocouple error after extended use is thought to be inevitable and impossible to correct [23]. However, recent studies have discovered new methods for decreasing the influence of the error due to acquired inhomogeneity on measurements of temperature using thermocouples.

According to mentioned [28], the error due to acquired inhomogeneity appears when the temperature field along the thermocouple legs is changeable. The main idea of the method proposed in [26] is as follows: in a stable temperature field along the legs, error due to inhomogeneity cannot manifest itself. Thus, the temperature field along the legs of the main thermocouple has to be purposefully and compulsory stabilized. Additional temperature control subsystems have been used to achieve this stabilization. These control subsystems are located along the legs of the main thermocouple. Each such subsystem consists of both an additional heater and an additional thermocouple.

The proposed sensor is called a thermocouple with controlled profile of temperature field (TCPTF) [26]. Its pattern is given in Fig. 3. In such a sensor the main thermocouple (MTC) measures the temperature of an object. The temperature control subsystems located along the electrodes of the MTC to develop and maintain a temperature field independent from that of the measured object. Each section of temperature field control system consists of a heater (H1...Hn) and a thermocouple (TC1...TCn). All the reference junctions of all the thermocouples are

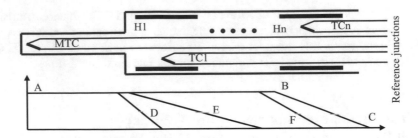

Fig. 3 The pattern of the TCPTF

wired to measure and control system (absent in Fig. 3). The temperature field marked ABC in the bottom part of Fig. 3, is set and maintained by the heaters H1… Hn of the temperature control system. Thus, even if the temperature field of the object changes within the borders from D to F, the profile of the temperature field ABC remains constant and this fact does not allow the error due to acquired inhomogeneity of the MTC to appear.

The emf developed by the MTC has to be measured with the highest possible accuracy using the measure and control system of the TCPTF. An emf of the section sensors TC1…TCn can be measured with usual accuracy because the effect of their error can be neglected while their error due to acquired inhomogeneity is negligibly small because they operate in stabilized temperature field. Because of this reason, it is reasonable to use a multipoint measurement system. The measuring channel is the same for all the sensors and each thermocouple is wired to it in a sequential order by a switchboard [29].

5 The Furnace to Control the Temperature Field Profile Control of the TCPTF

The sensor described in [26], whose pattern is given in Fig. 3, is designed as an integral sensor. However, considerable drawbacks are prone to such a design. These drawbacks are as follows:

1. The sensor design is complex.
2. It is not compatible with existing standard types of thermocouples.
3. It is difficult to replace or substitute the MTC when necessary (when significant degradation of the legs occurs).
4. Lack in standard metrological procedures for this, completely new type of sensor.

These drawbacks obstruct widespread use of the TCPTF. However, the proposed method of decreasing the impact of changes in the temperature field on the temperature measurement error is not prone to these drawbacks. The drawbacks appear because the TCPTF was designed in [26] as an integral sensor. To get rid of these drawbacks the design of the TCPTF should be divided into two parts: (i) the sensor itself that is a standard thermocouple being produced in industry should be used as the MTC; (ii) the means for stabilizing the temperature field along the MTC. In this case none of the abovementioned drawbacks can appear.

There is a pattern of the proposed furnace for the TCPTF [30] on Fig. 4. It consists of two tubular shells, one of which is outer (1) and the second one is inner (2). The outer diameter of a standard thermocouple corresponds to the diameter of the inner shell. The plugs (3) and (4) are welded to the shells (1) and (2). There is an insulation (ceramic) cylinder between the shell (2) and the heater (6). The heater (6) may be either wound or sprayed on the cylinder (5). The lead-outs (7) of the heater (6) extend through the insulating sleeve (8). To ensure better adhesion of the insulating sleeve (8) its shape is mushroom-like. The space between the shells (1) and (2) is filled with thermal insulation (9).

The main requirement for the design of the furnace is that the number of lead-outs wired to the left-hand side plug (see Fig. 4; the right-hand side plug is at high temperature) of the furnace must be minimal. To achieve it, the heaters H1... Hn and their corresponding thermocouples TC1...TCn are combined. As it was mentioned above, the heaters (6) are either wound or sprayed with the material of one thermocouple leg (e.g. chromel). The lead-outs (10) of the heaters (6) are made of the material of the other thermocouple leg (e.g. alumel). Therefore, each heater together with its lead-out forms a thermocouple to measure the temperature of an edge of the corresponding zone. The temperatures in the centers of zones are approximated as average temperatures of the temperatures at the edges of their neighbor zones. This superposition of the lead-outs and the heaters allows reducing the required number of lead-outs. The total number of lead-outs if the heaters H1... Hn and the thermocouples TC1...TCn are separate for n zones equals $4n$. The superposition of the lead-outs and the heaters described above requires the total number of lead-outs equal $n+1$.

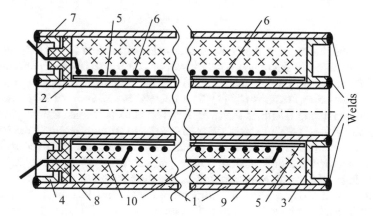

Fig. 4 The design of the proposed furnace

6   Computation of the Furnace Heaters' Power

To roughly estimate the maximum required power of heaters H1...Hn we assume that the zone of the TCPTF is an infinite cylinder with stable internal temperature t_1 (this temperature is by the heaters) and external temperature t_2 (environment temperature) on its surfaces. Thus, according to [31], it follows:

$$Q = -F\lambda \frac{dt}{dr} = -2\pi r l \lambda \frac{dt}{dr}, \tag{1}$$

where Q is the heat flux through the lateral area of the cylinder; F is the lateral area of the cylinder per unit length l; r is the radius of the cylinder; λ is the thermal conductivity of the cylinder material. The cylinder for the TCPTF is made of a layer of thermal insulation covering the heaters H1...Hn. The relation between the temperatures t_1 and t_2 is as follows

$$t_1 - t_2 = \Delta t, \tag{2}$$

where Δt is the maximum temperature change, which the heaters of the TCPTF must ensure.

According to [31], the solution of the differential Eq. (1) is dependence of the heat flux per unit length (the ratio of Q to l) through the lateral area of the cylinder on the ratio of internal to external diameters of the cylinder, that can be written as follows

$$\frac{Q}{l} = \frac{\pi(t_1 - t_2)}{\frac{1}{2\lambda} ln \frac{d_1}{d_2}} = \frac{2\pi\lambda\Delta t}{ln \frac{d_1}{d_2}}, \tag{3}$$

where d_1 and d_2 are internal and external diameters of the cylinder respectively.

From Fig. 4 it can be derived that

$$d_1 - d_2 = 2h, \tag{4}$$

where h is the thickness of the layer of thermal insulation of the TCPTF heaters.

The calculations of the required heat flow per unit length (the ratio of Q to l) for the TCPTF heaters to provide the required changes in temperature along the MTC legs are based on expressions (2)–(4). The heaters' power is equal to the heat flux on the corresponding length of the cylinder. The plots of the maximum power per unit length of the TCPTF heaters (W) versus the maximum required temperature change (°C) for different thickness of the insulation layer (m) are given in Fig. 5.

Figure 5 shows that high heating power (greater than 1 kW/m) is required only when the layer of thermal insulation is thin, and the changes of the temperature field along the thermocouple legs are comparable with the measured temperature. The TCPTF is not designed to operate in such conditions. It is designed to operate in massive thermal units, for instance, furnaces of thermal power plants, Changes of

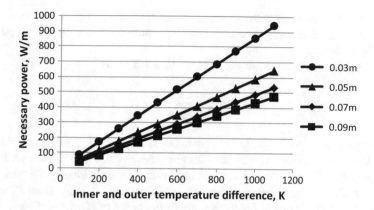

Fig. 5 Dependencies of the maximum power per unit length of TCPTF heaters on the maximum required temperature changes and thickness of the thermal insulation layer

temperature field in such objects are caused by changes in generated power, season and outdoor weather. These reasons cannot change the profile of the temperature field of the thermal unit more than 50 °C. Additionally, the length of the heater section should not exceed 0.1 m to ensure effective control of the temperature field profile. Thus, the maximum power developed of the TCPTF heater depending on the thickness of the thermal insulation, should not exceed 20–50 W. It is obvious that these figures of power are quite acceptable and do not distort operation conditions of massive thermal units.

7 Computation of the Parameters of the TCPTF Heaters

It is required to fit the parameters of the heaters (such as material and diameter of wires) to the required power (i.e. supply voltage and maximum current) in order to design the TCPTF. The resistance of a conductor R is as follows

$$R = \rho \frac{L}{S},\tag{5}$$

where ρ is the specific resistance of the conductor, for heaters, chromel should be used, whose $\rho \approx 0.65 \, \Omega \, mm^2/m$ [32]; L is the length of wire; S is the cross-sectional area of the conductor.

In case with the TCPTF, it can be written that

$$L \approx n \cdot \pi \cdot d_2, S = \pi \cdot d_P^2/4,\tag{6}$$

where n is the number of winds of the heater, n can be defined according to the formula $n = l/(d_P + w)$; d_P is the diameter of heater conductor; l is the length of one zone of the heater; w is the thickness of the conductor insulation (for continuously wound heater) or the distance between the conductors (for heater wound with compulsory step).

Taking into account (5) we can rewrite (6) as follows

$$R = 4\rho \frac{l \cdot d_2}{d_P^2 (d_P + w)}, \tag{7}$$

For the prototype in [26], we have: heaters made of chromel $d_2 = 0.02$ m, $l = 0.05$ m, $d_P = 0.5$ mm, $w = 1$ mm. Plugging these values into (7), we get $R \approx 7\,\Omega$. The thickness of the prototype thermal insulation is approximately 10 mm. In this case, according to (3), to ensure temperature variation in the range from 30 to 50 °C we need a maximum power for the heaters in the range of 20–50 W. Consequently, the supply voltage of the heaters is from 18 to 19 V and the maximum current of the

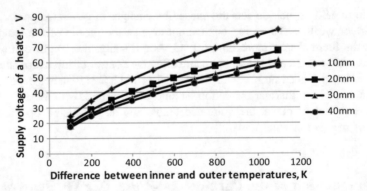

Fig. 6 The dependencies of the required supply voltage of a heater on thickness of the thermal insulation layer

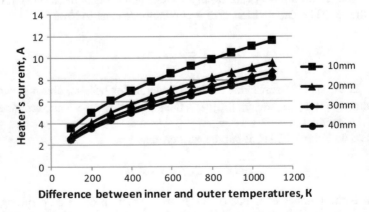

Fig. 7 The dependencies of required heater current on thickness of the thermal insulation layer

heater is from 2.6 to 2.7 A. These values are acceptable for practical use. Considerable reduction of the heater voltage and current (hence its power) can be got by increasing the thickness of the layer of thermal insulation. The corresponding plots are given in Figs. 6 and 7. It is shown in the section Conclusions: there exist no specific requirements for the materials from the furnace, for their processing and the hermeticity of the furnace.

8 Controlling the Temperature Field of the TCPTF with a System of Linear Equations

The significant thermal connection between the zones of the TCPTF leads to generation of temperature waves and instability of the temperature field along the TCPTF legs (this phenomenon can be also called self-excitation). Control of temperature fields in such multi-zone objects with mutual thermal influence requires complex methods of control and thus significant computational resources. However, there is a quite simple method for controlling of temperature field [33]. It was implemented on an 8-bit microcontroller in [33].

The assumptions below are the base of the method:

(1) There is a linear dependence between changes of temperature and power.
(2) Instead of the summation of heat flows, the summation of temperatures is carried out for computation of changes of the heater power.

These assumptions contradict to thermodynamics, however, the method works properly to compute approximately necessary changes of heaters when deviations of the temperature field from the preset one are relatively small. To calculate necessary power changes for establishing a preset temperature field the system of linear equations has to be solved. To solve it, any known algorithm can be used. Gaussian elimination is used in [33]:

$$\begin{cases} \Delta T_1 = k_{11}\Delta P_1 + k_{12}\Delta P_2 + \cdots + k_{1n}\Delta P_n \\ \cdots\cdots\cdots\cdots\cdots\cdots\cdots\cdots\cdots\cdots\cdots\cdots\cdots\cdots\cdots, \\ \Delta T_n = k_{n1}\Delta P_1 + k_{n2}\Delta P_2 + \cdots + k_{nn}\Delta P_n \end{cases} \quad (8)$$

where $\Delta T_1 \ldots \Delta T_n$ are the required changes in the temperature of each zone heater from 1 to n, $\Delta P_1 \ldots \Delta P_n$ are the required changes of the corresponding heater power $k_{11} \ldots k_{nn,}$ and the coefficients describe the effect of individual heaters on the temperature of other zones.

Each step of control (iteration) consists of: (i) solution to the system (8), (ii) changes in powers of heaters (iii) and the delay until the thermal transition ends. The next step of control (change of heaters powers), based on the new measurements of temperatures in all zones, is executed after a lag for three time constants of the TCPTF furnace.

Due to the method of control, the system turns into almost open one. The system is robust to self-excitation generation of thermal waves. Therefore, the profile of the temperature field always approaches to the preset one. However, the method is effective only when changes of the temperature field are both slow and quite small. The coefficients $k_{11} \ldots k_{nn}$ are defined from experiments for each operating temperature and zone [33].

The main drawback of the method is a long time required to establish the temperature field. Adjustment lasts for a few iterations even in the case of small deviations of the temperature field from the preset one.

9   Using a Neural Network for Temperature Field Profile Control

It makes sense to apply means of artificial intelligence to ease application of the abovementioned method of temperature control as well as to lessen the time required to establish the preset profile of the temperature field. ANN can take into consideration nonlinear nature of dependencies of dependencies of heat capacity, heat conductivity and heat transfer on temperature, as in [30, 34]. In this case, the required changes of powers for each iteration are defined with considerably higher accuracy. This method computes the necessary changes of heater powers better than the method from Chap. 6. This improvement of accuracy of the changes of the powers of the heaters leads to a decrease of the required iterations for establishing the preset profile of the temperature field. The main idea of the improvement is to apply the ANN to control the temperature field along the MTC legs. The ANN is trained under various temperatures of operation and temperature fields of the TCPTF. The number of inputs of the ANN is the same as a number of heaters H1... Hn. This number of inputs is two times bigger than the number of outputs to take into consideration the required temperature changes as well as corresponding temperatures of each zone.

The design of the proposed system to control the temperature field of the MTC is given in Fig. 8. It contains the TCPTF (in Fig. 8 it is given as a set of sensors (S) and corresponding heaters (H)), a multichannel measurement subsystem (MMS), furnace temperature set unit (TSU), subtract unit (SU), neural network (NN) and control unit (CU). The TCPTF is placed into the thermal unit, which is controlled by the thermal unit controller (TUC). The MMS consists of: (i) switchboard (SW), (ii) analog to digital converter (ADC), (iii) microcontroller (MC) and (iv) interface unit (IF).

The information about the temperature differences for each zone from the SU and their current temperatures from the MMS is sent to the NN. Therefore, the ANN gathers data about the current temperatures of each zone and the preset change of it. This information allows the ANN to take into consideration the dependencies of heat capacity and heat conductivity on temperature. That is why it

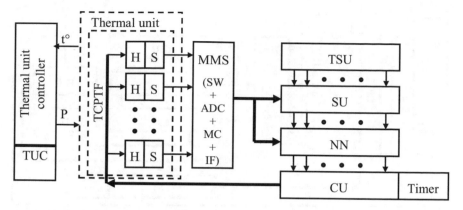

Fig. 8 A block diagram of the measure and control system in the mode of maintenance for the temperature field

is possible to calculate the required changes of powers of the heaters properly by the CU using the ANN. The CU stops any other actions until the end of the thermal transition process in all zones.

There are two the most widespread methods to train the ANN: (i) the one using the model of the object of control, and (ii) the one with parallel linking to the controller of the object (so called "teacher"). An accurate model of the controlled object, whose error is considerably less than the permissible error, is necessary for the first method. In general, it is not a simple task to design a model for a multi-zone object. It is necessary to carry out elaborate experiments to solve it. The method with a "teacher" needs regulators that have already provided appropriate control of the temperature field of the object. However, these regulators are able to solve the problem of the temperature field control without using an ANN.

The authors offer to train the ANN in situ. No model needed in this method [30, 34]. To apply the method, the structure of the control system has been changed while training the ANN according to the scheme in Fig. 9. The outputs of the ANN are wired to the training unit (TU) of the ANN and the inputs of the CU are wired to the power set unit (PSU).

During the training, the PSU gives a signal to the PSU to save the current temperatures of the zones and generates random power changes of the heaters (H) for the CU. These power changes may be either positive or negative. After the end of either heating or cooling the zones, the MMS measures all their temperatures and sends data to the SU as well as to the NN. The trained ANN generates the changes of powers for the heaters necessary to return the temperatures of all zones to the initial values. Corresponding changes are generated with the PSU. If the temperatures of the zones are not equal to the preset temperatures, the TU adjusts neuronal weights to approach the ANN closer to the changes generated with the PSU.

The PSU generates a test set of power changes at various temperatures to train the ANN from the experimental data. Each test set consists of 25–30 random test

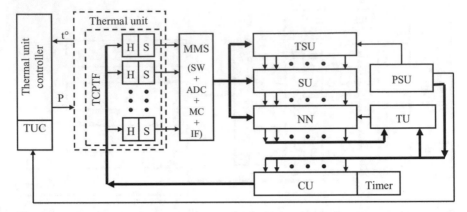

Fig. 9 A block diagram of the measure and control system during training of the neural network

power changes. If the ANN has not been trained properly using this set, one can continue training the ANN without changing the heater powers in reality. This method of training does not need any model of the measured object because the ANN has been trained during experimental studies.

The transition to new powers is carried out for all the heaters simultaneously. This transition is performed when all the transition processes (either heating or cooling) in all zones are finished completely (the same applies in [30, 34]). This explains why the control system can be considered an open one. The proposed method is not prone to self-excitation (auto-generation of thermal waves) [31, 34].

The computational complexity of this method is not greater than that of the method described in Chap. 6 and in [34]. The total amount of the computations performed by the ANN is approximately the same as the total amount of computations to solve the system of linear equations. However, the ANN collects and takes into consideration dependence of the mentioned above thermal parameters, such as thermal conductivity and heat capacity, of the TCPTF on temperature. Therefore, the ANN can take into consideration nonlinearity of the TCPTF as an object of control. On the other hand, the use of ANNs needs quite a big amount of experimental researchers which is not possible to decrease because the three-layer perceptron (the kind of ANNs used in the TCPTF) is not suitable to train on small samples.

That is why it is reasonable to use additionally other means of artificial intelligence to decrease laboriousness of ANN training.

10 Essential Features of Proposed Puzzle Methods

Logical Data Science applications are rapidly developing in various fields. Their success relies on the usage of enhanced modeling tools, which allow a deep interconnected (evolutionary) statistical and logical processing of the accumulated data or knowledge patterns. One of the featured characteristics of Data Science applications is their transparent (open-system), and data-/knowledge-driven character. This is the reason to avoid traditional algorithmic descriptions in this section, except few essential definition and application examples.

Contemporary data-driven approaches frequently use ANNs, for example, RBFs or spiking neurons. Here a different by nature data-driven method is considered. It is logically oriented and uses different novel types of constraints to narrow the field of possible solutions, and to infer new conclusions. The advantages rely on better modeling possibilities.

In this chapter a Data Science research named Puzzle method is considered aiming to improve the quality of the deep learning analysis, and of knowledge acquisition. For more details, please refer to the book [35]. The intelligent software plays more and more important role in contemporary machine learning systems and in their realizations: from industry to security systems. The Puzzle Method have been elaborated aiming to eliminate the well-known drawbacks of every ANN application: the learning experiment is very long, and expensive, and every ANN copies well-trained parts, but is almost useless in realistic, unknown conditions. Since the eighties the industry is controlled by flexible systems, and the pure algorithmic type of control couldn't assure the desired results. At first, neuro-fuzzy systems had been introduced. At the contemporary stage, data-driven and knowledge-driven methods are mainly used. Suggested innovations serve the more effective application of Data Science, Advanced Analytics, (Deep) data/web mining and/or collective evolutionary components. The latter should be used to combine logical and statistical results in one system, as considered in [35].

In industry, many innovations use traditional design approaches like TRIZ [36], brainstorming methods, etc. Unlike them, the Puzzle method is functioning in both manual and autonomous modes to improve the existing technical solutions and to elaborate the novel ones. In any case, TRIZ, ant other logical- or statistical-based methods are easily combinable with the below considered Puzzle method standards.

The research on traditional, syntactically produced puzzle methods revealed that their algorithmic complexity is rather high and this does not allow automatic processing of large enough sudokus, crosswords or puzzles. In this case, an emphasis is on the fact that the studied machine-based procedure takes the words from a predefined set using a random principle where only the length of the word is of importance. When statistical applications are used alone they are not so effective. On the contrary to the quoted syntactical realizations, a semantic variant of Puzzle method is considered below. It is based on the logical analysis of the existing data/ knowledge interconnections. Three types of relations have been used aiming at narrowing the set of possible solutions to the problem. An analogy with living creatures could be applied: the system using the Puzzle standard could focus attention on most significant things.

Briefly, every Puzzle method aims to discover new or hidden knowledge by connecting the unknown, the sought solutions with previous experience accumulated in knowledge bases (KBs). Let the constraints of the considered problem form a curve in space as depicted in Fig. 10. The main goal of the Puzzle method is to

Fig. 10 Binding, crossword
and classical sets of
constraints

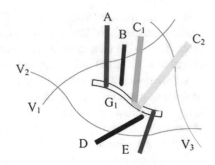

reduce the multidimensional search space for the solution. For this purpose, several limitations have been used. In [35] the study is focused on the case of using ontologies instead of a set of nonlinear constraints.

Furthermore, in certain cases the research of process dynamics aimed at reducing the field allows the derivation of a new knowledge in the form of rules as discussed in next subsections. This inference process and the usage of different constraints gives way to significant simplification of formal and evidence material in any research or lecture. Also it attracts attention on important details and simultaneously increases the activity of learners. Showing the process of connecting known to unknown improves the understanding and retention of the presented material.

For example, in this book chapter we use thermocouples in a furnace, and both mentioned decisions restrict the search space of the presented research via two lines, (nonlinear) curves. In this way a lot of unnecessary [re]search is avoided. It is also possible to inspect a case when the constraint is defined in the form of a surface but, as a result, a more general solution is obtained where a special interest is provoked by the boundary case of the crossing two or more surfaces. When the common case is inspected in details then in the majority of cases the problem is reduced to the exploration of the lines instead of complicated curves obtained as a result of crossing surfaces. Below the usage of constraints is investigated by using curves of first or higher orders.

Type V_i constraints of a linear/nonlinear form are the classic case borrowed from constraint satisfaction methods. The B&D binding constraints, dot-type variants are also possible, does not intercept the unknown goal but are located close to it, they reveal a kind of a *neighboring area* around the target. The A&C&E constraints are named the crossword constraints because their interception with G_1 gives a part of the searched goal where Ci form larger interconnected areas in G_1. The usage of the considered set of Ci-s is much more effective than of the classical set of V_i-s. The pointing and ontological constraints are not depicted because of similar images.

Through *binding* constraints (B or D on Fig. 10), it is convenient to implement many non-classical causal relations of the type "A is linked to B but the connection between them is non-classical, not implicative/provable". The same could be presented through heuristics, which is ineffective, hence not recommendable. It must be pointed out that through this type of constraints the location of the searched solutions is fixed in a way that is best combined with **fuzzy** methods.

Crossword constraints [A, C_1, C_2, E] offer new ways of assessment (outlook) for the searched unknown solutions on the basis of the accumulated so far knowledge.

The next group of newly designed constraints used in Puzzle methods is named **pointing constraints**. It can be classified as an exotic type of binding constraints, but constructs a new constraint type because of additions of new features: when at least one element of pointing constraints is reached, it shows the direction to the solution [to goal G_1]. In comparison, binding constraints show that the goal is somewhere near, let's say, with higher probability, but doesn't point the exact direction. For example, the mineral pyrite signs that the gold *may be* somewhere nearby. Other examples: if an infrared or thermal camera shows an image of a higher temperature zone inside the furnace, we know **where** to find this zone [*pointing constrains*], but if we are told that this zone should be explored in our project, we couldn't guarantee that it still exists and is usable. The latter analogy concerns just the *binding* constraint example.

On the other hand, many well-known statistical methods could effectively work with pointing/binding constraints and can improve the effectiveness of crossword constraint applications by calculating the probability value and the preferable direction of the goal location. Same for possibility, fuzzy or uncertain or even heuristic values.

The following example demonstrates how the proposed search process can be reduced using ontologies. The search space is presented on Fig. 11 where statistical data have been depicted on good practice methods how to use the furnace. The depicted data/knowledge sets had been generalized about different regions

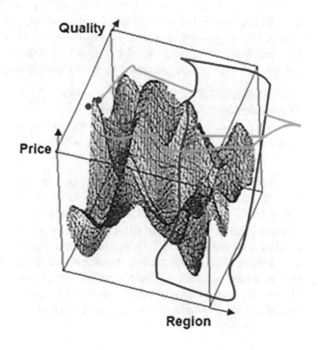

Fig. 11 Temperature zone application example

depending on their parameters and quality. Let the goal is: what zone is acceptable to our project. Let the right vertical, blue-colored (dark-grey), subset of feasible solutions is chosen: 'excluding the non-Eurasian experience'. Then the space of feasible solutions is to the left of the separating surface which is depicted on the figure in the rightmost corner.

In Fig. 11 another horizontal surface, depicted in the high corner in green (light-gray color), is shown. It delimits the search space of the solutions. In our case, it means 'frequently used applications'. It is accepted that in the scope there is no clear distinction related to the presented criteria. Hence, the search of the feasible solutions is nonlinear, of high dimensionality, and practically it cannot be solved using traditional methods. Nevertheless, by applying ontologies the problem is solvable via the proposed Puzzle method. There are two red dots (two brighter gray color dots) depicted in the left corner on the same Fig. 11. Each of them also represents a kind of constraint but of another type, the **binding constraint**. In this case its semantics is following: it doesn't point to a solution, but it resides close to the searched solution. We can say, that the two (or more) binding dots can fix up the solution surface. This good practice is recommendable as the most effective one.

10.1   Different Types of Binding Constraints

The following section discusses the introduction of following three types of binding constraints.

Every knowledge used in the Puzzle method can be presented as a part of information (atoms, linked by different relations). Usually, these relations have been obtained by some logical processing of information like structuring, extracting the meaning from information blocks or other knowledge processing. In this terminology, one rule can be presented in the following way:

The conjunctions of antecedents are A_1, A_2 .... A_z.
The conclusion/consequent is marked as B.
Let all z number of conjunctions are proved to be true/confirmed, then B is true, whereby the goal/problem of checking whether B is true has been solved (see Fig. 12).

When significance of the atoms has not been pointed out, as in the case shown in Fig. 12, then the significance of the conjunctions is considered equal ($1/z$). The above bows show that each atom A (conjunctival) has its individual *significance relation, mainly invisible and often informal, in proving the conclusion*. It is a real number from the interval [0, 1]. But in the common case, some atoms are of greater significance while others are less important, and the quoted numbers depend on the current situation. Part of these invisible links between the atoms of the type atom-conclusion or other parts of knowledge can be defeated (**torn**) in different conditions, for example, when additional information appears. During this process

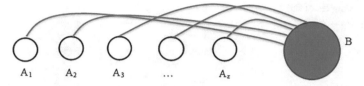

Fig. 12 Informal relations inside a rule

called *defeat*, the truthfulness of each atom can be changed, and its significance reduced even to zero in proving the conclusion/goal. Also, the whole rule could be changed completely, for example, by replacing it with another one.

The considered process of defeat is started by special knowledge forms named *exceptions to rules*, knowledge of the type E(C, Ap), (cf. Figure 13) where a prerequisite for the defeat is the argument C of the exception, which must be true in order to start this process of defeat. Therefore, both arguments of the exception enter a causal relation, which is not implicative, non-classical one. In life, we use so many similar relations and most of them are difficult to be formalized because they are not included in the classical formal logic.

The accumulation of such knowledge atoms, the compound with different classical and non-classical relations allows us to use new opportunities for realization of goals from the set.

For example, let a goal X had been proved via classical or non-classical means as shown in the diagram from Fig. 14 and let X and Y contain a non-classical causal relation, for example, X defeats Y; or for example, let X, and Y be statistically linked variables. In this case, the fact that there is a large volume of information linked to X, as shown in the figure, leads to imposing informal constraints over the choice of the condition of Y, regardless of the fact that in a formal sense, X and Y are not linked. Informally, proving X non-monotonically leads to the solution of the goal Y. In other analogous situations, solving X does not lead to proving Y. However, they are linked through the following non-classical relation: proving X shows that we are *close* to the solution of Y, the more rules and facts prove the truthfulness of X, the higher the confidence/sureness/certainty or belief in the hypothesis that Y is true. The described process will be called *binding*, as interpreted in Fig. 14.

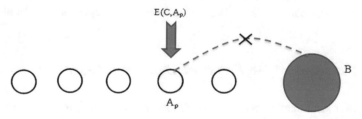

Fig. 13 The defeat of implicative connections

Fig. 14 Examples of the
binding process scheme

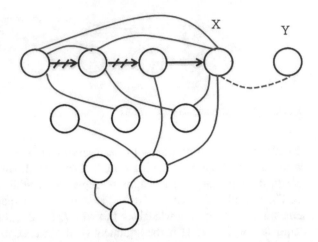

Very often using methods like fuzzy logic helps the binding process. For example, this can occur in situations where indefinite notions are used or notions that change their meaning depending on the situation and the context.

The following rule interprets the meaning of one of frequent binding processes:

If N is a high quality furnace, but N is a too expensive model, then N is not the best choice variant.

In the described example, there exist some notions that require further clarification: high quality, price, good buying practice, etc. Here it is important to clarify what is the desired type of the furnace, in what conditions it will be explored, and so on. It must be considered the fact that high quality is quite a subjective judgment and it can be greatly reduced due to a number of subjective reasons. In situations where there is doubt about the judgment of the quality, it is better to use binding relations, including the non-referable parts of the used rules. In this case, what is left is the relation that if something is a cheap option, it could be bought. Here the link is not an implicative one, we don't need a garbage in our offices. Obviously being a high quality device isn't enough for our decision making. Additional knowledge should be drawn here, increasing the confidence that if it is suitable for us, the other important facts can be drawn from the technical description, or from other sources. In the example, in fact, one or more atoms are defeated from the antecedent of the rule.

Defeating the link of the second conjunction with the conclusion actually removes the implicative link on the whole rule, and is replaced by a binding.

10.2 Essentials of the Proposed Defeasible Reasoning

One of the most effective ways of knowledge exploration is to add a new knowledge by appending/defeating the old one in a new situation. It is shown that almost every relation in a rule using classical implications may be defeated. Eleven main defeasible schemes have been introduced, and they describe the introductory notes to the proposed defeasible inference research. The represented formulas are the most usable ones.

Let the unity of classes K be comprised by the subsets $S_1, S_2, \ldots$ where $S \in K$. Every subset of type S includes elements $x_{s:1}, x_{s:2}, \ldots$, that form a new model. The original set S is related to one of the classes $S_i \in K$. The final result from the analysis of S is idenitified with one of the classes S_i from K. The output is an answer of the type $K_s = (T; F;?)$ with three truth values: "true", "false" and "uncertainty" (?, don't know). In the case with an answer $K_s = ?$ or $K_s = F$, the set S may be identified with more than a single known class $S_{i1}, S_{i2}, \ldots$ ($i1 \neq i2\ldots$). The answer $K_s = T$ is obtained if and only if the examined class S coincides with S_i.

Amongst the classes S_i, there exists an interdependence of the type "ancestor—successor", (is-a, e.g. S_i—an ancestor of S_{i1}). Thus, it is possible to form simple types of semantic nets—with one type of a relation. It is necessary to note that the elements $x_{si1;1}, x_{si1;2}, \ldots$ produce the differences between the class S_{i1} and the other successors of the common ancestor S_i. All differences that appear in the comparison process of S_{i1} with other classes that are not direct successors of S_i are determined after the application of the inheritance (heredity) mechanism.

The conclusion (response) $K_s = T$ is formed when the corresponding conjunction terms for all ancestors S_i and also for S_{i1}, are of the following form: $A_1' \wedge A_2' \wedge \ldots A_n'$, where A_k is x_k or $\neg x_k$; A_k' may coincide with A_k or it may include (using a disjunction) A_k and analogical terms for other variables.

Let's assume that a set of rules of Horn type describes the desired domain:

$$B \leftarrow \bigwedge_{i \in I} A_i. \tag{9}$$

During the usage of a truth/binary or even Boolean-like logic in the quoted rules, if at least a single variable A_i is different from 'true', then the truth of B is indefinite, i.e. the result is "?". In the case when the corresponding exclusion from the conjunction (9) of the rule is based on the inclusion of a term with any $A_k (k \in I)$ then the inference procedure changes. In the case when the exclusion $E(C, A_k)$ and C is true and A_k is false, then the right side of rule B may be true *as an exception*. The extended inference models with exclusions are introduced and generalized in following forms.

$$\frac{B \leftarrow \bigwedge_{i=1}^{z} A_i, C, E(C, A_k), \neg A_k \leftarrow C}{B \leftarrow A_1 \wedge A_2 \wedge \ldots A_{k-1} \wedge \neg A_k \wedge \ldots A_z}, \tag{10}$$

$$\frac{C, B \leftarrow \bigwedge_{i=1}^{z} A_i, E(C, A_k)}{B \leftarrow A_1 \wedge \ldots A_{k-1} \wedge A_{k+1} \wedge \wedge \ldots A_z} \tag{11}$$

$$\frac{C, B \leftarrow \bigwedge_{i=1}^{z} A_i, E(C, A_k)}{B \leftarrow A_1 \wedge \ldots A_{k-1} \wedge (A_k \vee C) \wedge A_{k+1} \ldots A_z} \tag{12}$$

It follows from formulas (10)–(12) that the proposed exclusions are a form of special, non-implicative causal relations. The interpretation of the formula (10) is based on the following facts. If an exclusion $E(C, A_k)$ exists and is related to one of the rules with conclusion B, and A_k is its effect, then the truth value of the conjunct A_k must be replaced to the opposite value "$\neg A_k$". In the case when C is not 'true' the corresponding replacement is blocked. Further applications of Modus Ponens (MP) rules lead to the fact that the defeat of the informal relation between B and $\neg A_k$ leads to a formal logical contradiction.

Therefore, the formation of exclusions of the type $E(C, A_k)$ may lead to a contradictory result provoked by an incompleteness in domain descriptions. In case when C is true the exclusion $E(C, A_k)$ includes this meaning in the conjunct A_k to defeat the meaning of the last conclusions. The result is that A_k is replaced by C because the test of its meaning does not influence the output. In the case when C is true, the corresponding conjunct A_k is directly replaced by C, this leads to various schemes like the one from (12).

Rules of type (9) are united in following systems where inter-relations between atoms from different rules are not explored:

$$\begin{cases} B_1 \leftarrow \bigwedge_{i \in I} A_{1i}. \\ B_2 \leftarrow \bigwedge_{j \in I} A_{2j}. \\ \ldots \end{cases} \tag{13}$$

In the general case the causal-effective relation may be applied using non-classical operations of successions that are denoted '<-' in the paper and B_i may be presented as combinations of sophisticated logical relations like in formula (14).

The usage of exclusions (10) up to (12) could be also applied in systems (13) or (14); in the general case it reflects the interrelations between different parts of the causal-effective relations influenced by a new information, an exclusion that is attached to one or another group of relations. The new information may influence mutual relations between the atoms, elements of the rule (9) or of systems (13); (14).

$$\begin{cases} B_1 < - \bigwedge_{i \in I} A_{1i}. \\ B_2 < - \bigwedge_{j \in I} A_{2j}. \\ \ldots \end{cases} \tag{14}$$

As a result, the relations of a causal type are defeated or they are strengthened due to an additional information contained in the exclusions or shown by them. The rest of the paper does not include versions (13) and (14) because in the majority of practical applications it is sufficient to confine ourselves to rules type (9), in this way the algorithmic complexity of the used combination of methods is significantly lowered. The presented exclusions by their nature are an enlarged version of defeasible inferences that are widely used in the intelligent systems. The considered research differs from classical inference schemes using exclusions because it is possible not only to exclude the atom A_k that is contained in and tailored to the rule, but also because a totally new formula may be included in the rule, e.g. $\neg A_k$ in formula (10) or an interrelation between A_k and C in (12). The research also includes versions of formulas using a weak non-classical negation $\sim$, versions with exclusions of implications influenced by exclusions, and so on:

$$\frac{B \leftarrow \overset{z}{\underset{i=1}{\wedge}} A_i C, E(C, A_k), \sim A_k \leftarrow C}{B \leftarrow A_1 \wedge A_2 \wedge \ldots A_{k-1} \wedge \sim A_k \wedge A_{k1} \wedge A_{k+1} \wedge \ldots A_z}, \tag{15}$$

$$\frac{C, B \leftarrow \overset{z}{\underset{i=1}{\wedge}} A_i, E(C, A_k)}{A_1 \wedge \ldots A_{k-1} \wedge A_{k+1} \wedge \ldots A_z}, \tag{16}$$

$$\frac{C, B \leftarrow \overset{z}{\underset{i=1}{\wedge}} A_i, E(C, A_k)}{A_1 \wedge \ldots A_{k-1} \wedge A_k \wedge A_{k+1} \wedge \ldots A_z}, \tag{17}$$

where A_{k1} is an additional condition for transitions from $\sim A_k$ to $\neg A_k$. The investigation includes schemes with multi-argument exclusions $E(C, A_k, A_l, \ldots A_s)$ that lead to the simultaneous change of several parts of the rule. The introduced defeasible method leads to three basic results. The truth of parts of the rule is altered when influenced by the exclusion if the conditions for activation of the exclusion are enabled; formulas are included in or excluded out of the rule or the rule itself is defeated as it is shown in (15) or (17). The results from the research led to a large number of inference versions with exclusions; a part of them is included in the bibliography list.

Hence a generalized concept of defeating is introduced that is based on the following facts. Object scope modeling is a dynamic process. In the act of scope-field completion by the system the old relations between separate parts of the knowledge and/or between different knowledge pieces may be eliminated, changed or their effect may be *redirected*. This is accomplished by the influence of the new knowledge that completes or corrects the primary existing knowledge or its inter-relations. The described processes are formalized in the following way.

We did a research of the situations that appear after the addition of the new knowledge to the existing experience/knowledge base, and we divided them into 11 basic groups. Let P be the part of the new knowledge that influences one or more formulas like in (9) up to (12).

I. P 'nullifies' A_k: it breaks its relation to the conclusion B. As a result of this process, the significance of A_k has a meaning of 0, and no matter whether it is true or false, the truth value of the conclusion does not change:

$$\frac{B \leftarrow \overset{z}{\underset{i=1}{\wedge}} A_i, P}{B \leftarrow A_1 \wedge A_2 \wedge \ldots A_{k-1} \wedge A_{k+1} \wedge \ldots A_z, \neg \left(B \leftarrow \overset{z}{\underset{i=1}{\wedge}} A_i \right),} \tag{18}$$

where in difference with above mentioned defeasible schemes, the original rule format existing before the appearance of P becomes false.

II. This is an extreme version of the situation from group I when all the atoms in the antecedent are defeated. After that the rule (9) turns into a fact: B ←.

$$\frac{B \leftarrow \overset{z}{\underset{i=1}{\wedge}} A_i, P}{B, \neg \left(B \leftarrow \overset{z}{\underset{i=1}{\wedge}} A_i \right)}, \tag{19}$$

III. P changes the truth value of A_k from true to false or v.v.

$$\frac{B \leftarrow \overset{z}{\underset{i=1}{\wedge}} A_i, P}{B \leftarrow A_1 \wedge A_2 \wedge \ldots A_{k-1} \wedge \neg A_k \wedge \ldots A_z, \neg \left(B \leftarrow \overset{z}{\underset{i=1}{\wedge}} A_i \right),} \tag{20}$$

IV. P defeats the existing significance of A_k and increases it to 1. The significance factors of other parts of the antecedent from (9) duly drops down to 0. Independently of the way (conjunctively or disjunctively connected atoms), all of them are related to A_k and in this situation they are defeated by the antecedent of the rule (9):

$$\frac{B \leftarrow \overset{z}{\underset{i=1}{\wedge}} A_i, P}{B \leftarrow A_k, \neg \left(B \leftarrow \overset{z}{\underset{i=1}{\wedge}} A_i \right)}, \tag{21}$$

V. P redirects all relations between the rule and the other knowledge in the domain.

The causal relations are not exhausted by the classical implication and the next consideration will show that even by formal means it is possible to present different forms of non-classical causal relations. Let following two rules are introduced:

$$R_1: B \leftarrow A; \quad R_2: N \leftarrow M. \tag{22}$$

Let both rules initially be related to the object X. Let also after the appearance of the new set of conclusions P R_1 be related to Y and R_2 to the former object X. In this case, the first rule is preserved but its effect is redirected to another object.

In medical domain it is known that by nature a disease is provoked either by a virus or by bacteria. However, let us have a case when a patient manifests simultaneous symptoms of an illness both from a virus and from bacteria. The sequent investigation (P) shows that the symptoms of a virus-provoked disease are related to the patient's throat and that the bacterial symptoms are related to the patient's lungs. The redirecting of the conclusion that contradicts the rule from the example and the discovery of the second disease gives the solution to the problem. It is possible to redirect whole rules by analogy to the above presented case.

VI. P breaks or amplifies the relation between the considered rule and other knowledge in the domain.

The difference with the previous situation V now is either the elimination of the existing relations or the addition of new relations between the existing rules. The very rules are preserved with that.

For example, every chess-player must have a good physical condition so that he/she can present himself/herself well in the tournaments. If however the 'examined' chess-player is a computer program—this is the effect from the new information P—then the already said does not at all concern this program.

VII. P influences the conclusion from one or from a group of rules: from R_1: B $\leftarrow$ A into R'_1: $B^* \leftarrow$ A. In this way, the old conclusion P is defeated or it is replaced by the new one B^*.

$$\frac{B \leftarrow \overset{z}{\underset{i=1}{\Lambda}} A_i, P}{B^* \leftarrow \overset{z}{\underset{i=1}{\Lambda}} A_i, \neg\left(B \leftarrow \overset{z}{\underset{i=1}{\Lambda}} A_i\right)}, \tag{23}$$

VIII. The appearance of P changes the antecedent of the examined rule (9). It imports a new atom on the place of A_k, before or after the chosen one A_k. In the last two cases, the new atom is conjunctively or disjunctively related to A_k, e.g.

$$\frac{B \leftarrow \overset{z}{\underset{i=1}{\Lambda}} A_i, N(P, J)}{B \leftarrow A_1 \wedge A_2 \wedge \ldots A_{k-1} \wedge J \wedge \ldots A_z, \neg\left(B \leftarrow \overset{z}{\underset{i=1}{\Lambda}} A_i\right)}, \tag{24}$$

This situation can be named by specifying the antecedent as a result from the new information P.

IX. R_1 is replaced by R_2 and influenced by P:

$$R_1: B \leftarrow A; \quad R_2: N \leftarrow Q. \tag{25}$$

The difference from the previous situation here is in the complete replacement of the rule provoked by P in accordance with the a priori defined concepts.

$$\frac{B \leftarrow \overset{z}{\underset{i=1}{\Lambda}} A_i, P}{N \leftarrow Q, \neg \left(B \leftarrow \overset{z}{\underset{i=1}{\Lambda}} A_i \right)}, \tag{26}$$

X. We have a situation from I to IX, but the obtained consequences may not be used in the antecedents of the other rules. The reasons for similar constraints are different, e.g. limiting an insecure information along long chains of rules, etc.

XI. The atoms of the investigated rule (9) remain the same, but some of the logical operations are changed affected by P, e.g.

$$\frac{B \leftarrow A_1 \wedge A_2 \wedge \ldots A_{k-1} \wedge \neg A_k \wedge \ldots A_z, N(P, J)}{B \leftarrow A_1 \wedge A_2 \wedge \ldots A_{k-1} \wedge \sim A_k \wedge \ldots A_z, \neg \left(B \leftarrow \overset{z}{\underset{i=1}{\Lambda}} A_i \right)}, \tag{27}$$

A characteristic example of a similar situation is the transformation of the strong classical negation '¬' into a weak paraconsistent negation '~'.

Let's discuss the following illustrative example. In principle, it is not possible that the same person is a teacher and a student at the same time. Let's denote that 'John is a teacher' by the variable Q. Then it will not be an error if we denote that 'John is a student' by ¬Q.

This is valid in the prevailing number of situations, but it is inapplicable in condition (P) that John is a student in one subject in one school, but he is a teacher in another subject in other e.g. sports school or at the same school. After the advent of the new information P, it is not possible to say that 'John is a student' is ¬Q; now it is correct to use the weak negation and ~Q will lead to a contradiction only in the cases when definite conditions hold—in the example the conditions are the subject for teaching and also the location for teaching.

The described situations from I to XI present a research for the influence of the new information P over different parts and relations between existing conclusions. In the majority of the discussed situations, contemporary mechanisms for defeasible inference may be used. The difference is just in the fact that P totally changes the situation existing a priori. However, if P replaces the literal in the first argument in the exception $E(C, A_k)$, then the exclusion does not change the action progress for the existing up to the advent of P things and it adds a new scheme to them that is activated if and only if P is false. The present chapter does not contain formal descriptions of all the possible variants of the situations from I to XI because the number of their combinations in all the possible applications is too large.

In the end, the whole rule may be replaced by totally different rule(s) using one or more exclusions type $E(C, A_k, A_l, \dots A_s)$. The total change isn't the recommended one because the proposed MAS environment isn't revolutionary but evolutionary one. Total changes should be carefully checked if acceptable.

The proposed defeasible reasoning could be widely applied in Puzzle method applications.

10.3 On the Usage of Crossword Constraints

This section discusses the introduction of one type of constraint called *crossword*.

Let's assume that KB contains deeply structured knowledge, for example, ontology on the problem. In this case, if other knowledge is discovered, and is related to the ontology, but badly structured, for example, written in another language, or written via pictograms, incomprehensible to us, or information lacking in text or noised or encoded, i.e. in situations when the meaning of just a fragment of the information is comprehensible, and out of which only part of the information can be drawn, and if this part of the information is new, in a sense that it complements the ontology, then the mentioned new knowledge could added to the ontology regardless of the missing parts of the knowledge. In this case, invisible rule/clause relations are being used, similar to the one in Fig. 12 and linking the non-structured knowledge with the structured one, ontology, as well as with the atoms of the non-structured knowledge through the used crossword constraints. Thus, new knowledge extraction is carried out. In this case most often searching in the constrained area of options is avoided, and the solution is obtained straightforward and non-alternatively. For example, we assume as a known fact that there was an incident in a factory in Bulgaria. The event relates to the ontology, furnace' with relations to a broken furnace, fire, and Bulgaria. Here as another column we can add to the ontology *when* the incident took place, etc. Then, if someone watches a TV show on the Albanian television and does not understand the local language, they can nevertheless determine that the news is about the well-known incident in Bulgaria for example by the written time of the incident and so on, but when in this context of unfamiliar words appears the known word BULGARIA..., our ontology of the incident is complemented with a new relation: a fire has been exterminated during the incident. In this case, the context of the event described in an unfamiliar language characterizes the unknown, and only the word Bulgaria and some other specific information units are understood/known or link the unknown to the known from us. Regardless of the lack of information, by using the crossword constraints here we will obtain an extension of the existing ontology solution: we have obtained an improved with new knowledge ontology.

By introducing new constraints or, sometimes, ontologies our goal is to show that it is possible to use causal links different from implications and that they help us search the goals in a more effective manner.

In the tutoring/industrial education case, the usage of crossword constraints as depicted in Fig. 10 also helps us to reveal the dynamics of the problem-solving process. There the resolution process is sometimes more important than the decision/solution itself. The usage of the classic cases, Vi-constraints one by one doesn't always form the necessary closed area. Revealing type-Vi and other constraints one by one helps represent the resolution process in its dynamics and hence make a deep inference to the problem. Even in the worst case, when the accumulated knowledge is incomplete for the problem resolution, the dynamics of what is represented analogically to Fig. 10 will show what should be the resolution. The binding/crossword constraints are the form of non-implicative rule-based relations. Their application effects are discussed in the next sections.

10.4   Essentials of the Usage of Pointing-Type Constraints

The pointing constraints are best illustrated by different gradients. Their essential principle is depicted in Fig. 15. The pointing nature of the phenomenon is that the best way to find optimal solutions (min/max value) from the depicted point x, result

Fig. 15 Interpretations to the specifics of gradient methods

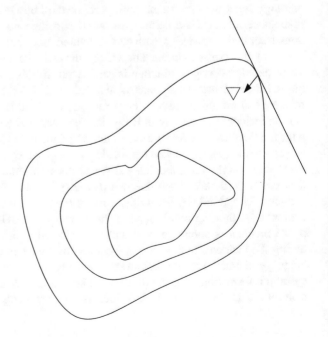

of an intersection of the line and the isosurface line is via (anti-)gradient. Nabla symbol is used to depict the gradient direction which is perpendicular to the depicted line.

Every gradient illustrates a restricted case of the proposed pointing constraints. The universal, pointing cases show not only the local/global optimal direction, but also the informal notion what is the most preferable (where is it?), what should be executed first using the precedence operators, or, for example, what is the simplest or most logical solution. Beautiful, effective, perspective, and other informal notions cause no problems for pointing constraint operations. In this sense, some forms of pointing constraints are similar to ontological constraints. This question should be further investigated.

A set of [mutually] connected pointing constraints may form an analog to an algorithm describing the resolution to the goal.

One of the actual forms for pointing constraints uses the logical analysis of the situation. There exist many situations, when there is only one solution to the problem, or their number is fixed and no problem to be found. Then just the a priori given pointing constraints set helps us to satisfy the goals concerning them. For example, let the goal is to find the direction of the temperature drift in the thermocouple. As an answer, we use one formula to show the temperature drift in conditions of increasing or stable temperatures, and another formula for decreasing cases. Let as a result the drift is in a positive direction in the first case and in a negative direction in the second case. Both directions consider just one of the cases of above mentioned pointing constraints.

11 Puzzle Method Applications for Temperature Field Profile Control

The Puzzle method rules may seemingly look like too formal and rather ineffective ones, but the following application example should show their strength and flexibility. During the training of the ANN (RBF-ANN or other type) using the thermocouple data, many drawbacks may appear. One of the frequent issues is the unstable temperature field in the furnace during the training process. In these conditions, the learning process is prone to errors. It is resolved by the introduction of above described set of constraints V_i. As depicted on Fig. 10, only one constraint does not resolve the problem, but may diminish it. For the substantial restriction of the research area we used a closed area of constraints V_1, V_2 and V_3. In the proposed research, these constraints are 'materialized' in the form of a thermo-proof shell around the thermocouple. After that, the ANN training process will be improved and stable.

The ANN learning process concerns exact examination of the proposed data patterns. In the considered furnace, such stable conditions can't be fulfilled. For example, signals may be temporary lost, the thermocouple could be changed with a new one, etc. It is impossible to start a machine learning procedures each time when any changes appear. The described Puzzle principles reinforce the adjustment of the learning process in the following manner. The following three constraint conditions are recommended for modelling and usage during the analysis.

- The pattern/form of the investigated object;
- Location;
- Main model parameters.

It is very important to make a closed constrained area as shown in Fig. 16. Otherwise, the method applications will produce much lower effects.

For example, if at the learning stage the temperature drift line is inclined at 3° for temperatures below 900 °C and at 2.5° for temperatures above 900 °C, then the change of the incline is an additional factor to secure that the temperature is near to reach the 900 °C threshold. This revealed fact isn't a complex and very innovative one, but when it is used in an autonomous mode, its usage is very productive.

If the form of the temperature drift pattern is analyzed and memorized, and later, the same form appears in the image of the temperature evaluation, it is no problem to calculate the error deviation because of the temperature drift. Analogically, the measurement error will reveal its specific form.

When two pieces of knowledge have been put into a constrained area, they can easily form new information pieces in the form of rule-based knowledge. First of all, this process should be schematically described.

These natural language or formally described fragments of knowledge can be presented in plain text for learners, but further it is shown that by the Puzzle method this is executed much more naturally and better. For example, Fig. 17 shows another example of Puzzle method application revealing certain processes and relationships between objects. Concentrating the attention in the closed area of Fig. 17 and narrowing the set of analyzed elements supports the revealing of new implicative relation M-N referred to in Fig. 18. Informal (sometimes the formal one) causal relations are elaborated that are associated with the implication 'from M it follows that N'.

Fig. 16 The main constraint elements

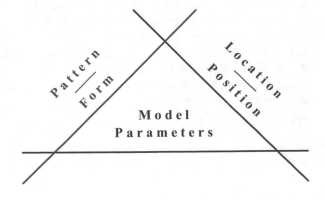

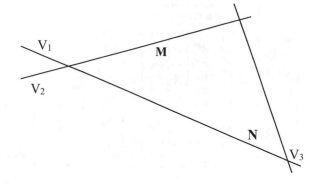

Fig. 17 M and N are unrelated but belong to same confined area

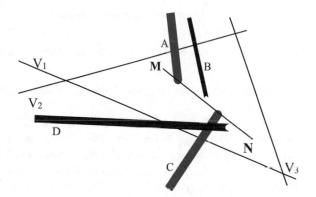

Fig. 18 Detection of the rule connection between M and N

In Fig. 18 the system of three different constraint sets helps detect a causal relationship between M and N, with no evident correlation between them in Fig. 15. The connection occurs in imposing additional constraints in Fig. 18. The process is dynamic in nature and it is almost impossible to be properly explained only with words. Here the set of the pictures has a role of the replacement of the intuition process.

The process of finding new rules using Puzzle methods gives the trained people the necessary deep knowledge. This form of knowledge can be sometimes represented in the form of ontologies (in any words, an information transferred by sense), but for the sake of simplicity this case isn't described.

The image (it could be a media) from Figs. 17 to 18 is convenient and efficient to visualize, transform and use different dynamic processes through sense-based explanations. Examples and figures can lead to the conclusion that ontologies are introduced, used and dynamically changed applying the Puzzle methods. Hence they can be improved using Puzzle methods.

The block diagram from Fig. 19 concerns the machine learning cycle. The Puzzle method processing is included into the learning cycle. Its inclusion allows deep knowledge processing and more, novel solutions to intelligent measurements.

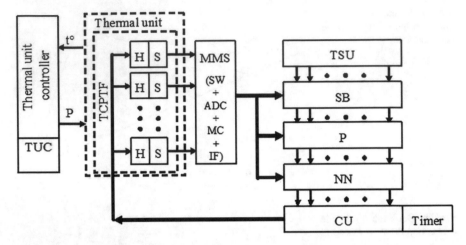

Fig. 19 Block diagram for the machine learning scheme using proposed Puzzle methods (P)

The Puzzle block (B) may control the ANN or function alone. All variants have been explored.

The proposed experiment includes the investigation of temperature drift parameters. The diagram from Fig. 20 contains a schematic (numbers excluded) view of temperature values given by the investigated thermocouple in same conditions during different machine learning sessions. The maximal values are included in the set depicted by the red curve. The dotted line shows the real, average values of the temperature conditions in the moment. In most cases, it is impossible to obtain the exact temperature values at any time, in any point of the oven because of different thermal currents in the furnace, too many factors, influencing the temperature processes, and so on. Another example of the same case is the thermocouple 'inertia' error in the case of a temperature spike or burst or during significant temperature changes. Its influence is measured just after the change in the dotted line at T = 900 °C. In the regional pattern just above T = 900 °C a situation

Fig. 20 Main temperature drift parameters

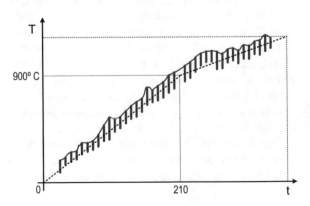

occurred where the minimum measured value is significantly higher than the real one.

Of course, the above quoted ANN applications may be applied aiming to catch the above-mentioned problems, but this is an expensive and labor-consuming process. The proposed applications make analogical processes easier and faster.

A situation in Fig. 21 with the same process parameters used in slightly different conditions includes an addition of a higher temperature value in accordance with the set depicted in a form of a dashed line at the bottom of Fig. 21. It is used as a gradient of additional temperature changes. Considering the data from Fig. 20, the system is capable to autonomously calculate the direction of the temperature changes, signed by 'D in a circle' from Fig. 21: this is one of the discussed pointing types of constraints applied in the experiment. Also, without any expert help, the system calculates the most possible location of 'B in a circle', the binding constraint showing that in this region the temperature shifting point is located. When $t > 170$, and not $t > 210$ from Fig. 20, another 'D in a circle' is applied showing the change/fluctuation in the main temperature direction. All three constraints have been calculated using classical statistical methods, and obviously all three of them are interconnected: using the first direction D_1, we go to the point B* where the direction shifts to D_2.

The exact calculation of the coordinates of the binding constraint 'B in a circle', and of its range, and of both pointing constraints 'D in a circle' depends on concrete task parameters. To date, best solutions have been obtained by using different clustering algorithms. Actually, the direction of D may be calculated via the formula with arguments—mean values of multiple temperature measures at the same time. They are depicted as vertical set of lines on Figs. 20 or 22.

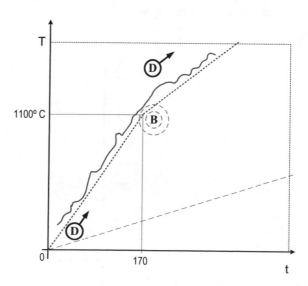

Fig. 21 Recognition of temperature drift parameters

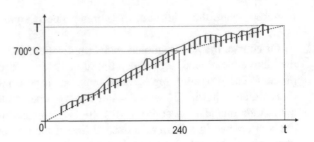

Fig. 22 Recognition of stretched patterns similar to the learned ones

The next experiment considers the same situation with an analogical pattern, where the curve is a stretched variant of the one from Fig. 20. In this case many more clustering algorithms could be effectively applied.

In some complicated cases with many temperature change points and/or only one part of the learned pattern applied, the experimental data are similar to ones from Figs. 23 or 24. In these cases, more crossword/binding/pointing constraints are welcome together with the classical linear, discrete or non-linear constraint satisfaction ones.

On the other hand, the strengths of the Puzzle methods are same in all above considered formalized experimental descriptions. Actually all described types of the constraints are included in one logically guided system of constraints. In Fig. 20 the pointing constraints set the way to the binding constraint. The position of the second pointing constraint depends on the position of the binding constraint. This interconnection strategy is helpful in case-based applications: how to calculate the unknown parameters using the well-known ones in the same or similar cases.

The proposed set of Puzzle methods is not a heuristic by nature. Frequently its motto is 'computing with words', in other conditions statistical formulas are applied. It provides a good basis to evolutionary applications of logical and statistical methods in same systems.

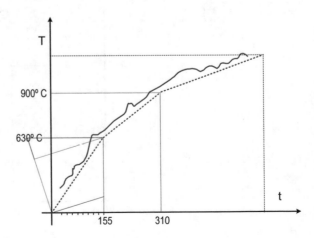

Fig. 23 Recognition of patterns similar to the explored ones

Fig. 24 Identification of similar and different parts of the patterns

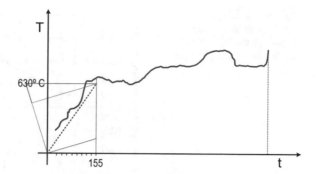

12 Experimental Studies of the Control Methods

The method of temperature control (described in Sect. 6) requires determination of the coefficients $k_{11} \ldots k_{nn}$. To determine these coefficients it is required to create the system of n^2 linear equations (like (8)) and solve it for $k_{11} \ldots k_{nn}$. It is not a complex problem to solve the system like this. However, the problem is in creating the system of linear equations, since the set of experimental data should consist of n^2 profiles of temperature field. Therefore, the creation of the system of linear equations is very laborious. The second drawback of this method is low sensitivity and accuracy because large number of different temperature field, which leads to small differences between these temperature fields. That is why the measurement errors as well as possible incomplete thermal transient processes will affect the determination of the coefficients $k_{11} \ldots k_{nn}$.

To simplify coefficient determination the simplified technique has been suggested. Its idea is to run the experiment with only one heater turned on. Say it is the j-th heater. In this case, the system of Eq. (8) turns into a column. Having solved such a system it is possible to calculate the coefficients $k_{j1} \ldots k_{jn}$. The temperature field for each experiment can be given as a plot temperature versus number of zone, as in Fig. 25. The relative temperature is referred to as a ratio of the change of the temperature in a particular zone of the furnace developed by the j-th heater to the maximum change of the temperature developed by the heater in its zone. Coefficients $k_{j1} \ldots k_{jn}$ are to be calculated from the plots. In this technique the total amount of experiments is n, instead of n^2, where n is the number of zones and thus the number of heaters.

This technique stipulates the equality of powers for all heaters. However, in a real prototype of the TCPTF the heaters' powers are not equal. To take the inequalities of power into account between the individual zones, the heaters let us express the heater power changes $\Delta P_1 \ldots \Delta P_n$ in terms of the right hand side of (8). To take into consideration these differences of heaters let us express the power changes $\Delta P_1 \ldots \Delta P_n$ in terms of the right hand side of (8)

126   V. Jotsov et al.

Fig. 25 The relative
temperatures versus number
of zone of the TCPTF with
just one heater on (the fifth
one)

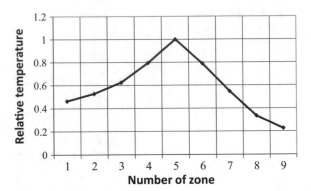

$$\Delta P_i = \Delta P_{MIN} \cdot k_i^P, \tag{28}$$

where k_i^P is the coefficient of excess. This coefficient describes the excess of the power of the i-th heater over the heater developing the least temperature change in its zone. The coefficients $k_1^P \ldots k_n^P$ can be computed as follows

$$k_i^P = \Delta T_i / \Delta T_{MIN}, \tag{29}$$

where ΔT_{MIN} is the minimum temperature change of a heater among the other heaters; ΔT_i is the temperature change with its i-th heater turned on.

The experimental study of establishing the temperature field using the method of linear Eq. (8) is given in Fig. 26. The units of axis x are the units of relative time. The unit of relative time corresponds to three time constants of the TCPTF. One unit of relative time is not greater than 0.5 h when the TCPTF compensates the change of a temperature field in large object operating in the vicinity of a sole temperature. As is clear from Fig. 24, the differences between the real and the preset temperatures of the zones of the TCPTF gradually decrease. It is evident from Fig. 20 that the temperatures of the TCPTF zones gradually approach to the preset values. However, it takes quite a long time to establish the preset profile of a temperature field. Even for a relatively small difference between the real temperature field of the TCPTF and the preset one (the maximum difference for a single zone does not exceed 2.5 °C in Fig. 26) 6 iterations needed.

Fig. 26 Establishment of the
temperature field profile using
the control method based on a
system of linear equations

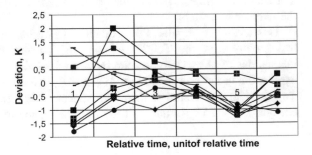

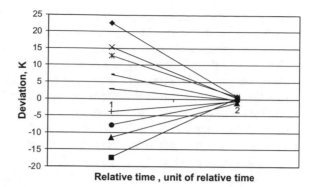

Fig. 27 Establishment of the temperature field profile using the control method based on the neural network

The experimental study of establishing the temperature field using the ANN method is given in Fig. 27. As seen from Fig. 27, the process of establishing the preset temperature field takes only one time constant, thus, it takes one iteration. In addition, the method works properly with larger (greater than 20 °C) differences between the real and the preset temperature fields.

13　Conclusions

The chapter researches and develops a method to reduce the error due to acquired thermoelectric inhomogeneity of thermocouple legs. The main idea of the method is to stabilize the temperature field along the main thermocouple that measures a temperature of an object. This stabilization of the temperature field significantly reduces the influence of the changes in the temperature field of the measured object on the emf developed by the thermocouple measuring its temperature. One of the problems is to create an appropriate method to control the heaters while stabilizing the temperature field profile. The problem occurs because of likely self-excitation. One of the possible and simple solutions to this problem is the proposed method of control based on a system of linear equation and on an ANN. It allows solving the problem in a simple way. The computational resources of an 8-bit microcontroller are enough to carry out calculations of the required power changes of the heaters. However, a personal computer is needed to adjust the control subsystem and train the ANN before the operation.

The main computer power had been used for the presented data science application processing. The Puzzle method is a good addition to ANN-based training approaches because it reveals, models and used a sense-type knowledge and such a system has some understanding of the situation, and not only a training element which is expensive-to-be-obtained and frequently useless in our constantly changing world. The theoretical study showed possibility of significant improvement in the accuracy of the temperature measurements using the proposed TCPTF. The error due to acquired inhomogeneity of the MTC legs is no greater than 0.13 °C (if the temperature field of the object changes to 20…25 °C along its legs). The total measurement error of the TCPTF is no greater than 1.4 °C [26]. To get such accuracy a reference thermocouple with the error no greater than 0.6 °C needed.

The requirement for accuracy and stability of the temperature field are not strict. This is because the influence of the error in temperature field maintenance can be neglected. We can presume that the dependence between the error due to inhomogeneity and the change of the temperature field is proportional. Therefore, a change of the temperature field reduced in 20–30 times leads to the proportional reduction of the error due to the inhomogeneity of the MTC.

That is why changes of the temperature field of the TCPTF due to instability of the additional thermocouples (6, 7 and 10 in Fig. 4) almost do not affect the error of temperature measurement (it is decreased in 20–30 times as well). In addition, it should be noted that the additional thermocouples work in stabilized temperature field and this stabilization damps possible instability of the temperature field on the measured object. Initial differences between the temperature field (developed by the furnace of the TCPTF) and the preset one minimally affect the error of temperature measurements because they are taken into consideration when calibrating the MTC using a reference thermocouple. It should be noted, that instability of the temperature field caused by inhomogeneity of the additional thermocouples with respect to the error of temperature measurement is infinitesimal of the third order. The change in the profile of the temperature field of the furnace due to its imperfection leads to the error due to residual inhomogeneity of the MTC. This error is infinitesimal of the second order. The error due to drift of the additional TCs with respect to imperfection of the furnace is small, so the error due to it in relation to the changes in the profile of the temperature field is infinitesimal of the second order. In addition, it should be noted that the additional thermocouples work in stabilized temperature field and this stabilization damps possible instability of the temperature field on the measured object. The latter makes the problem of the error due to imperfection of the additional thermocouples rather theoretical.

Experimental studies of the proposed sensor prototype proved the theoretical statements. The proposed sensor is neither very complex nor expensive. The accuracy of temperature measurements can be significantly improved when using the sensor instead of the standard thermocouples. It ensures high accuracy despite the fact that an inhomogeneous thermocouple is used.

References

1. Derde, M.P., Massart, D.L.: Supervised pattern recognition: the ideal method? Analytica Chimica Acta **191**, 1–16 (1986)
2. Turchenko, I., Osolinsky, O., Kochan, V., Sachenko, A., Tkachenko, R., Svyatnyy, V., Komar, M.: Approach to neural-based identification of multisensor conversion characteristic. In: Proceedings of IEEE International Workshop on Intelligent Data Acquisition and Advanced Computing Systems: Technology and Applications, 21–23 Sept 2009, Rende, Italy, pp. 27–31

3. Voschinin, A., Skibitski, N.: Interval calibration model of multisensor system. Computing **2** (2), 82–86 (2003)
4. Michie, D., Spiegelhalter, D.J., Taylor, C.C.: Machine Learning, Neural and Statistical Classification. Ellis Horwood, New York (1994)
5. Kroese, B.: An Introduction to Neural Networks. University of Amsterdam, Amsterdam (1996)
6. Turchenko, V., Kochan, V., Sachenko, A.: Estimation of computational complexity of sensor accuracy improvement algorithm based on neural networks. In: Dorffner, G., Bischof, H., Hornik, K. (eds.) Lecture Notes in Computing Science, vol. 2130, pp. 743–748. Springer (2001)
7. Scervini, M., Rae, C.: An improved nickel based mims thermocouple for high temperature gas turbine applications. J. Eng. Gas Turbines Power **135**, 091601-1–091601-6 (2013)
8. Scervini, M., Rae, C., Lindley, B.: Transmutation of thermocouples in thermal and fast nuclear reactors. In: 3rd International Conference on Advancements in Nuclear Instrumentation, Measurement Methods and their Applications (ANIMMA), Marseille, France, pp. 10–17 (2013)
9. Roshchupkin, O., Smid, R., Kochan, V., Sachenko, A.: Reducing the calibration points of multisensors. In: Proceedings of the 9th IEEE International Multi-conference on Systems, Signals and Devices (SSD'2012), Chemnitz, Germany, 20–23 March 2012, pp. 1–6. Digital Object Identifier. https://doi.org/10.1109/ssd.2012.6197987
10. Webster, J.: Measurement, instrumentation, and sensors handbook. CRCnetBase (1999). http://www.crcnetbase.com/isbn/9780415876179
11. International Electrotechnical Commission: Thermocouples. Part 2: Tolerances. International standard IEC 584-2, Geneve (1989)
12. Glowacz, A., Glowacz, A., Korohoda, P.: Recognition of monochrome thermal images of synchronous motor with the application of binarization and nearest mean classifier. Arch. Metall. Mater. **59**(1), 31–34 (2014)
13. Maruda, R.W., Krolczyk, G.M., Feldshtein, E., Pusavec, F., Szydlowski, M., Legutko, S., Sobczak-Kupiec, A.: A study on droplets sizes, their distribution and heat exchange for minimum quantity cooling lubrication (MQCL). Int. J. Mach. Tools Manuf. **100**, 81–92 (2016)
14. Glowacz, A., Glowacz, A., Glowacz, Z.: Recognition of monochrome thermal images of synchronous motor with the application of quadtree decomposition and backprop-agation neural network. Maint. Reliab. **16**(1), 92–96 (2014)
15. Temperature sensor solutions. www.thermo-electra.com
16. Kortvelyessy, L.: Thermoelement Praxis, 3rd edn. Vulkan-Verlag, Essen (1998)
17. Sloneker, K.C.: Life expectancy study of small diameter type E, K, and N mineral-insulated thermocouples above 1000 °C in air. Int. J. Thermophys. **32**(1–2), 537–547 (2011)
18. Southworth, D.J.: Temperature calibration with Isotech block baths. Handbook of Isothermal Corporation Limited (1999)
19. Sloneker, K.C.: Thermocouple inhomogeneity. Ceram. Ind. Mag. **159**(4), 13–18 (2009)
20. Holmsten, M., Ivarsson, J., Falk, R., Lidbeck, M., Josefson, L.-E.: Inhomogeneity measurements of long thermocouples using a short movable heating zone. Int. J. Thermophys. **29**(3), 915–925 (2008)
21. Temperature web portal. http://temperatures.ru/pages/termoelektricheskie_termometry
22. Brignell, J.E.: Digital compensation of sensors. J. Phys. E: Sci. Instrum. **20**(9), 1097–1102 (1987)
23. Kirenkov, I.: Some laws of the thermoelectric inhomogeneity. Research in the field of temperature measurements, pp. 11–15. VNIIM, Moscow (1976). (in Russian)
24. Reference materials. http://www.omega.com/temperature/Z/pdf/z016.pdf
25. Zvizdic, D., Sestan, D.: Zinc-filled multi-entrance fixed point. Int. J. Thermophys. **36**(2), 336–346 (2015)
26. Kochan, O., Kochan, R., Bojko, O., Chyrka, M.: Temperature measurement system based on thermocouple with controlled temperature field. In: Proceedings of the IEEE International

Workshop on Intelligent Data Acquisition and Advancing Computing Systems (IDAACS'2007), Dortmund, Germany, pp. 47–51 (2007)

27. Sachenko, A., Kochan, V., Turchenko, V.: Instrumentation for gathering data. IEEE Instrum. Meas. Mag. **6**(3), 34–40 (2003)

28. Jun, S., Kochan, O.: The mechanism of the occurrence of acquired thermoelectric inhomogeneity of thermocouples and its effect on the result of temperature measurement. Meas. Tech. **57**(10), 1160–1166 (2015)

29. Kochan, R., Berezky, O., Karachka, A., Bojko, O., Maruschak, I.: Development of the integrating analogue to digital converter for distributive data acquisition systems with improved noise immunity. IEEE Trans. Instrum. Meas. **51**(1), 96–101 (2002)

30. Jun, S., Kochan, O., Kochan, V., Wang, C.: Development and investigation of the method for compensating thermoelectric inhomogeneity error. Int. J. Thermophys. **37**(1), 1–14 (2016)

31. Lienhard, J. IV, Lienhard, J. V: A heat transfer textbook. Phlogiston Press (2008)

32. http://conceptalloys.com/electrical-resistance-alloys-chromel-c113-alloy/

33. Vasyl'kiv, N., Kochan, O., Kochan, R., Chyrka, M.: The control system of the profile of temperature field. In: Proceedings of the IEEE International Workshop (Rende (Cosenza)) Intelligent Data Acquisition and Advancing Computing Systems, Rende, Cosenza, Italy, pp. 201–206 (2009)

34. Kochan, O., Jun, S., Kochan, V.: Decreasing of thermocouple inhomogeneity impact on temperature measurement error. In: Proceedings of the 13th IMEKO TC10 Workshop on Technical Diagnostics Advanced Measurement Tools in Technical Diagnostics for Systems' Reliability and Safety 2014, Warsaw, Poland, pp. 105–110 (2014)

35. Jotsov, V.: New proposals for knowledge and data driven applications in security systems. In: Sgurev, V., Yager, R., Kacprzyk, J., Jotsov, V. (eds.) Innovative Issues in Intelligent Systems, Studies in Computational Intelligence, vol. 623, pp. 231–294. Springer, Berlin, Heidelberg (2016)

36. TRIZ Innovation Portal. http://www.innovation-portal.info/resources/triz/

Personal Assistants in a Virtual Education Space

Jordan Todorov, Vladimir Valkanov, Stanimir Stoyanov,
Borislav Daskalov, Ivan Popchev and Daniela Orozova

Abstract This chapter introduces personal assistants operating in a Virtual Education Space (VES). VES is built as an Internet-of-Things ecosystem consisting of autonomous intelligent components displaying a context-aware behavior. A personal assistant plays as an "entry point" to the space operating as a kind of intelligent user interface. The life cycle and the architecture of a generic personal assistant are presented in the chapter. Some suggestions for provision the assistant with learning capabilities are also given. A prototype implementation known as LISSA is discussed in more detail. Currently, a second version is developed known as IoT LISSA where the personal assistant will be able to account for the reality of the surrounding physical world in planning and executing its actions.

J. Todorov (✉) · V. Valkanov (✉) · S. Stoyanov
Department of Computer Systems, Plovdiv University "Paisii Hilendarski",
Plovdiv, Bulgaria
e-mail: jtodoorv@uni-plovdiv.bg; jtodorov@leet-soft.com

V. Valkanov
e-mail: vvalkanov@uni-plovdiv.net

S. Stoyanov
e-mail: stani@uni-plovdiv.net

B. Daskalov · I. Popchev
Institute of Information and Communication Technologies, BAS, Sofia, Bulgaria
e-mail: bdaskalov07@gmail.com

I. Popchev
e-mail: ipopchev@iit.bas.bg

D. Orozova
Burgas Free University, Burgas, Bulgaria
e-mail: orozova@bfu.bg

© Springer International Publishing AG, part of Springer Nature 2018   131
V. Sgurev et al. (eds.), *Practical Issues of Intelligent Innovations*,
Studies in Systems, Decision and Control 140,
https://doi.org/10.1007/978-3-319-78437-3_6

1 Introduction

In recent years the interest towards eLearning has been growing stronger. Besides the most well-known open-source systems (e.g. Moodle [1]) and corporate systems (e.g. Microsoft Classroom [2]), a lot of universities develop their own eLearning systems. In line with this trend a Distributed eLearning Centre (DeLC) project was implemented in the Faculty of Mathematics and Informatics at the University of Plovdiv aiming at the development of an infrastructure for delivery of electronic education services and teaching content [3]. The DeLC architecture is designed as a network, which consists of separate nodes, called eLearning Nodes. Nodes model real units (laboratories, departments, faculties, colleges, and universities), which offer a complete or partial educational cycle. Each eLearning Node is an autonomous host of a set of electronic services [4]. The eLearning Nodes can be isolated or integrated in more complex virtual structures, called clusters. Remote eService activation and integration is possible only within a cluster. In the network model we can easily create new clusters, reorganize or remove existing clusters, because the reorganization is done on a virtual level and it does not affect the real organization. DeLC provides mobile access to services and content over an extended local network called InfoStations [5].

One serious disadvantage is that the DeLC implemented as a virtual environment does not account for the physical world in which the learning process is performed. Building an infrastructure where the virtual world is integrated in a natural way within the surrounding physical environment would open new opportunities for delivering education services and learning content in a personalized and context-aware way. For achieving this integration in recent years we started a transformation of DeLC in a new, cyber-physical infrastructure known as Virtual Education Space (VES). VES is developed as an Internet-of-Things ecosystem, consisting of autonomous intelligent components displaying a context-aware behavior [6]. Furthermore, the space is enhanced by approaches using semantic models, mainly in the form of ontologies.

In working with complex systems such as VES, it is advisable for the users to be supported by specialized interface components in an intuitive and intelligent manner. One possible solution is the use of personal assistants. In the space architecture personal assistants play an important role.

This paper presents the creation of a personal assistant, which will aid students in their work with the space. The rest of the chapter is organized as follows: Sect. 2 presents a state of the art of various related works. Section 3 briefly presents the VES infrastructure. Section 4 describes the students' personal assistant known as LISSA (Learning Intelligent System for Student Assistance). Section 5 discusses some details related to the implementation of the personal assistant. Finally, Sect. 6 concludes the chapter.

2   Related Works

Introduced in the 1990s, the concept of using personal assistants for aiding people in their everyday business and personal affairs is being developed for many years. As is shown in [7], technologies from the field of AI present opportunities for building intelligent machines, which autonomously perform tasks on the user's behalf. In [8] we discover that the applications of agents are divided in two main groups—distributed systems and personal software assistants where agents play the role of proactive assistants to the users in working with some application. Personal assistants (PA) can help the everyday as well as long-term management, execution and control of different types of tasks such as those related to planning, reservations, shopping, payments. Nowadays these assistants are usually put on mobile devices, they can use the resources of social networks and can learn. Two useful reviews, the first one for the use of PA in the context of IoT [9] and the second one of Intelligent Pedagogical Agents (IPA) for personalized learning and increasing the students' motivation [10], demonstrate the expanding domains of personal assistants.

Several big projects can be shown, which outline the prospects and stimulate the research in that area. PAL (Personalized Assistant that Learns) [11] is one of the first broad-range programs for scientific research in the field of cognitive systems. The program's goal is to radically improve the way in which computers interact with humans. One of the most interesting systems developed within PAL is the personal assistant CALO (Cognitive Assistant that Learns and Organizes) [12]. The assistant, capable of self educating, supports the users in making decisions mainly for military problems. Of great interest are the user interface and database, which is built as a semantic application frame including a platform for military education as well. The frame offers different services to the users such as views, context navigation, calendar, web- and file browser, e-mail client, instant user messages. Another project for creating intelligent assistants is COMPANIONS [13], which aims to change the generally acknowledged way of interaction between humans and computers. A virtual colloquial "Companion" is an agent, which "co-exists" with the user for long periods of time during which it "builds" a friendship and "studies" the preferences and desires of its owner. The agents communicate with the users mainly by use and understanding of speech, but they can also use other means such as sensor displays and detectors. The project PAL (Personal Assistant for healthy Lifestyle) [14] predicts the development of a social real NAO robot, its mobile avatar and an expandable set of mobile applications in the field of healthcare, which use a common database and conclusion mechanism. With the system's aid health specialists can set goals and problems to children and monitor their development through their achievements and solutions. Thus PAL can be an assistant and teacher to the children.

A number of corporate developments are well-known among a broad range of users. Generally, these personal assistants implement user interfaces in a natural language to interact with users. Usually these assistants (such as Siri [15], Microsoft Cortana [16], Google Now [17], LG Voice Mate [18]) are used to support the performing of common everyday activities. For instance, Apple was the first large technological company, which in 2010 integrated the intelligent assistant Siri in its operating system. In its newer versions Siri supports different techniques for self-education. A large group of personal assistants are applicable in healthcare (e.g. HealthPal [19], BeWell [20], PPCare [21]).

Interesting researches for the creation of personal assistants are conducted in many universities. Within those projects predominantly are combined the results of researches on the Internet of Things, robotics and machine learning. Over the last years, significant efforts have been made to create personal assistants for aiding people with different needs. Realizing the necessity for developing multimedia technologies for people with disabilities, the Center for Cognitive Ubiquitous Computing (CUbiC) at Arizona State University devoted its efforts in overcoming this obstacle. One exemplary project of this center is the planning and developing of an assistant for social interaction and improving the possibilities for interacting with people with ocular disabilities [22]. This assistant's goal is to improve the accessibility of social non-verbal signals for people who are blind or with impaired vision. It consists of a pair of glasses equipped with an appropriate camera. The incoming video stream is analyzed with the help of algorithms for machine learning and computer vision, which allows the extraction of the respective non-verbal signs. Therefore, this information is delivered to the user. Over the last years, a lot of efforts have been made for creating robots able to help taking care of old people. There are significant difficulties in the personalization of these robots. Therefore, they are capable of responding to the specific needs of the people they are taking care of. Even robots with abilities to self-learn take a long time to understand the preferences of the old people. The paper [23] presents an intelligent agent, which provides personalized care for old people, which is realized as an integrated help, combining different intelligent services. The service's common architecture consists of three main components—Virtual caretaker, Virtual Care Personalizer and Template Care. For the effective functioning of the integrated service, the personalizer controls and generates the needed actions personalization, which provide the care in a special Care Cloud. The appearance of intelligent personal assistants is an example of offering new types of services using electronic devices with big potential, the achievements of AI, as well as the evolution of information and communication technologies and the Internet. Combining the concept for personal assistance and the paradigm of IoT presents new opportunities for presenting context-aware services to the end-user. In [24] is presented a health scenario where a mobile gateway is integrated in PA. With the help of this gateway, individual information on the patient's physical condition, acquired from a body sensor net, is transferred in real time to PA.

With the rapid development of technologies, a shaping tendency is the increase in traditional educational environments with intelligent components or the development of educational environments as an integral part of intelligent infrastructures.

Some systems which operate as PAs have a new application through promoting instant messaging technologies, which allows the users to select available and accessible online recipients for receiving messages. In the research presented in [25], a platform (type of personal assistant) is developed, which is specialized for students and aids them in their academic and everyday tasks. The personal assistant provides useful interactive functions for control of school curricula, diaries, finances and chat systems. The assistant possesses a simple interactive interface and is proven to be reliable, effective and user-friendly, when is used in learning and working environments. The paper [26] researches the importance of introducing a learning environment as an educational infrastructure in an intelligent settlement. The demands of an intelligent settlement are reviewed and a two-way relation is proposed between the physical and virtual environments. Based on previous papers for creating intelligent pedagogical agents, a non-conventional intelligent interface is proposed for intelligent urban services for satisfying educational needs. The proposed architecture allows the integration of different new application scenarios and potential new services, including serving students with special needs.

Another tendency is the introduction of personal assistants in aiding the users' personal mobility. IRMA [27] researches personal mobility in a possible near-future scenario, which is oriented towards a green, mutual and public transport. The project's goal is to propose an adaptable, easy-to-execute and stable modular platform, which uses a combination of different information sources and propositions with added value, each serving a group of interested users, such as a municipality, citizens or transport service providers. Thus, IRMA aids the users in the entire lifecycle of controlling mobility. The platform is developed as a SOA/EDA (Service Oriented Architecture/Event Driven Architecture) infrastructure presenting a hierarchy of re-usable services that can be put on mobile- as well as web devices. With the increase of the population, people's mobile behavior in big cities is about to change faster than ever. The new opportunities for transportation, along with the comfort and an improved ecological awareness lead to the ever more common behavior of intermodal mobility. The broad use of mobile devices and the presence of a high-speed Internet in transport vehicles allow users to stay informed about the immense multitude of possibilities for mobility. What could be discomforting is the vast quantity of information that the users receive. Digital Mobility Assistants can alleviate the difficulty of choice for the best option of mobility for a given user by accounting for their habits and preferences as well as delivering the respective information at the right time. Personalized models for mobility are proposed in [28] to provide such intelligent assistance that include not only information on common trips and destinations, but also allow accounting for the preferred means of transportation. The proposed system is specially designed for the use of limited sensory data from mobile devices in order to adequately balance the battery life and information quality.

Using modern information and communication technologies, many useful services are presented to users every day. Still, in some situations the users are unable to quickly find sufficient data. In [29] is presented a cyber-physical space, which consists of physical space, digital space and agent space, which acts as a mediator. The agent space consists of three types of agents—PAs, service agents and resource agents. PAs have the ability to dynamically track the users' activities, which allows the space to operate as a context-aware informational system intended for the user.

3 Overview of the Virtual Education Space

3.1 Basic Characteristics

The Virtual Education Space is developed as an Internet of Things ecosystem with the following characteristics [30]:

- *Distribution and Autonomy*—the space is designed as a distributed infrastructure comprised of autonomous components.
- *Smartness*—the space exposes an intelligent behavior [31, 32] monitoring what's happening inside itself (as well as in the physical world), interacting with the components it is built of, making decisions and acting towards their execution.
- *Context-awareness*—within the space is accepted the definition of context and context-awareness from [33, 34]. In our case the space has the ability to find, localize and identify the changes (events), which occur in itself and operate while accounting for them.
- *Accessibility*—the access to information resources in the space is possible only through the so-called "entry points". For registered users the role of such entry points is played by personal assistants. The non-registered users have free access to certain resources through a specialized portal.
- *Adaptability*—the space is independent from any specific learning process. It can adapt to different forms of learning. Our future intentions are to make it adaptable to different domains.
- *Personalization*—in the space a personal use of resources is supported.

A major challenge is to provide an effective interaction among autonomous intelligent components, operating in the space. Our approach includes the following three steps:

- *Building a unified integrated technology*—the integrated technology has to ensure syntactical and communicational interoperability among different types of components used in the space and located on different architectural levels. Intelligent components of the space are the assistants, realized as rational agents. The agents are software components that operate accounting for the dynamic in the surrounding environment. However, they are unsuitable for delivering a business functionality. Contrarily, the services are a good solution for realizing functionality; however, they are not flexible, neither proactive and cannot be separate components in the space. Therefore, the agents include in their internal architecture suitable interfaces to services. Thus, the space operates as an ecosystem for electronic education, open to expansion with new education services. Furthermore, the assistants have to be able to communicate with the physical world. For that purpose, suitable agent-sensory interfaces are used.

- *Providing semantic interoperability*—the proposed integrated technology is not sufficient in itself for providing the conditions for intelligent interaction. It needs to be supplemented by approaches, methods and means aiding semantic aspects of the interaction. Approaches used in the space are the following:

  - Agent-oriented approach—a powerful approach for building autonomous intelligent software components with a mentality, which also includes a language for interaction (ACL [35]).
  - Using standards—the two standards for electronic education SCORM 2004 and QTI 2.1. use and specify structures (e.g. LOM) with clear syntax and semantics, which allows a unified interpretation of the different types of space components.
  - Semantic modeling—an important aspect of intelligence is the degree of formalization and automated interpretation of the used data. An agent-oriented approach and using standards can solve this problem only partially and with limitations. This is why we additionally use semantic modeling of information in the form of ontologies.
  - Integrated unified model of events for the space.

- *Proactive and learning assistants*—an important requirement for providing satisfactory intelligence in the VES is the presence of intelligent assistants. Under "intelligence" here is meant assistants, which show context-aware, reactive, proactive and social behavior in dependence with the condition of the space. With that regard, it is essential to build proactive and learning agents.

3.2 Architecture of the Space

As an IoT ecosystem [6] the components of the space can be grouped in three architectural levels—access level, operative and analytic level and sensory level (Fig. 1).

For registered users the access to information resources and services in the space is realized mainly through personal assistants (PA). The main designation of PAs is to aid users (in this case students and teachers) in their work in the space. They operate as specific access points of the VES. During the initial registration in the space users are provided with their own PA. A genetic PA is supported for that purpose, which by interacting with the education portal and the module for registration generates a specific for the particular user PA. Personal assistants are

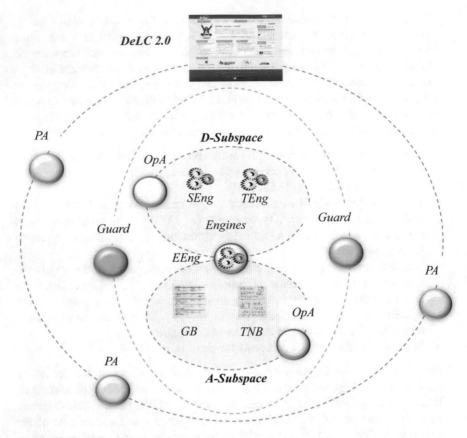

Fig. 1 VES architecture

developed as BDI rational agents [36]. In the current VES version, a PA prototype for students, known as LISSA, is created. In LISSA is integrated a simplified interface with the ability to comprehend and generate phrases in a natural language (English).

Non-registered users may have access to the space through the education portal DeLC 2.0. [37]. The current version of the education portal supports two forms of electronic learning—blended learning and lifelong learning. For the blended form of learning more than 20 lecture courses are provided. Examining students is made through a system for electronic testing integrated in the portal [38].

Operation and analytical level (A-Subspace in Fig. 1) is the level with extreme importance in the degree of intelligence of the space because this is the place where the sensory information is collected in order to support a solution-making, related to the operation and control of VES. The components operating in this part of the space, known as operative assistants (OpA), support two models in the current version:

- Gradebook (GB)—repositories and solutions for improving students' results;
- Teacher's notebook (TNB)—repositories and solutions for supplementing the gradebook and improving the production of the teacher's activity.

The main functions of the sensory level are gathering, registering, transforming and transferring different types of data, relevant to the operating and managing of the space. Generally, in VES two types of sensors are supported—virtual and physical. The physical sensory information is received and initially processed by the guards. For the space the physical world is a collection of physical sensors accessible to the guards.

A specific feature of guards is that they can operate anywhere within the space. Accounting for that circumstance we distinguish different types of guards (Fig. 1). The role of the so called *logical guards* (LGs) is very important, which task is to gather raw data from single or a group of physical sensors. This data can be initially processed, transformed and transferred to the other components in the space. Thus, logical guards operate as a unique interface between the physical and virtual world in VES.

The second type, virtual guards (VGs), operate entirely in the virtual space. The two types of guards can interact among themselves as well as with other components of the space. Unlike the physical ones, virtual sensors are abstractions. Typical sources of virtual sensory data in the space are the three machines situated in D-Subspace (Fig. 2):

- SCORM 2004 Engine (SEng)—sensory information for the students' homeworks;
- QTI 2.1. Engine (TEng)—sensory information for examining students;
- Event Engine (EEng)—sensory information for events that originated within the space.

3.3 Event Model

To ensure an effective control and interoperability among different components, the concept of events is created. In our model an event is defined as *Event = (e_id, e_type, e_pars)*, where:

- *e_id:* event identifier;
- *e_type:* type of event;
- *e_pars:* event parameters which are usually different characteristics of the events. Those parameters can be common (such as duration, period, uniqueness) or specific for the particular type of event. Events can also be parameters.

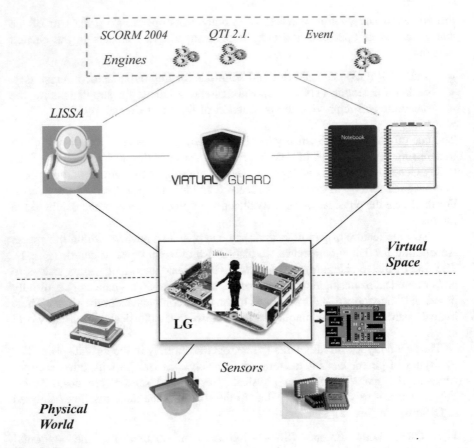

Fig. 2 Guards

The following main types of events are defined in the model:

- *Basic events*—these are atomic events (events without arguments), usually occurring in the physical world. Date, time and location are the three main events that play an important role in the work of the personal assistant.
- *System events*—usually these events identify the condition of the infrastructure, which supports the space. Typical examples of such events are generating/removing components in the space or sending/receiving messages among the components of the space.
- *Domain events*—these events are specific for the domain in use (in this case electronic learning). Typical examples are lectures, exercises, examinations, homework, schedules, curricula.
- *Emergency events*—special types of events related to emergency situations, which impact the learning process.

Usually different assistants react differently to specific events. For instance, personalized assistants intensively use basic events, while operative agents react to domain events. Emergency events are processed by guards. Therefore, by event we mean mainly whatever has happened from the applied aspect, i.e. the event is a term related to the domain in use.

Regarding the place of occurrence events can be:

- *Real*—occurring in the real world, such as lectures, exercises, examinations. These events need to have some representation in the virtual space.
- *Virtual*—occurring in the virtual world, such as schedules, charts, curricula.

Different connections (relations) can exist between the two types of events.

4 LISSA

Personal assistants are main components which task is to aid the users in working with the VES. The personal assistant known as LISSA (Learning Intelligent System for Student Assistance) aims in assisting student participation in the learning process by monitoring the performance of tasks related to an individual curriculum.

4.1 Life Cycle

The LISSA's life cycle (Fig. 3) is the life cycle adaptation of a practical reasoning agent. Unlike the suggested in [8], the LISSA's life cycle has its own characteristics such as:

- The initialization phase in LISSA is more complex;
- Desires are limited within the frame of the current curriculum at the university.

The separate phases of the LISSA's life cycle are presented in this paragraph.

Registration. To be able to receive a personal assistant, new users must undergo registration. The registration's goal is to generate a user profile (Fig. 4), which is used for providing personalized help. In the current version the profile there is information for the student's personal identification, background (e.g. GB, where information for the student's grades is saved), types of events to which the personal assistant has to react (incl. type of event with maximum reaction time). In working with the system, the profile is updated constantly. To generate the profile, LISSA interacts with the DeLC 2.0 educational portal (register). The educational portal, through its interface to the integrated university system, acquires the necessary personal information for the profile (faculty number, major, course, etc.). Furthermore, the portal checks what information is present in its repositories for the student.

```
/* Registration */
B₀ ← get_percept;
profile ← register(B₀, student_id);
/* Initialization */
Desires ← create_PC(profile, B₀);
B ← B₀;
I ← I₀;
/* Deliberation */
while true do
      percept ← get_percept;
      B ← update(B, percept);
      D ← identify_goal(B, Desires);
      I ← compose_goal(B, D, I);
      /* Planning */
      π ← plan(B, I, Ac);
      while not (empty(π) or succeeded(I, B) or impossible(I, B)) do
            α ← head(π); execute(α); π ← tail(π);
                  percept ← get_percept;
                  B ← update(B, percept);
                  if reconsider(I, B) then
                        Desires ← update(B, I, Desires);
                        if needed then update(profile);
                        break;  // go to external while cycle
                  end-if
      end-while
end-while.
```

Fig. 3 Life cycle of LISSA

```
profile:
      identification: name, faculty_nr, ... ;
      background: GB_reference, ...;
      events: event_types, max_reaction_time, ...
      guard: G_reference;
end-profile.
```

Fig. 4 Profile

The personal assistant can maintain its own guard, which provides interaction with the physical world. This is a useful option, particularly when the LISSA user has some physical problems (e.g. requires a wheelchair).

Initialization. During the initialization *Desires* of the personal agent are generated (create_PC). In this case, the role of *Desires* is played by a personal calendar presenting the student's participation in the learning process for a certain period (semester, study year). The records in the personal calendar are separate domain events in accordance with the VES event model. Identically to the registration, the personal calendar is generated through an interaction with the educational portal DeLC 2.0. The educational portal, upon receiving data from the student's profile, extracts the necessary information from the faculty's curriculum. The initialization is done periodically, e.g. at the start of each new academic year.

Deliberation. In this phase of the life cycle, the immediate goal has to be determined (I) to which meeting the student will be currently assisted. The goal, presented as a domain event, is identified through conducting a search (identify_goal) in the personal calendar depending on the actual beliefs (B) of LISSA (Fig. 5). In certain cases, it is necessary for the assistant to react at some point to more than one goals, i.e. composite goal (compose_goal).

Planning. After the current goal is determined, the agent needs to prepare a plan (plan) for its achievement. In the current version, LISSA usually generates reminders and warnings to the student. Furthermore, the agent possesses an elementary word processor. In some cases, it is necessary to review the goal "outside the plan" (reconsider). This usually requires communication with the LISSA environment, which aims to find whether and what actualization is needed in the personal calendar and/or profile before identifying a new goal. A typical example: the student failed an exam in some subject. Then communication with Grade Book (GB) is needed to ascertain that fact and afterwards LISSA interacts with the DeLC

```
function identify_goal(B) returns Goals, a list of next goals
    persistent: Desires, the student's personal calendar
                profile, student's profile
    input: B, the agent's current beliefs using as search parameters
    output: Goals
    local variable: PC_slice
        Goals ← ∅;
        PC_slice ← Desires(B.date, profile.max_reaction_time);
        for each PC_slice[i] do
            if PC_slice[i].type ∈ profile.event_types
            then Goals ← extend(Goals, PC_slice[i]);
        end-for
        return Goals
end identify_goal
```

Fig. 5 Function *identify_goal*

2.0 portal to acquire the date of the supplementary exam. Finally, the agent updates the personal calendar.

4.2 Architecture

The architecture of LISSA is developed in accordance with the proposed life cycle (Fig. 6) which includes Personal Assistant (PA), Generic Personal Assistant (GPA) and Generic Dispatcher Agent (GDA).

Personal Assistant (PA). Upon registration for each user, the agent's instance is configured and placed on the user's mobile device. Each PA operates as an "entry point" to the VES operating as a kind of intelligent user interface. After the configuration, the actual PA is activated where it assumes the life cycle control If needed, for the execution of certain activities, the PA interacts with the other components of LISSA. The agent's plans usually include actions for realizing interaction with the student, presenting recommendations and warnings. The current prototype offers basic language means of communication.

Generic Personal Assistant (GPA). The main goal of GPA is to register new users and configure respective instances of PA. The current instance is configured with the following steps:

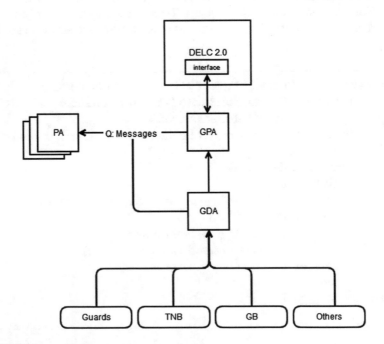

Fig. 6 LISSA's architecture

- Generating a newly registered student's profile—GPA interacts with the educational portal DeLC 2.0 to deliver the necessary information which is present in the portal's storage and in the university information system (only the portal has an interface to that system);
- Initiating mental states—at this step are generated and initialized the structures presenting the mental states of the PA. In our case *Beliefs* are basic events (according to the event model), i.e. date, time, location. The current values of Beliefs are given by the get_percept function. *Desires* are presented as the student's personal calendar, which is initially created as a personalized extract of current study schedule. Generating the personal calendar (create_PC) happens through interacting with the portal, which has access to the university's curricula. To achieve that, GPA gives the portal the necessary parameters from the student's profile such as major, course, semester, type of program. The separate entries in the calendar are domain events. In defining *Desires* of great importance are the type of event and its parameters such as date of event, repetitiveness and uniqueness. The values of these parameters have an important influence on the PA planning mechanism. The student can complement the personal calendar with personal events (e.g. birthdays, meetings, visitations, participations), which should be monitored by the PA.

Once configured, the life cycle control is submitted to the new PA instance.

For existing users, the GPA (interacting with the respective PA) is used when necessary from periodic initializations; for instance, the student's personal calendar (*Desires*) has to be initialized at the beginning of each new semester or each new academic year depending on the assumed approach.

Generic Dispatcher Assistant (GDA). GDA serves as an interface agent for receiving and transferring messages between PA and GPA, on the one hand, and other assistants in the space (GB, TNB, Guards, et al.), on the other hand.

4.3 Learning

Creating a personal learning assistant, which can effectively be used in real applications, is a complex task. In VES, the evolution of LISSA into learning LISSA is made in two steps:

- Creating a basic learning model—a type of preliminary survey, prototyping and experimenting for determining the goal, task specification and determining the learning approach.
- Creating a complete learning model—experimenting with the prototype we expect to be able to offer a second evolved version of the model, which can be used into accomplish learning LISSA.

In the article, we will shortly present our work on creating the basic model. In the basic model, certain simplifications have been made from real practice. In theory, there are different approaches that we could use to create the model. To determine our approach, we are looking for answers to two questions:

- Is it possible to learn from examples or do we need to use some of the methods of reinforcement learning?—our analysis shows that it is possible to define examples that can be used for training.
- Is it possible for the learning model to be defined as a mathematical function?—at this stage it is difficult to define learning through a mathematical function and therefore we are using a constructive analytical approach.

In our case, the learning is based on the results of the educational process where the personal assistant attempts to have a deeper understanding of the student's learning behavior. Therefore, the main problem is finding the right definition for the term "*learning behavior*". Generally, learning behavior is determined by the following factors:

- The student's self-study—in which way the student masters the teaching content and how he copes with the homework.
- Participation in tutorials—to what extent the student completes lab exercises, his scores on tests and intermediate grades.
- Participation in lectures—regular attendance, early comprehension of the subject matter, participation in discussions.
- Examination scores—what are the final grades in the separate subjects.

The main question is whether we can define and present well-structured examples, which can be used to educate the personal assistant. In the model we offer the following definition of an "example": an example ex = (la, r), where la = <spl, tut, lec> is a structure that consists of the following three elements:

- spl—set of parameters describing the student's self-study;
- tut—set of parameters describing the student's participation in lab exercises;
- lec—set of parameters describing the student's participation in lectures.

r is a structure presenting the student's results in the respective subject which in the simplest example is the grade from the examination. It is possible to use more complex grading models.

All of the model's elements are presented as attribute/value pairs. Besides, we define the trained set LBT = $\{ex_1, ex_2, ..., ex_n\}$. Roughly, the learning of LISSA happens according to the model presented in Fig. 7.

We assume that during the self-study the student is working with textbooks developed in accordance with the SCORM 2004 standard and that the exams are performed in the system for electronic testing in accordance with the QTI 2.1 standard. VES provides those opportunities. SCORM2004-Engine gathers and delivers different data describing the process of self-study for the student. This data is used to generate spl. Identically, QTI-Engine provides result data from the exam, which is used to generate r. The data for the student's participation in lab exams and lectures are delivered by the panel of teachers through suitable user interfaces.

1. **Step (Initialization of the Training Set):**
   LBT ← ∅;

2. **Step (Generation of a new example):**
   - spl ← SCORM2004_Engine(student,discipline);
   - tut ← Tutor(student,discipline);
   - lec ← Lecturer(student,discipline);
   - r ← QTI_Engine(student,discipline);
   - ex ← Integrate(spl, tut, lec, r);
   - LBT ← LBT + ex.

3. **Step (Evaluation of Consistence):**

   if ¬LearningBehavior_H(ex) **and** evaluate(r)
   then LearningBehavior_H ← globalization(LearningBehavior_H)

   become consistent
   else if LearningBehavior_H(ex) **and** ¬evaluate(r)
   then LearningBehavior_H ←
   specification(LearningBehavior_H)

Fig. 7 Learning approach

From the separate structures is generated a new example with which the training set is expanded. The grade for consistency of the new example (ex) and hypothesis (LearningBehavior_H) are made from an adapted version of the method presented in [39].

The training set and the current hypotheses are stored in the space's GB. In fact, work hypotheses can be defined for different aspects and degrees of understanding of the subject matter. One possibility is to grade the student's abilities according to the Bloom taxonomy [40]. Another possibility is on the basis of statistical data to define working hypotheses for assessing success on some grading scale (in our example, a five-step scale).

A significant disadvantage of the assumed approach is the small and slowly growing training set—the exams that students have to take during the year are few for the used methods. The following improvements are possible:

- Using examples from all available GBs in order to propose a classification of different types of "learning behavior". In generating a given student's PA, we are trying to determine to what type of learning behavior the student belongs. Thus, we have some preliminary idea about what we can expect from this student's performance.
- The used analytical approach can be improved through using background knowledge. The GBs can operate as repository for background knowledge.

5 LISSA's Prototype

The LISSA's prototype PA is implemented as an Android app for two reasons:

- Android is the most widely-used operation system for mobile devices and can be used on a large group of different devices such as: mobile phones, tablets, smart watches, TVs, et al.
- The second important reason is that it uses the JAVA programming language which will retain the system's homogeneity.

The prototype for LISSA was designed as BDI rational agents. For its implementation is used the development environment JADEX [41] an extension of JADE [42] adding BDI mental states support. Using JADEX provides a large number of service interfaces, e.g. RESTful, High-Level Streaming, Search. LISSA intensively interacts with the DeLC 2.0 portal, which operates as a special entry point in the space. DeLC 2.0 is implemented as a dynamic web application distributed in two main areas—an educational portal operating as a specialized user interface and server side. Both areas communicate by using pure HTTP requests, RESTful

Fig. 8 Login in LISSA

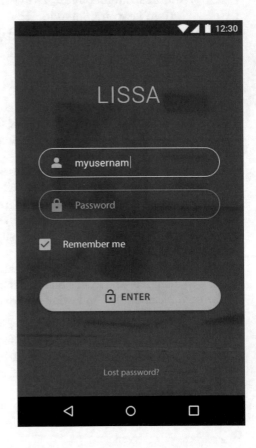

services and Web Sockets. At the server side, we implemented an agent-oriented interface, known as JADE & Jadex Container [43], which makes possible the creation of agents that can operate in the portal and in this way to ensure interaction with the personal assistant LISSA and the operative assistants of the space.

In building the prototype, we faced several problems that we were able to overcome with the help of JADEX developers at the University of Hamburg. Besides, the implementation of a simple language interface presented the following problems:

- The process of voice recognition is very complex and is an entire scientific area.
- Recognizing voice commands requires lots of resources from the device and the most common mobile devices do not possess large hardware resources.
- The device's battery life is a valuable resource and we had to find a way to save it as much as possible.
- In order for the assistant to be useful even in the absence of connection with the network, the voice recognition has to be able to work in offline mode.

Creating a library for voice recognition from the beginning would take a lot of time and resources and that is why we decided to use an existing one. For dealing with

Fig. 9 Operative state of LISSA

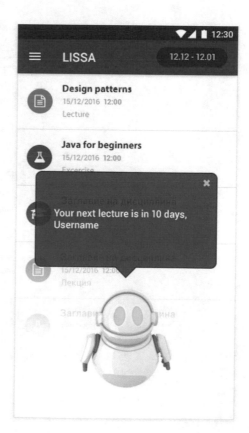

the second problem, in the current version of the prototype we are using a limited set of commands which the assistant can recognize. Thus, we are significantly reducing the search area and therefore the resources necessary for completing the operation. For maximum battery life the voice commands are only working when the user is actively manipulating the phone; otherwise, if the device is in sleep mode, it is not listening for commands. To further decrease the battery's use and recognize commands more easily, we included the keyword "LISSA", which has to be said at the beginning of each command. In that way, before the start of the command, only one word is being looked up and that significantly decreases the resources necessary for the voice recognition process, and thus reduces the need for additional operations, which would lead to a larger consumption of the device's battery.

For voice recognition in offline mode, we use the CMU Sphinx library [44]. Compared to the Google voice recognition library [45] this one offers more limited functional abilities but can work in offline mode.

We will briefly demonstrate working with LISSA. To be able to work in the space, the user has to initially log in or register in the system (Fig. 8). If the user is

Fig. 10 Selected event details

in registration mode GPA, in interaction with DeLC 2.0, delivers data for the creation of a profile and desires of the personal assistant.

In normal working mode, on the screen is visualized the LISSA avatar, and above it is a list with pending events (Fig. 9). The most recent events are on top. For each event there is an event name, date and time, and below it is the type of event. In the upper right corner, we see a range, which is used for reviewing events and depends on the types of events in the calendar. For range is used the event with the most distant occurrence. When a reminder is necessary, the assistant simultaneously verbalizes and displays the message on the screen.

The normal working process of the prototype runs through the BDI architecture as described in the previous chapter by using the date and time to factor in the environment, and the user location is found with the help of the GPS sensor.

If the user wishes to review additional information for the current or future event (Fig. 10), by pressing on the screen or on command "LISSA, show me information about next lecture" will be visualized detailed information about the particular event, which includes: type of event, name, date, time, description, lecturers.

6 Conclusion

In this paper, the personal assistant LISSA operating in the VES is presented. The VES is implemented as an IoT ecosystem. A second version is currently developed, known as IoT LISSA, where the personal assistant will be able to account for the reality of the surrounding physical world in planning and executing its actions.

With these capabilities, the personal assistant will offer more effective support to its users. Besides, the new version will be able to effectively adapt to a group of users needing different additional help in their everyday activities. (e.g. people with different physical disabilities). To achieve this goal, IoT LISSA will be actively supported by the guards in the space.

Currently, two important problems are being solved in relation with creating the second version of LISSA:

- Refinement of the life cycle of LISSA to include abilities to account for the physical world.
- Building a suitable interface between LISSA and guards in the space using the OSGi model [46] and the existing in JADE interface to OSGi components [47].

References

1. Moodle. https://moodle.com/, Sept 2016
2. Microsoft Classroom. https://classroom.microsoft.com/, Sept 2016

3. Stoyanov, S., et al.: From CBT to e-Learning. J. Inf. Technol. Control **3**(4), 2–10 (2005)
4. Stoyanov, S., Zedan, H., Doychev, E., Valkanov, V., Popchev, I., Cholakov, G., Sandalski, M.: Intelligent distributed eLearning architecture. In: Koleshko, V.M. (ed.) Intelligent Systems. InTech, Mar 2012, Hard cover, 366 pages, pp. 185–218. ISBN: 978-953-51-0054-6
5. Stoyanov, S., Ganchev, I., O'Droma, M., Zedan, H., Meere, D., Valkanova, V.: Semantic multi-agent mLearning system. In: Elci, A., Kone, M.T., Orgun, M.A. (eds.) Semantic Agent Systems: Foundations and Applications. Studies in Computational Intelligence, vol. 344. Springer (2011). ISBN: 978-3-642-18307-2
6. Stoyanov, S.: A virtual space supporting eLearning. In: Proceedings of the Forty Fifth Spring Conference of the Union of Bulgarian Mathematicians Pleven, pp. 72–82, 6–10 Apr 2016
7. Maes, P.: Agents that reduce work and information overload. Commun. ACM **37**(7), 30–40 (1994)
8. Wooldridge, M.: An Introduction to MultiAgent Systems. Wiley (2009)
9. Santos, J., Rodrigues, J., Casal, J., Saleem, K., Denisov, V.: Intelligent personal assistants based on internet of things approaches. IEEE Syst. J. **99**, 1–10 (2015)
10. Soliman, M., Guetl, C.: Intelligent pedagogical agents in immersive virtual learning environments: a review. In: The 33th International Convention on Information and Communication Technology, Electronics and Microelectronics (MIPRO 2010), Opatija, Croatia, pp. 827–832, 24–28 May 2010
11. PAL, The PAL Framework. https://pal.sri.com/, Sept 2016
12. CALO, Cognitive Assistant that Learns and Organizes. http://www.ai.sri.com/project/CALO, Sept 2016
13. Companions-project. http://www.companions-project.org/, Sept 2016
14. PAL, Personal Assistant for a healthy Lifestyle. http://www.pal4u.eu/, Oct 2016
15. Siri web page. http://www.apple.com/ios/siri/, Oct 2016
16. Microsoft Cortana web page. https://www.microsoft.com/en-us/cloud-platform/cortana-intelligence-suite, Oct 2016
17. Google Now web page. http://www.digitaltrends.com/mobile/how-to-use-google-now/, Oct 2016
18. LG voice mate webpage. http://www.lg.com/us/mobile-phones/VS985/Userguide/388.html, Oct 2016
19. Komninos, A., Stamou, S.: HealthPal: an intelligent personal medical assistant for supporting the self-monitoring of healthcare in the ageing society. In: Proceedings of the 4th International Workshop Ubiquitous Computing for Pervasive Healthcare Applications, 17–21 Sept 2006
20. Chen, Z., Lin, M., Chen, F., Lane, N.D., Cardone, G., Wang, R., Li, T., Chen, Y., Choudhury, T., Campbell, A.T.: Unobtrusive sleep monitoring using smartphones. In: Proceedings of the 7th International Conference on Pervasive Computing Technologies for Healthcare, Venice, Italy, pp. 145–152, 5–8 May 2013
21. Tang, Y., Wang, S., Chen, Y., Chen, Z.: PPCare: a personal and pervasive health care system for the elderly. In: Proceedings of the 9th Conference on Ubiquitous Intelligence and Computing. 9th International Conference on Autonomic and Trusted Computing, Fukuoka, Japan, pp. 935–939, 4–7 Sept 2012
22. Panchanathan, S., Chakraborty, S., McDaniel, T.: Social interaction assistant: a person-centered, approach to enrich social interactions for individuals with visual impairments. IEEE J. Sel. Top. Signal Process. **10**(5), 942–951 (2016)
23. Tokunaga, S., Horiuchi, H., Tamamizu, K., Saiki, S., Nakamura, M., Yasuda, K.: Deploying service integration agent for personalized smart elderly care. In: 15th IEEE/ACIS International Conference on Computer and Information Science (ICIS 2016), Okayama, Japan, pp. 897–902, 26–29 June 2016
24. Santos, J., Silva, B., Rodrigues, J., Casal, J., Saleem, K.: Internet of things mobile gateway services for intelligent personal assistants. In: 17th International Conference on E-health Networking, Application and Services (HealthCom), Boston, MA, USA, pp. 311–316, 14–17 Oct 2015

25. Zhang, Y., Hu, Y., Zhang, P., Zhang, W.: Design of personal assistant robot with interactive interface of learning and working. In: Fifth International Conference on Intelligent Systems, Modelling and Simulation, Langkawi, Malaysia, pp. 502–506, 27–29 Jan 2014
26. Soliman, M., Elsaadany, A.: Smart immersive education for smart cities with support via intelligent pedagogical agents. In: The 39th International Convention on Information and Communication Technology, Electronics and Microelectronics, vol. 10
27. Motta, G., Sacco, D., Ma, T., You, L., Liu, K.: Personal mobility service system in urban areas: the IRMA project. In: 2015 IEEE Symposium on Service-Oriented System Engineering (SOSE), San Francisco, CA, USA, pp. 88–97, 30 Mar–3 Apr 2015
28. Nack, L., Roor, R., Karg, M., Birth, O., Kirsch, A., Leibe, B., Strassberger, M.: Acquisition and use of mobility habits for personal assistants. In: 2015 IEEE 18th International Conference on Intelligent Transportation Systems, Las Palmas de Gran Canaria, Spain, pp. 1500–1505, 15–18 Sept 2015
29. Sugawara, K., Moulin, C., Kaeri, Y., Barthès, J.P., Manabe, Y.: An application of an agent space connecting real space and digital space. In: Patel, S., Wang, Y., Kinsner, W., Patel, D., Fariello, G., Zadeh, L.A. (eds.) Proceedings of the 2014 IEEE 13th International Conference on Cognitive Informatics and Cognitive Computing (ICCI*CC'14), pp. 230–235
30. Valkanov, V., Stoyanov, S., Valkanova, V.: Virtual education space. J. Commun. Comput. **13** (2), pp. 64–76 (2016). ISSN: 1548-7709 (Print) 1930-1553 (Online) (Serial Number 124)
31. Liu, B., et al.: Intelligent spaces: an overview. In: IEEE International Conference on Vehicular Electronics and Safety, Beijing, 13–15 Dec 2007. ISBN: 978-1-4244-1266-2
32. Wang, F.-Y.: Driving into the future with ITS. IEEE Intell. Syst. **21**(3), 94–95 (2006)
33. Dey, A.K.: Understanding and using context. Pers. Ubiquitous Comput. J. **5**(1), 4–7 (2001)
34. Dey, A.K., Abowd, G.D.: Towards a better understanding of context and context-awareness. In: Proceedings of the Workshop on the What, Who, Where, When and How of Context-Awareness. ACM Press, New York (2000)
35. Agent Communication Language Specifications. http://www.fipa.org/repository/aclspecs.html, Oct 2016
36. Rao, A.S., Georgeff, M., BDI Agents: from theory to practice. In: Proceedings of the 1st International Conference on Multi-Agent Systems, San Francisco, CA, pp. 312–319 (1995)
37. Stoyanov, S., Popchev, I., Doychev, E., Mitev, D., Valkanov, V., Stoyanova-Doycheva, A., Valkanova, V., Minov, I.: DeLC educational portal, cybernetics and information technologies (CIT). Bulg. Acad. Sci. **10**(3), 49–69 (2010)
38. Gramatova, K., Stoyanov, S., Doychev, E., Valkanov, V.: Integration of eTesting in an IoT eLearning ecosystem—virtual eLearning space. In: BCI '15, 02–04 Sept 2015, Craiova, Romania, © 2015. ACM, ISBN: 978-1-4503-3335-1/15/09. http://dx.doi.org/10.1145/2801081.2801086, Art. 14
39. Russell, S., Norvig, P.: Artificial Intelligence: A Modern Approach. Prentice Hall (2011)
40. Bloom, B.S.: Taxonomy of Educational Objectives, Handbook I: The Cognitive Domain. David McKay Co Inc., New York (1956)
41. Pokahr, A., Braubach, L., Lamersdorf, W.: Jadex: implementing a BDI-infrastructure for JADE Agents. Search Innov. (Special Issue on JADE) **3**(3), 76–85 (2003)
42. Bellifemine, F.L., Caire, G., Greenwood, D.: Developing Multi-Agent Systems with JADE. Wiley (2007)
43. Doychev, E., Stoyanova-Doycheva, A., Stoyanov, S., Ivanova, V.: Agent-based support of a virtual eLearning space. In: Nguyen, N.T., Manalopoulos, Y., Iliadis, L., Trawinski, B. (eds.) Proceedings of 8th International Conference ICCCI 2016, Part II, Halkidiki, Greece, Computational Collective Intelligence, pp. 35–44, 28–30 Sept 2016. Springer
44. CMU Sphinx. http://cmusphinx.sourceforge.net/, Nov 2016
45. Google voice recognition. https://cloud.google.com/speech/docs, Nov 2016
46. The Dynamic Module System for Java. https://www.osgi.org/
47. JADE. http://jade.tilab.com/download/add-ons/

Practical Guidelines for Design of Human-in-the-Loop Systems: Lessons Learned

Vasily Moshnyaga

Abstract Technology has entered the age of smart systems, which not only implement traditional human functions of analyzing real-world data and making decisions but also employ humans within the feedback control loop. The development of such systems is not trivial, however, involving several challenges. In this chapter, we share our experience of building human-in-the-loop systems, such as a user-aware computer display, a viewer-conscious TV, a smart door, a smart carpet, a medication adherence control system and smart in-home system for monitoring people with cognitive impairment. We identify problems related to incorporating humans into the control loop and present guidelines for hardware and software designers.

1 Introduction

Advances in ICT and electronics engineering technologies have led to significant increase in smart devices and systems that can perceive environment and create actions similar to humans. According to IDC forecast [1], there will be 80 billion smart internet-connected devices in the year 2025, five times more than in 2012. Unlike existing systems, which aid their users with information processing, communication, industrial and environmental control, future systems are expected to have humans in the loop (HIL), controlling both environment and humans and making decisions. Such systems present exciting opportunities for a wide range of applications such as energy management, healthcare, automotive systems, environment monitoring, etc. [2].

A typical autonomous smart system keeps humans out of the control loop, allowing him/her to intervene if necessary, as shown in Fig. 1a. It generates actuation decisions and control commands based on environmental sensors. An HIL system

V. Moshnyaga (✉)
Department of Electronics Engineering and Computer Science, Fukuoka University,
8-19-1 Nanakuma, Jonan-Ku, Fukuoka 814-0180, Japan
e-mail: vasily@fukuoka-u.ac.jp

© Springer International Publishing AG, part of Springer Nature 2018
V. Sgurev et al. (eds.), *Practical Issues of Intelligent Innovations*,
Studies in Systems, Decision and Control 140,
https://doi.org/10.1007/978-3-319-78437-3_7

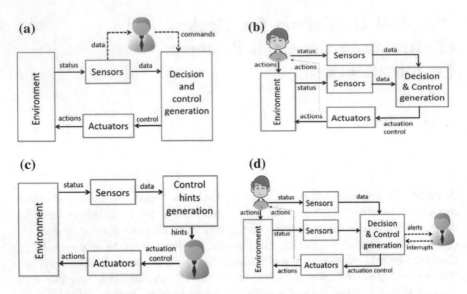

Fig. 1 Smart system organization: **a** conventional with human out of the control loop; **b** HIL system with passive human involvement in control; **c** HIL system with active human control; HIL system with hybrid control

includes humans within the control loop. In contrast to conventional one, it "infers the user's intent by measuring human activities through sensors and translates the intent into control signals to interact with the environment via actuators" [3].

The human involvement in the system control can be passive, active and hybrid. Passive involvement occurs when the system monitors the human status (Fig. 1b) and takes appropriate actions to adjust the environment and sometimes regulate the human's health [4]. Active involvement means that the human performs some of the main control decisions, while the system generates hints to simplify the decision-making process for him or her, as shown in Fig. 1c. Finally, the hybrid involvement means that system not only monitors the human but also accepts human input for decision making and control generation (see Fig. 1d).

Developing an HIL system is a complex and very challenging process [3, 4]. Unlike the conventional one, which involves exploring a variety of design choices and alternatives in the conceptual design phase, making a system prototype and validating its operation, building a smart HIL system involves incorporating humans into the system control and hence requires not only understanding, modeling and realization of functions to be performed by the system but also understanding peculiarities of human behavior. Building an HIL system demands knowledge across diverse domains, such as engineering, computer science, design, social sciences and relevant health-oriented sciences (e.g. gerontology). Recent research from Fujitsu found 38% of European companies already have a shortage of skilful designers [5]. As the industry moves towards smart HIL technologies, the lack of skilful talents capable of carrying out their implementation will become even more prevalent [5].

Since 2007, our group has been involved in the development of a number of HIL systems, which allow the monitoring of humans and take appropriate actions depending on the human behavior. Through this experience, we have learned a number lessons, which we think could be useful to others to cope with the difficulties of smart system design and make better design choices.

In this paper, we discuss lessons learned from building HIL smart systems. In contrast to related studies [6–9], which concentrate on either the team management, the user's acceptance or technology deployment, we focus here on problems associated with the design. Namely, we identify challenges that software and hardware developers can face and outline design guidelines to increase the system efficiency and acceptability. The objective of the guidelines is twofold:

 (i) to prevent design errors that might have profound consequences and
(ii) to shorten the design phase, which proved to be tedious and expensive.

We believe that our experience and the guidelines presented in the paper will help young and inexperienced researchers to overcome problems in developing their HIL systems.

The chapter is organized as follows. The next section outlines the smart systems we developed. Section 3 describes the lessons learned. Section 4 presents our conclusions.

2 The Systems We Developed

The developed HIL systems can be divided into two groups:

(1) *Systems for enhancing the energy efficiency of general purpose electronic devices.*

   • User-aware computer display,
   • Viewer-conscious TV set.

(2) *Systems for assisting independent living of people with cognitive impairment:*

   • Mobile face recognition system.
   • Smart door
   • Smart carpet
   • In-home patient monitoring system
   • Medication adherence monitoring system

The first two systems of the second group are dedicated to assisting people with cognitive deficiency with face recognition and opening/closing of doors without keys. The other systems of the group are intended to help caregivers to remotely

monitor the patient's activity and medication adherence. Below we briefly describe
the developed systems.

2.1 User-Aware Computer Display [10]

Our goal in designing this system was to increase the energy efficiency of the
computer display through user "monitoring". The system contains a camera and
controller (see Fig. 2a) which detects the user's gaze from the camera readings and
based on this activates the appropriate power mode. Namely, when the user looks at
the screen, the display is kept in the "power-up" mode to provide the best visibility.
If the user shifts his/her attention from the screen, the controller dims the backlight
luminance to decrease energy consumption. Finally, if the user has not been looking
at the screen for a long time or has disappeared from the camera's view, the display
is turned off. With this protocol, the human is engaged in the system control loop
passively. The controller takes the camera images as inputs, determines the user's
gaze and adjusts the display's operating voltage and the screen brightness as outputs. The control has to maintain a consistent user experience while saving the
display energy when user attention is distracted. Given that computer users may
have a large variation of face illumination and face inclination/rotation, providing
accurate and real-time unobtrusive detection of the user's gaze with a low power
overhead was quite challenging.

The system implements closed-loop control accurately responding to the user in
real time. To reduce energy overhead, it is fully realized in the hardware, composed
of a FPGA-based gaze detector and analog voltage controller embedded into the
display as shown in Fig. 2b. Experimental evaluation of the system prototype

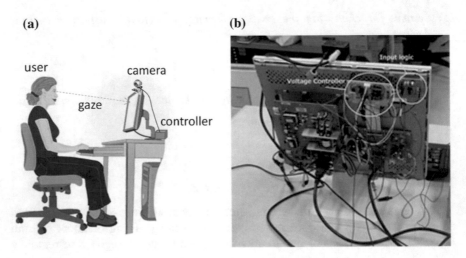

Fig. 2 Illustration of a user-aware display (**a**) and its implementation (**b**)

showed that it can reduce display energy consumption by up to 40% on typical usage patterns in comparison with adaptive power management that activates low-power display modes after fixed periods of inactivity [10].

2.2 Viewer-Conscious TV Energy Management System [11]

This system monitors the TV viewers in order to dim the TV screen when nobody is looking at it. The idea is similar to the user-aware computer display (see Sect. 2.1) with the difference that the viewer-conscious TV management system employs a camera for face rather than gaze detection. When turned on, the TV has a pre-set brightness to provide good picture quality. For every image frame captured by the camera, the system runs face detection and dims the screen brightness if no human face is detected in N consecutive frames. If the screen brightness cannot be dimmed anymore and the human face has not been detected in N consecutive frames, the screen is turned to the sleep mode to save energy while the sound still remains ON. Any detected face reactivates the initial screen brightness by changing the TV onto the normal mode.

The biggest challenge in developing this system was similar to the previous one. Namely, achieving accurate, real-time and low-power face detection in conditions of unrestricted room brightness and face rotation at a distance of up to 3 m from the camera. The developed prototype executes close-loop feedback control with passive human interaction. The system's face-detection module was implemented in software on the ARM-based "Beagle-board", while the screen power controller was realized in specific hardware. Figure 3 shows the system's block diagram and screenshots of the TV and the face detection in "normal", "low-power" and "screen-off" operational modes, in which the total power of TV system was 193 W, 39 W and 10 W, respectively [11].

2.3 Smart Door [12]

The smart door is a keyless HIL control system that automatically opens a door for predefined people only. When someone pushes the outside handle of the door, the system microcontroller (see Fig. 4) receives the request to open the door and activates an electronic circuit to flash the lamp and take a picture of the person at the door. The captured image is compared by face recognition against people who have permission to enter (i.e. whose faces are preset in the database in advance). If there is a match, the system opens the door, welcoming the person by name, and recording the time and visitor's name in a database. Otherwise, the door remains closed. Pushing the lock handle from the inside opens the door automatically.

Our goal was to make the door operation process easier, better and more secure for the person with cognitive impairment. To achieve that goal, we employed

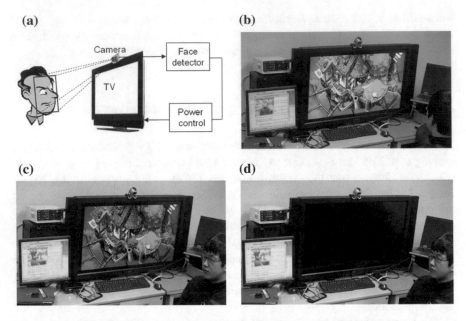

Fig. 3 Block-diagram of the viewer conscious TV (**a**) and screenshots of its operation in normal mode (**b**), low-power mode (**c**) and the screen-off-mode (**d**)

Fig. 4 An illustration of the smart-door system

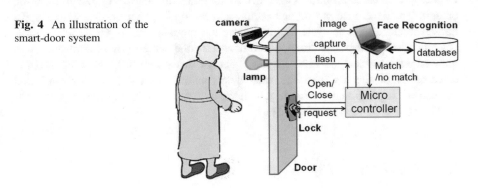

close-loop feedback control with passive human involvement. Providing very accurate face recognition was central to this control [12].

2.4 Smart Carpet [13]

This system was developed to assist caregivers in remote patient monitoring within a room, automatically detecting the location of a person on the carpet, his/her status such as walking, staying, sitting on the floor and lying on the floor, identifying falls, and alerting the caregiver in an emergency via the internet. The smart carpet contains an array of mats, each incorporating a pressure sensor located under an expanse of carpeting.

Figure 5 (left) illustrates the carpet structure on an example of 4 × 4 mats. If no pressure is made over the mat, the sensor idles. Otherwise, the sensor sends a signal through horizontal and vertical lines to a microcontroller which computes coordinates of the active mats and wirelessly transfers them (via XBee transmitter) to a server. The server decodes the signals received, evaluates the pressure pattern and if it corresponds to a fall sends an alert to the caregiver via the internet. Because each mat has a rigid location within the carpet, the position of a person on the carpet is simply determined from the reading the mat sensors. Walking, staying, sitting or lying over the carpet is modeled by a proper signal pattern.

Figure 5 (right) exemplifies the patterns displayed on the server PC and smartphone when a person lies on the carpet. The mats that sensed the pressure are shown in bright color. As the test revealed [13], the carpet sensors efficiently detect gait characteristics and are not perceptible to the users.

2.5 In-home Patient Monitoring System [14]

This system automatically monitors a person with cognitive impairment at home, determines his/her current status (e.g. lying in bed/on the sofa/floor, sitting on the chair/sofa/bed/floor, walking, exiting the room, visiting the toilet, bath, attempting to open home doors, etc.), detects falls, assesses an emergency for assistance, alerts the caregiver in case of emergency by text message, records the patient's movement, generated alerts, displays the results of patient monitoring on the caregiver's phone or PC, enables real-time video and voice communication with the patient in a case of emergency, provides a graphic interface to display the history of patient's

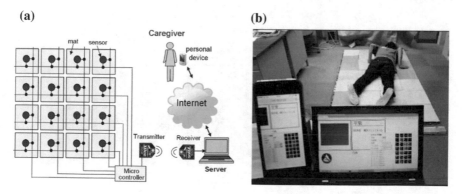

Fig. 5 Organization of the smart carpet having 4 × 4 mats (**a**) and a snapshot of the system testing at smartphone and PC (**b**)

activity, frequency of using facilities, and frequency of alerts. Unlike the above-mentioned systems, this one utilizes a hybrid loop control with the passive involvement of the monitored person and active interaction of the caregiver. The caregiver, for example, can modify the parameters of monitoring and activity assessment, and initiate real-time audio and video communication with the monitored person if necessary.

Structurally the system contains a network of heterogeneous sensors installed in the apartment of a dementia patient and a server, as shown in Fig. 6. The sensor network includes a Smart Carpet (SC), Bed Sensor (BS), Door Sensors (DS), Motion Sensors (MS), video cameras (or video sensors) and microphones. The data from the sensors are wirelessly transmitted to the server that fuses the data and uses artificial intelligence to assess the activity of the person, needs for assistance, and generates alerts to the caregiver. The results of activity monitoring and assessment are recorded in the database and can be viewed online from the caregiver's personal device or PC. In the design of this system, we faced several challenges related to building a network of inexpensive unobtrusive heterogeneous sensors, fusing their data and accurate assessment of the human behavior. Although the system prototype still has a long list for improvement, it has already shown performance superior to related designs in the accuracy of real-time activity recognition [15].

2.6   Smart Medication Adherence Monitoring System [16]

This hybrid-type HIL system was developed to assist a memory-impaired person with medication adherence. Unlike existing smart pill-dispensers, it not only provides the prescribed medication dosages at predefined times but also unobtrusively monitors the actual medical adherence of the user. The system consists of two main units: the dosage unit and the PC-based controller as shown in Fig. 7. The dosage unit stores medication doses preset by the caregiver. The controller is programmed to implement open-loop control of the following functions:

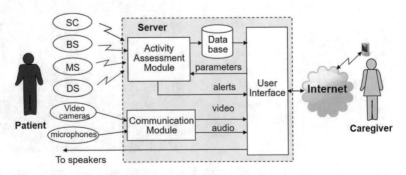

Fig. 6 Organization of caregiver assisting system

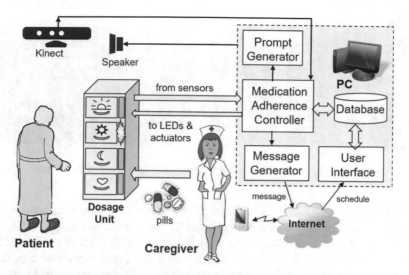

Fig. 7 Organization of the medication adherence monitoring system

- To assist the caregiver to set both the medication doses and the intake schedule.
- To remind the patient of the time for medication intake.
- To prevent the patient from incorrect use of prescribed medications. At the time of medication intake, the patient has access to the correct medication dose only, while all the other doses are locked.
- To guide the patient through steps of medication intake by vocal prompts.
- To monitor the correctness of each step.
- To record in a database the results of monitoring in terms of completeness/incompleteness of the steps and problems observed if any.
- To send the caregiver a text message on the results of monitoring.
- To display the patient's medication adherence history for a given period.

In order to assess and supervise the patient's activity, the system contains a visual sensor (Kinect), speakers, and an array of simple Reed-switch sensors, micro-actuators, and LEDs. The Reed-switches, sensors, actuators and LEDs are built into the dosage unit, while Kinect and the speakers are directly connected to a PC. The results of monitoring are stored in the system database and can be viewed online from a PC or online from the caregiver's personal device.

The system operates as follows. By using the system interface, the caregiver unlocks the dosage unit, fills medication doses into compartments and sets up the intake schedule. As the schedule is saved in the database, the compartments are automatically locked. When the time comes, the medication adherence controller unlocks the compartment with the correct dose, activates the prompt generator to vocally remind the patient to take medication, and initiates the patient's monitoring by Kinect. To help the patient find the required dose, only the LED of the unlocked compartment is switched ON (see Fig. 7), while all the other compartments are

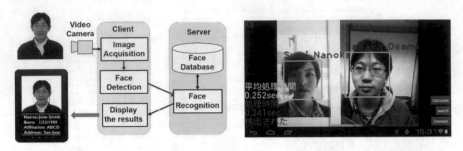

Fig. 8 The face recognition process (left) and the snapshot of the recognition results (right)

locked and have their LEDs OFF. As the patient opens the compartment, the system vocally prompts him/her on the correct steps for medication intake. If it recognizes that the patient completed the intake activity successfully, it sends an informative message to the caregiver. Otherwise, it prompts the patient again to repeat the uncompleted actions (e.g. intake the medication; close the compartment, etc.) If after several reminders the action remains still incomplete, the system alerts the caregiver about the problem by a corresponding message, switches the LEDs OFF and returns to the scheduled monitoring. Robust real-time recognition of patient's activity was the most challenging issue in the development of this HIL system.

2.7 Mobile Face-Recognition System [17]

The mobile face-recognition system is a standalone software application that was developed to help a forgetful person identify the names of people. With the system, a user needs only to direct the camera of his/her mobile device to a person of interest and view the results on the display, as shown in Fig. 8 (left). The face detection/recognition tasks are performed in real time using computational resources and the database of a high-performance network server (i.e. cloud). If there is a match, the results in terms of a face rectangle and data identifying the person of interest are displayed on the screen. Otherwise, the person is identified as "unknown". Figure 8 (right) illustrates the result of mobile recognition of multiple faces at the same time. Due to new offloading architecture, the system was faster than related designs by an order of magnitude [17].

3   Lessons Learned

From building the systems we learned a number of lessons which can be summarized as follows.

3.1 Correct Modeling of Human Behavior Is Extremely Difficult

An HIL system literally integrates a human with the physical or engineered environment. Humans cannot be engineered and hence must be in the HIL system design by corresponding mathematical models. In addition to models, designers need algorithms to interpret human intent, imitate decision-making and simulate the system-human interaction. Obviously, these models and algorithms must be robust, reliable and efficient. Even though our systems were relatively simple, the correct modeling of human behavior in the systems was quite hard due to complex physiological, psychological and behavioral aspects of human beings. This task was particularly challenging in the design of the medication adherence system, whose user was assumed to be at home, unrestricted in movement and actions, and having mild or moderate memory loss. Due to the cognitive impairment, the user might interrupt his medication intake with another activity, such as wandering, moving around or sitting while keeping the medicine in his or her hand. The person might either put the medicine somewhere thus absolutely forgetting about intake or complete the intake activity later after recovering the medicine. Modeling the behavior of cognitively impaired person was paramount to design.

To model a human as a generalized element of feedback control, we adopted the loop structure shown in Fig. 9. The human observes signals S(t), generated by the system in terms of vocal prompts and LED flashing, and on the basis of these observations takes action A(t) (e.g. opens a compartment in the dosage unit, takes the medication dose, etc.). The system monitors both the human (through Kinect) and the environment (through sensors) taking the sensor readings as inputs, X(t), for decision making and control. The system task is to generate proper outputs Z(t), which will enforce humans to act "close" to the desired sequence of actions A'(t) as time evolves. Generally speaking, there are random and/or bias inputs that disturb humans from the desired activity. The cause of disturbances can be external (environmental) or internal; that is related to a physiological, psychological, cognitive and behavioral condition of the person. Clearly, these superfluous input deviations

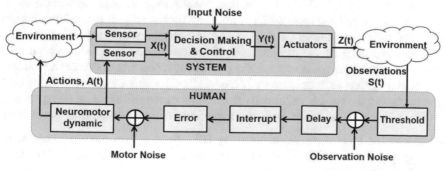

Fig. 9 Human as a generalized feedback control element

must be reflected by the model. The basic question is how should the model count the internal and external variations inherent in dynamic human behavior?

To cope with the psychophysical variations in human behavior, the model includes specific operators: Threshold, Delay, Interrupt, Error, Neuro-motor dynamic and noise operators (see Fig. 9). The *Threshold* operator specifies discrepancies in human perception capability. Due to visual and hearing deficiencies, people may not react to vocal prompts or visual signals of small amplitude. Alternatively, some may not understand the meaning of the prompt and consequently choose not to react. This threshold phenomena are nonlinear and tend to worsen with the human's age. The *Delay* operator determines human delays associated with audio, visual, processing and neuro-motor pathways. It also depends on the personal perceptual and cognitive ability and age. The *Interrupt* operator describes the possibility of discontinuous human behavior and activity interruption. The *Error* operator predicts random inaccuracy in human response characteristics. The *Neuro-motor* operator models lag between the "intended" action and the actual action executed by the human. Because of central processing and neuro-muscular dynamics, the action cannot be done immediately. Furthermore, as people get older, they lose the ability to control their movements (tremor) and frequently make errors in executing the desired action. Moreover, due to forgetfulness, old people can change their activity "on the flight" and thus do not complete the intended task at all. To specify the neuro-motor dynamic in time, we used the hidden Markov model, in which hidden states correspond to human poses and/or actions (e.g. raising a hand, lowering a hand, etc.) and observations (or activities) are defined by a chain of stochastic hidden states. The anomalies in human perception and action generation are modeled by observation and motor noises, respectively.

Despite the level of modeling differs by application, there are common challenges associated with the:

- Large number of user-specific thresholds and parameters,
- Wide variation of human behavior over the time, and
- Technology capable of sensing the appropriate aspects of human behavior.

The accuracy of modeling strongly depends on collected data. Obtaining necessary data however may not always be possible as people usually hesitate to play roles of "guinea pigs". Collecting the data usually requires multiple repetitions of the same behavioral routines, which is not easy for old people. In addition, human behavior varies dramatically with age and the physiological, psychological, and cognitive condition of the person. Even if we can collect data for a person with cognitive impairment (or dementia), the data become obsolete as dementia worsens. Potential solutions such as learning behavioral models from medical professionals and extracting the models statistically from big data are still far from being solved.

Other observations drawn from our experience can be summarized as follows:

- Models of individuals work better in HIL systems than models obtained as aggregated summaries of isolated functional variables. Especially if an individual suffers from memory loss.

- Models of functional interaction between human and system work better than generic behavioral models of humans.
- Traditional laboratory methods are not applicable for real task simulation. For instance, data accumulated from behavioral patterns of university students was inadequate for modeling an old person.
- Human behavior is difficult to formalize and model. Due to a lack of theory, existing human modeling techniques shortfall with confidence indicators.
- Contrary to data- or goal-oriented methods, human modeling in HIL systems demands predictive and stochastic formulations.
- Human modeling is a tradeoff between simulation accuracy and speed. Accurate modeling requires a large number of states, equations and transition probabilities, and therefore takes a long time to develop and process. On the other hand, simplified models are fast but inaccurate.

Next, HIL systems require advanced technologies for human identification and activity recognition. While advances in human gaze/face/pose recognition are astonishing (see [18] for example), available technologies are still inefficient in real-life environments with various face/pose rotations, face occlusion, face illumination (light variation), face blur from weather conditions, etc. For example, a viewer may watch TV from a distance while laying on sofa at night with face rotated and possibly occluded by glasses, hair or (in Japan) mask. Robust recognition of rotated, inclined and occluded faces from a far distance in a dark room with partial or almost no face illumination remains very difficult. Even for simple settings, the recognition might be unsatisfactory [19].

Furthermore, employing face recognition in a decision-making system (e.g. smart door) must be done with caution, because even the robust techniques fail to distinguish a photo from a real person (see Fig. 10) and thus may grant permission to a wrong person. Smart HIL systems for human behavior recognition and support clearly will require not only more complex (3D) models but also predictive models capable of imitating different variations of human behavior and avoiding problems before they occur. Currently, state-of-the-art techniques that model certain aspects of human behavior are either general or very specific.

Fig. 10 An illustration of inability of an image-based system to distinguish a real person from a photo

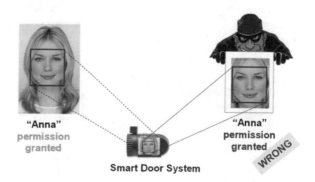

3.2 Design of HIL Smart Systems Requires not Only Technology Integration but Integrated System-Human Modeling

In spite of differences in implementation, existing smart systems utilize a general architecture, shown in Fig. 11. Sensing, power management, energy storage, signal processing, computing, communication, actuation and other technologies necessary to make a system smart to some extent already exist. The main challenge here goes beyond the design of components (an already difficult task in itself). It is in the integration of heterogeneous technologies and components into a system. As the system involves a human in a feedback control, the technologies must be coupled coherently with the human. Although there have been studies on methodologies and tools for automatic design flow from the architectural (board/module) level to the circuit and the physical design levels [20], they are still at the research stage. Also, currently, human and technical models are developed independently. While there are many approaches for modeling and analysis of human motion, motion prediction, and algorithm optimization [20, 21], there are no tools yet capable of supporting simultaneous modeling and virtual human-system simulation at the system design phase.

To engineer an HIL system, one must first identify design criteria (e.g. cost, robustness, response time, integration complexity, installation and maintenance difficulty, etc.) and based on these choose the technology and hardware components to be integrated in the system. It is important that the selection is evaluated from both the human's point of view and the system perspective. The system components are usually acquired as modules (e.g. sensors, RF modules, analog macros, etc.) or implemented using specific methods and CAD tools, while their integration is done from scratch by using ad hoc techniques and high-level modeling tools such as MATLAB-SIMULINK, UPAAL [22] and Transaction-Level Modeling tools [23]. Our experience showed that combining technologies from the analog and digital worlds, embedding them with pre-defined hardware and software modules to implement a prototype of a smart system typically does not require scientific innovation. It is mainly an engineering task which involves time-consuming and costly iterations. At the same time, modeling the human within the control loop, optimizing the system-human interaction, finding solutions which improve the performance or reduce cost or energy consumption usually demands lots of creativity, invention and research.

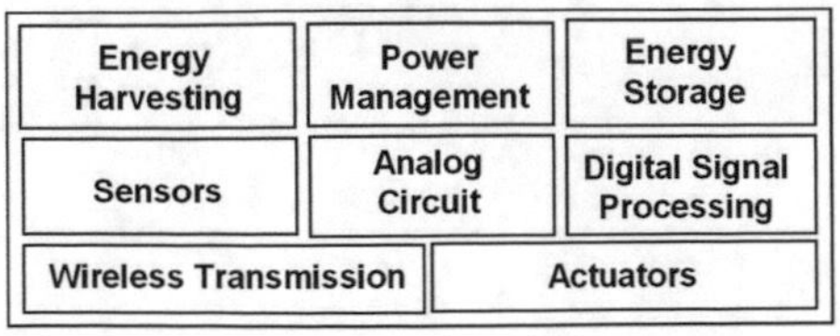

Fig. 11 Typical organization of a smart system

Energy Harvesting	Power Management	Energy Storage
Sensors	Analog Circuit	Digital Signal Processing
Wireless Transmission	Actuators	

The problem is that general-purpose computer-based system simulators such as MATLAB-SIMULINK, UPAAL, etc., are actually not well-suited to model a human in the control loop. Although a system simulator can be used as modeling environment to study system interaction with a human, it assumes that the human is out-of-the-loop by default. In particular, in the case of the medication adherence system, we tried to incorporate the above-mentioned model of a human with cognitive impairment into MATLAB-SIMULINK and apply it to a set of initial parameters. However, the simulation results were quite far from the results obtained from a real experimental data. While simulation results can easily be replicated in time by simply providing identical parameters, humans are unpredictable in their behavior even if environmental conditions remain unchanged. The lack of a design environment capable of virtual modeling of man-machine interaction was one of the main obstacles in assessing the functionality of our HIL systems for performance and reliability.

3.3 System Perception Depends on the Sensing Methods Used

A number of sensing methods can be used to implement system perception. However, all have limitations. For example, video-based methods might be inaccurate during to occlusion, brightness, etc., and also raise privacy concerns. Acoustics-based methods are quite susceptible to ambient noise. Video and acoustics based methods require expensive pre-installation. Pressure-based methods are area restricted. Wearable-based methods use a built-in gyroscope or accelerometer, which must be worn all the time. Besides, the user must know how to operate them, which is often challenging for old people. Despite a wide variety of sensors on the market, none of them alone can satisfy the perception necessary for human health monitoring or identifying complex daily activities, such as medicine intake, eating, etc. The only way to improve perception is to fuse data from multiple heterogeneous sensors, which sense both the environment (vision sensors, IR motion sensors, RFID, pressure sensors, magnetic (Reed) sensors, temperature sensors, moisture sensors, etc.) and the human health (blood-pressure sensor, Pulse-Oximetry Sensor, Temperature, sensor, ECG (heart rate) sensor, EMG sensor, blood glucose sensor, etc.).

When selecting sensors a number of constraints should be considered. Although energy might be the most limited resource in sensor nodes, bandwidth, cost, robustness, invasiveness, and ease of installation can affect the selection. For example, an optimal solution might use 6 sensors. However, due to bandwidth constraints, only 2 can be activated. These 2 sensors might not even be from the optimal set of 6 sensors.

A sensor network conventionally executes background data acquisition to support decision making and control. Periodic readings might be hierarchically assembled and memorized in the back-end with the controller sending actuation commands as necessary. But these protocols may not be well suited for embedded interaction with ordinary users or network extension. In applications with a large number of sensors, devices may automatically determine neighboring sensors by sending identification requests as it done in SenQ systems [24]. Such a request may include filtering criteria for device types or IDs and sensor modes to curtail redundant and unwanted replies.

To fuse data streams together, spatial sensor aggregation, data querying and sensor virtualization can be used. The sensor aggregation means that one sensor (aggregator) is functionally designated to collect low-level data from its neighbors, make a query and re-send it to k-hops in the networks. Sensor virtualization allows the system user to replace a group of sensors of the same type by a virtual entity (tag) transforming their data flow into a metadata stream for simplicity of data interpretation and processing. The virtualization can be done both spatially and functionally. In the former case, the user, for example, can combine all sensors inside a room into a virtual "*Room*" node of a home network to detect any activity within this particular room. In the latter case, the user may create, for example, a virtual sensor "*Fall*" that merges body network sensors to identify whether the wearer has fallen.

Another key element of choosing the sensing technology is justification that all sensors have to be used in the system. In other words, all sensors must satisfy a verifiable and quantifiable set of requirements (e.g. the total system cost, energy consumption, system response time, etc.) that will demonstrate their benefits in the system. As the sensors are allocated close to the source of potential problem phenomena and the data from sensors is processed by the system server, the decision on sensing technology must be done concurrently with technologies employed for data transmission (wired or wireless, e.g. Wi-Fi, Bluetooth, Zig-Bee, etc.), data conversion, data storage, energy supply and so on, as well as the decision on server architecture (centralized or distributed). A good strategy to explore the design alternatives is to start from a simple and inexpensive centralized solution with wired sensors, iteratively changing to decentralized architecture and wireless connections.

3.4 Energy Is Everything

Energy management (power sources, power delivery, operation modes, etc.) is a focus of any system development. Poor energy management affects the system's performance, durability reliability and maintainability. In contrast, effective energy management increases durability and reliability of devices and sensors, reduces energy consumption, heat and cost. If the smart system to be used in a residential environment, low energy consumption might be a crucial issue and an important argument for its user.

We learned that making the energy management event-driven and user-centric is a simple yet very effective approach to save energy. The event-driven means that the system responds and adapts its operation to variations in the environment. For example, many devices react to the change from AC to battery power by automatically dimming the display [25]. The user-centric energy management means that the system operation is conditional on its user in a feedback. In other words, the system neither computes nor delivers results when no-one needs them. For instance, not only display but also the corresponding video playing system can be put into power saving mode or even switched off to save energy when nobody is looking at the screen.

Making the system energy management event/user-driven may require some extra sensors for the physical environment and the ability to generate events or state changes. The system should respond to the event and take actions that will conserve energy—or at least offer options to the user. These options may include a power source (AC vs. battery), remaining charge on the battery, the status of energy consuming devices such as WiFi and Bluetooth radios and possible system reconfigurations, etc. [25]. Our experience shows that just simple data reuse and a sleep mode can save a lot of energy. Both hardware and software must activate sleep mode as much as possible, so the system consumes energy only when it is actually doing something.

To facilitate sleep mode operation the following guidelines can be considered:

- Remove (or at least deactivate) unused background tasks or services.
- Reduce frequency of device pooling and maintenance timers.
- Filter redundant application requests.

To save energy in active mode, several guidelines can be used. Some of them are:

- If there are several algorithms, select one which has less computing complexity and fewer memory accesses.
- Use only required resources.
- Trade off accuracy for energy if possible.
- Keep data and control logic as close to the associated devices as possible to minimize data transfers (it not only saves energy but also retains network scalability).
- Make applications context-aware, i.e. adaptive to the applications, available resources and the user's demands.
- Decrease frame rate, screen resolution, data movement, etc. For instance, a 2 MB image transmitted wirelessly consumes five times more energy than a 200 KB image.
- Displays are main sources of power consumption in any computing system. Because display power consumption is proportional to brightness, dimming the display, making brightness adaptive to the image content is energy efficient.
- Send/upload files less frequently. Use compression to transmit data sets with high compression ratios (>3.0) because it saves energy. However, for data sets with compression ratios less than 1.2x, it is better to transmit in original

(uncompressed) form due to the high compression overhead. A file compression/decompression takes usually 100% of CPU power while a file transfer takes only 4–7% [26].

- Reduce data translation between components. By simplifying the data formats and the translation steps, unnecessary energy consumption can be reduced.
- Refrain from redundant features. Anything that offers no value to the user consumes energy, which can be saved by omitting them. No chatty protocols, frivolous graphics, animations, assisted editing, suggestions for further navigation, etc.
- Optimize communications. Because energy is consumed by each party of the communication loop (e.g. sender, gateways, proxies, receiver, etc.), reduce the number of parties involved in the communication. Fewer communication actions mean less energy consumed. Avoid data broadcasting wherever possible. Replace pooling by published subscribe communication.
- Offload energy-consuming data processing tasks from the battery-operated devices (sensors, control system) to a server or cloud.
- Run multiple applications on shared servers. When two applications, for example, utilize their own servers, both servers will be doing nothing useful most of the time. But they have to be on and consume electricity all of the time. By letting applications share the same server, fewer servers and less energy will be needed to do the same work. Many applications can be easily made suitable to run on shared servers.
- Log less. Logs are used to diagnose application bugs. Storing logging data to disk requires both large space and energy.

Although the abovementioned guidelines are quite helpful in reducing system energy consumption to an extent, a really energy-efficient system can be achieved if and only if energy is considered at all design phases. The modified design process might be as follows. At the requirement engineering phase, the developers specify requirements for the system platform, the dynamic energy management, and test. At the design phase, they select the system platform, develop (or incorporate) a power management application and test interfaces to system components, choose a design environment, which supports power estimation, and develop an event-driven system architecture. At the implementation phase, the tasks include realization of energy-management interfaces and monitoring interfaces, profiling and computing the system energy budgets for the target applications. At the testing phase, the designers develop test tools and run tests to evaluate system operation, energy consumption for each application and the energy hotspots.

3.5 Keep It Simple, Small and Inexpensive

Albert Einstein once said: "Everything should be made as simple as possible but not simpler." In the same way, HIL systems should be made simple, easy-to-use and easy to maintain. The human is the main source of errors and faults in man-machine interaction. While some of the errors are due to stress, fatigue and work overload that users experience daily, a large number of errors can be attributed to poor user interface design [27]. If a system is difficult to operate, has an unreasonable user interface or responds to the human inputs in an unpredictable manner, user mistakes can more easily occur.

In general, the user interface in an HIL system can be either passive or active. The passive interface assumes that the user does not affect the system operation unless he or she induces an exceptional event such as a wire disconnection, a sensor breakup or other hardware failure. The active interface means that the user receives information from the system and can interrupt and modify the system operation by changing system settings. For example, a patient interfaces the caregiver-assisting system passively while the caregiver interfaces the system actively. The errors caused by an active interface are due to a wrong decision and/or an incorrect, mistaken, input. Unfortunately, no user interface can completely eliminate human errors. Nevertheless, it is very important that it alleviates the incidence of errors as well as their consequences as much as possible.

We suggest that user interface specifications are developed at the requirements engineering stage and included into requirements for HIL systems. The User-Centric Design approach [28] or Cognitive Work Analysis [29] can be used to specify human interactions in HIL systems. However, if neither usability specialists nor results of application-specific usability studies are available, International Standards (ISO 9241-110, ANSI/HSF 100) and design principles summarized in [30] could help to improve HIL interface design. Additionally, we found the following guidelines useful.

Guidelines for Improving the Passive User Interface:

- All system components (sensors, actuators, routers, computing elements, etc.) must be invisible and unobtrusive for passive users. They must be hidden or made unnoticeable (colored as background) or allocated in places which are difficult for the user to reach (e.g. placed up on walls or ceiling).
- Wired connections must be minimized, hidden to not obstruct the user.

Guidelines for Improving the Active User Interface:

- Users desire simplicity not complexity. Most users are ordinary people without much computer background. Nevertheless, they have to be able to understand how to operate the systems without reading manuals. For example, many caregivers of people with dementia in Japan are spouses, relatives, neighbors or friends who are 60 + years old. Clearly, devices designated for caregivers must have a user interface understandable for elderly people. Unlike computer applications designated for youngsters and mature computer users, the user interface for the elderly has peculiarities. To prevent an incorrect action, such as pushing a wrong button or selecting a wrong menu, the displayed information must be visible and easy to comprehend by an old person. For instance, consider Fig. 12 which shows two versions of a GUI, developed for our in-home patient monitoring system [15]. Although designers suggested the one shown in Fig. 12 (left) due to its simplicity and possibility to put more information within the screen frame, elderly users preferred the large illustrative icons depicted in Fig. 12 (right). Text-, voice-, and gesture-based user interfaces are quite unpopular among elders.

- The number of steps necessary to complete a task by a user must be minimized whenever possible. The system must not place too much load on the user's cognitive capabilities and short memory because they decrease with stress, fatigue and age. At each step, the system should guide its user by proper cues to prevent skipping of necessary steps. Designers should provide visible, unambiguous and easily accessible controls to recover the user from wrong actions.
- When making a critical decision, a user must have a clear display of the current state of the system and consequences of actions. For example, when setting the medication intake schedule into the medication adherence system [15], the user must have information about his or her current input and consequences of incorrect medication scheduling on the patient's health. Warnings (by active pop-ups and passive alert indicators) with clear instructions how to avoid an error are indispensable for the user in this situation. We also suggest using communication delivery indicators to be sure that the user has not overlooked the warning.

A good approach to improve the HIL user interface is to employ empirical user studies for identification and prevention of potential failures, as proposed in [31]. In comparison to the model-based interface design, it does not require modeling of user behavior, which is quite difficult due to the complexity of man-machine interaction. Including probabilities of certain user actions into the models provides some quantitative estimates of the user's behavior. However unexpected man-machine interactions are hard to predict.

Furthermore, it is not a secret that existing smart systems are expensive and unaffordable for the majority of people today. Developers should remember that

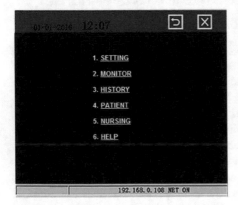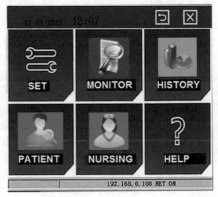

Fig. 12 An example of two user interfaces for a caregiver assisting system

Fig. 13 Two examples of personal communication system

customers (i.e. users) vote by their wallets and frequently select those devices which are less expensive. As an example, consider Fig. 13 which shows two systems capable of delivering remote communication with caregiver or relatives for an old person. Despite the fact that a robot-like system can follow an old person anywhere in the home supporting the communication on the move, most people will rather select a simple portable system, which might be not so advanced and has to be carried around but costs an order of magnitude less. The ground rule is that cost reduction must be an aim at every step of design.

3.6 Systems Must Be Easy to Deploy, Maintain and Extend

Deploying an HIL system containing any small sensors to sense both environment and humans is always a challenge. As more sensors are added to the system, it becomes complicated to handle, maintain, extend and repair. Just selecting an appropriate device from a long list can be quite troublesome.

In a home setting, wireless devices are preferable for users as they do not require long connections and are easy to maintain. Nevertheless, they must be easy to access. Otherwise changing the batteries will be a problem. In comparison to the use of a large number of small wireless sensors, installing a few highly accurate and inexpensive devices gives uniform coverage, unobtrusiveness, ease of deployment, low power consumption, better maintenance and, hence, greater acceptability by users.

Furthermore, system assembling and re-assembling must be as easy as Lego: simple and inexpensive. Figure 14 illustrates the idea of the smart-carpet assembling. Here each mat has a fixed size, shape and number of connectors. To prevent assembling errors (i.e. connecting inputs with inputs or inputs with wrong outputs),

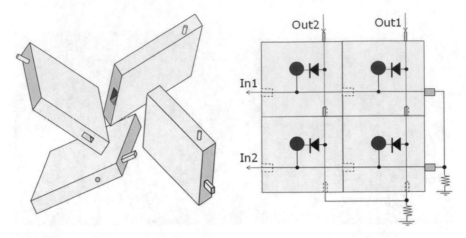

Fig. 14 An example of Lego-motivated assembling of 4 × 4 smart carpet. Mats used in the carpet assembling (**a**), the assembled circuit (**b**)

the connectors must have different shapes and sizes, for instance, cylindrical and rectangular. Also using colored connections to specific pins helps prevent misconnections.

In addition, system reliability and ease of repair have to be a concern in the design of a distributed smart system. As the number of system components grows, the reliability of the system decreases. If the components are distributed and hidden all over the place, detecting a faulty component or a connection becomes a hard task. Providing solutions to automatically distinguish hardware failures in real time could save a lot of time and work. A possible solution is to check signals on inputs/output ports of the system controller. Those ports which return the high-impedance values probably have broken wires.

3.7 Validate HIL Systems by Ordinary People and Common Cases

One important lesson is that all models have shortcomings. Without experimental validation, there is no reason to believe that the system works well. However, laboratory-based experimentation in the best case can only partially verify the target functionality. In a real setting, environmental conditions (room brightness, temperature, available space, etc.) might be quite different. System developers frequently have insufficient knowledge of the system' application environment as well as peculiarities of system users. For instance, using students as subjects in testing a system for monitoring activity of people with dementia is inappropriate, because old and sick people walk, think, and converse differently to young and healthy people. Figure 15 shows an example. In order to really validate the system, the tests

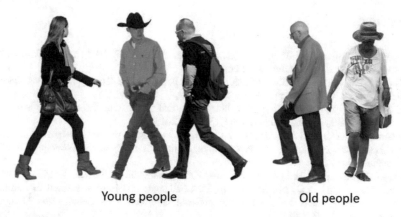

Young people   Old people

Fig. 15 Illustrative walking patterns of young and old people

must be conducted in a real environment and with people for whom the system is dedicated. Also, the test results have to be discussed with both ordinary people and technical experts (professionals) to figure out what is practically important for ordinary people, meaningful and feasible for specialists.

4   Conclusion

We discussed lessons learned from our experience of building HIL systems. Development of such systems requires accurate modeling of human behavior which is very difficult to achieve. Our main goal in writing this chapter was to provide basic information about ways to overcome problems which implicitly appear in the design and implementation of HIL systems and their prototyping. The presented guidelines and design choices are based on our own experience and of course, are not universal. However, as the system prototypes revealed, the guidelines are practical. They help developers to make well-informed choices in the design of systems that are not only smart but also inexpensive, easy to install and maintain. Though our lessons have been learned from application-specific systems, they are generalizable to a wide community of electronic system designers. Hopefully, future studies will help validate this. We invite other researchers to improve the presented guidelines with their own experience in the construction of HIL systems and in the use of specific technology, etc. We are sure each new experience in HIL system design is valuable additional information to the field.

Acknowledgements The author thanks all members of the Computer Systems Lab. of Fukuoka University for their hard work in implementing the smart systems. Without their contribution, this manuscript would not be possible.

References

1. Worldwide Internet of Things Forecast Update: 2016–2020, IDC, Doc.# US40755516. Accessed May 2016 http://www.idc.com
2. Nunes, D.S., Zhang, P., Silva, J.S.: A survey on human-in-the-loop applications towards an internet of all. IEEE Commun. Surv. Tutor **17**(2), 944–965 (2015)
3. Schirner, G., Erdogmus, D., Chowdhury, K., Padir, T.: The future of human-in-loop cyber-physical systems. Computer **46**(1), 36–45 (2013)
4. Munir, S., Stankovic, J.A., Liang, C.J.M., Lin, S.: New cyber physical system challenges for human-in-the loop control. In: Proceedings of the 8th Internationa Workshop Feedback Computer, June 2013
5. Reger, J.: Internet of Things—2017 trends, IOT Business News, 08 Dec 2016
6. Kawahara, Y., Minami, M., Saruwatari, S.: Challenges and lessons learned in building a practical smart space. In: Annual International Conference on Mobile and Ubiquitous Systems, pp. 213–222 (2004)
7. Linton, R.J., Schafeld, J., Padur, T.: Smart wheelchairs or not: lessons learned from discovery interviews. In: Annual International Conference of the IEEE Engineering in Medicine and Biology Society, pp. 5016–5019 (2015)
8. Bauer, L., et al.: Lessons learned from the deployment of a smart phone-based access-control system. In: SOUPS'07, pp. 64–75 (2007)
9. Bouchard, K., Bouchard, B., Bouzouane, A.: Practical guidelines to build smart homes: lessons learned. In: Opportunistic Networking, Smart Home, Smart City, Smart Systems, pp. 1–37. CRC press (2014)
10. Moshnyaga, V.G., Hashimoto, K., Suetsugu, T.: A camera-driven power management of computer display. IEEE Trans. CAS Video Technol. **22**(11), 1542–1553 (2012)
11. Lee, C., Moshnyaga, V.G.: Embedded system for camera-based TV power reduction. In: Euromicro Conference on DSD, pp. 764–768 (2011)
12. White, P., Sumi, N., Hayashida, A., Hashimoto, K., Moshnyaga, V.: Smart Door, Embedded System Symposium (2015)
13. Tanaka, O., Ryu, T., Hayashida, A., Moshnyaga, V.G., Hashimoto, K.: A smart carpet design for monitoring people with dementia. In: Salvaraj, H. et al. (eds.) Progress in System Engineering, Advances in Intelligent Systems, vol.330, pp. 653–659. Springer (2015)
14. Moshnyaga, V., Tanaka, O., Ryu, T., et al.: An intelligent system for assisting family caregivers of dementia people. In: 2014 IEEE International Symposium on Series on Computational Intelligence, pp. 1–5 (2014)
15. Moshnyaga, V., Tanaka, O., Ryu, T.: Identification of basic behavioral activities by heterogeneous sensors of in-home monitoring system, in human behavior. In: Salah, A.A., et al. (eds.) Lecture Notes in Computer Science, vol. 9277, pp. 160–174. Springer (2015)
16. Moshnyaga, V., Koyanagi, M., Hirayama, F., Takahama, A., Hashimoto, K.: A medication adherence monitoring system for people with dementia. In: Proceedings of the IEEE International Conference on Systems, Man and Cybernetics, pp. 194–199 (2016)
17. Imaizumi, K., Moshnyaga, V.: Network-based face recognition on mobile devices. In: IEEE ICCE-Berlin, pp. 406–409 (2013)
18. Taigman, Y., Yang, M., Ranzato, M.A., Wolf, L.: DeepFace: closing the gap to human-level performance in face verification. In: Proceedings of the International Conference on Computer Vision and Pattern Recognition (CVPR), pp. 1701–1708 (2014)
19. Bombieri, N., Drogoudis, D., Gangemi, G., et al.: Addressing the smart systems design challenge: The SMAC platform, Microprocessors and Microsystems, MICRO 2228, Elsevier B.V., 4 June 2015, pp. 1–17
20. Liu, Y., Feyen, R., Tsimhoni, O.: Queueing network-model human processor (QN-MHP): A computational architecture for multitask performance in human-machine systems. ACM Trans. Comput. Human Interact. **13**(1), 37–70 (2006)
21. Booher, H.R.: Handbook of Human Systems Integration. Wiley, Hoboken (2003)

22. Behrmann, G., David, A., larsen, K.G.:. A Tutorial on Uppaal. In: Formal Methods for design of real-time systems, vol.3185, Lecture Notes in Computer Science, pp. 200–236 (2004)
23. Cai, L., Gajski, D.: Transaction level modeling: and overview. In: International Conference on HW/SW Codesign & System Synthesis, Oct.2003, pp. 19–24
24. Wood, A., Selvo, L., Stankovic, J.A.: SenQ: an embedded query system for streaming data in heterogeneous interactive wireless sensor networks. In: Lecture Notes in Computer Science, Distributed Computing in Sensor Systems, vol.5067, pp 531–543 (2008)
25. Larsson, P.:. Energy-efficient software guidelines. Intel Software Solutions Group, Technical Report (2011)
26. Steigerwald, B., Chabukswar, R., Krishnan, K., De Vega, J.: Creating Energy—Efficient Software, Intel White Paper (2008)
27. Insup, L., Sokolsky, O.: Medical cyber physical systems. In: Proceedings of the 47th ACM/IEEE on Design Automation Conference (DAC), pp. 743–748 (2010)
28. Gulliksen, J., Goransson, B., Boivie, I., Blomkvist, S., Persson, J., Cajander, A.: Key principles for user-centered systems design. Behav. Inf. Technol. **22**(6), 397–409 (2003)
29. Rasmussen, J., et al.: Cognitive Systems Engineering. Wiley, NY
30. Rae, A.: Helping the operator in the loop: practical human machine iinterface principles for safe computer controlled systems. In: the Proceedings of the 12th Australian Workshop on Safety related Programmable Systems (SCS '07) (2007)
31. Cranor, L.F.: A framework for reasoning about the human in the loop, CMU-Cylab-08–001, 24 Jan 2008

NEO-Fuzzy Neural Networks for Knowledge Based Modeling and Control of Complex Dynamical Systems

Yancho Todorov and Margarita Terziyska

Abstract Capturing the dynamics and control of fast complex nonlinear systems often requires the application of computationally efficient modeling structures in order to track the system behavior without loss of accuracy and to provide reliable predictions on purpose to process control. An available approach is to employ fuzzy-neural networks, whose abilities to handle dynamical data streams and to build rule-based relationships makes them a flexible solution. A major drawback of the classical fuzzy-neural networks is the large number of parameters associated with the rules premises and consequents parts, which need to be adapted at each discrete time instant. Therefore, in this chapter several structures with reduced number of parameters lying in the framework of a NEO-Fuzzy neuron are proposed. To increase the robustness of the models when addressing to uncommon/ uncertain data variations, Type-2 and Intuitionistic fuzzy logic are introduced. An approach to design a simple NEO-Fuzzy state-space predictive controller shows the potential applicability of the proposed models for process control.

1 Introduction

Combining neural networks and fuzzy systems in one unified framework has become popular in the last decades. The fusion of the fuzzy logic with the neural networks allows to incorporate the learning and computational ability of neural networks with the human like IF-THEN reasoning of a fuzzy system [1, 2]. During the last decades, many Neuro-Fuzzy Networks have been adopted in the practice as:

Y. Todorov (✉)
Department of Chemical and Metallurgical Engineering, Lab. of Automation and Process Control, Aalto University, Kemistintie 1, 02100 Espoo, Finland
e-mail: yancho.todorov@aalto.fi; yancho.todorov@ieee.org

M. Terziyska
Department of Informatics and Statistics, University of Food Technologies, 26 Maritza blvd., Plovdiv, Bulgaria
e-mail: m.terziyska@uft-plovdiv.bg

© Springer International Publishing AG, part of Springer Nature 2018
V. Sgurev et al. (eds.), *Practical Issues of Intelligent Innovations*,
Studies in Systems, Decision and Control 140,
https://doi.org/10.1007/978-3-319-78437-3_8

ANFIS [3], DENFIS [4], NEFCON [5] and etc. The main advantage of these modeling concepts relies on their flexibility to employ data from the process and to adapt the parameters of the network quickly to changing nonlinear process behavior by using fairly standard optimization procedures. A serious drawback for their on-line application for dynamical modeling purposes is the number of parameters under adaptation at each sampling period, since it grows exponentially with the increasing level of nonlinearity. Another disadvantage of the classical neuro-fuzzy systems, especially when they operate in on-line mode is the slow convergence of the conventional gradient-based learning procedures and the computational complexity of second-order ones. In addition, such classical structures cannot handle major process uncertainties in many complex situations.

The idea for a NEO-Fuzzy Neuron (NFN) has been introduced in the early 90s by the works of Uchino and Yamakawa as a potential simpler approach for modeling of highly nonlinear systems. The concept allows to construct a network of artificial neurons, whose structure represents a MISO rule-based element with so-called nonlinear synapses. When the input of a NFN is fired through a vector signal, its output is defined by both the input membership functions and the tunable synaptic weights. Compared to the classical approaches the NFN is similar to a zero order Takagi-Sugeno system and a simple Radial Basis Functions neural network. The most important properties of the NFN's are their computational simplicity, the proven high approximation properties and the possibility of finding the global minimum of the learning criterion in real time [6, 7].

During the last years, different approaches for nonlinear system modeling using the NFN concept are reported in literature. For instance, in [8, 9] authors propose an approach for on-line linear system parameter estimation using a NFN algorithm and a universal approximator employing NFN's. In [10, 11] different NFN topologies are presented, while in [11, 12] respective learning approaches are reported. Practical applications to flux observation in induction motors, bearing condition prediction, stock exchange forecasting and bacteria foraging optimization are shown in [13–16]. An approach to classification task is discussed in [17] and evolving NFN structures are proposed in [18, 19].

Despite the proposed applications, the adoption of the NFN concept in various fields is still limited on purpose to dynamic identification of nonlinear plants and intelligent process control. On the other hand, beyond the numerous modeling structures discussed in the literature less attention is given to the problems related to input data variations and occurring uncertainties caused by undetermined conditions of the environment which may prevent the reliable on-line model operation.

The classical fuzzy-neural approaches have been extended with Type-2 fuzzy logic instead of the generally used Type-1 [20]. Originally, Type-2 fuzzy logic was proposed by Zadeh in response to continuing criticism that Type-1 fuzzy sets can't deal with uncertainties. His theory was subsequently developed by Karnik and Mendel [21, 22]. A Type-2 fuzzy set is characterized by a fuzzy membership function, unlike a Type-1 set where the membership grade is a crisp number in [0, 1]. The idea behind the Type-2 fuzzy logic is well accepted in academics and nowadays lot of scientific papers are dedicated to Type-2 Fuzzy Logic and Type-2 Neuro-Fuzzy Networks [23–31].

In the beginning of 80s Atanassov [32] extended Zadeh's fuzzy set concept to Intuitionistic Fuzzy Sets theory (IFS) by introducing an additional attribute parameter called non-membership degree. IFS has been shown to be superior to Zadeh's fuzzy set in e.g. semantic expression and inference ability. A relevant comparison between the concepts of interval Type-2 fuzzy sets and Intuitionistic Fuzzy Sets is given in [33].

The possibility for combination of ideas from neural networks and IFS are discussed in [34]. Several applications of IFS are reported in literature. For instance, in [35] the mathematical apparatus necessary to design an IFS Feed Forward Neural Network is given. An IFS Neural Network with triangular membership functions is described in [36]. An Adaptive Intuitionistic Fuzzy Inference Systems of Takagi-Sugeno Type is discussed in [37]. In [38], a max-min Intuitionistic fuzzy Hopfield neural network (IFHNN) is proposed.

The application of Type-2 and IFS fuzzy logic implies the realization of additional operations such that the overall number of parameters in a typical fuzzy-neural network will increase sharply, depending on the dimension of the input/output space. Therefore, in most cases it is needed to be made a tradeoff between the modeling accuracy and the computational simplicity, since the great number of parameters under on-line adaptation may limit the applicability of a fuzzy-neural network to capture the behavior of fast dynamical systems. Since, the NFN concept proposes a low level integration of the fuzzy rules into a set of simpler connectionist structures with smaller number of parameters, compared with the classical structures will jeopardize the extension of the concept to structures with Type-2 and IFS fuzzy logic.

In process control, a dynamical model is usually needed to capture and predict the system behavior in order to control a plant process or a system. In most situations, the industrial plants have a slower dynamics within a range of minutes to hours, while specific systems (e.g. chemical and automotive industries) require a response within seconds to a minute. Thus, it is needed to be designed and implemented more computationally efficient modeling tools with less parameters under adaptation, implying less computational time.

In recent years, the model based control systems have become popular in various applications. They use an available dynamical process model which captures and predicts the system response within a time interval ahead. Afterwards, an optimization algorithm on the basis of the obtained model predictions optimizes the future process response over a finite time horizon by minimizing an economic cost while assuming a set on imposed operating constraints. Usually, this control concept requires a substantial computational effort to be executed on-line. Therefore, the potential reduction of the adapted parameters will bring a significant impact on the control approach as a whole and broaden its potential applicability in control of various dynamical systems.

The potential of the NFN's to express knowledge and to build relationships upon dynamically changing input vector with less parameters make them a possible reliable solution to be included in model based control schemes.

In this chapter, several modeling structures and model based predictive controller lying of the NFN concept are proposed. The potential effects of application of Type-2 and IFS fuzzy logic, as well as an approach to design a multi dimension input/output NFN network are studied by numerical experiments to capture and estimate the dynamically changing behavior of several chaotic time series while uncertain conditions occur. To simplify the implementation of a possible model based predictive controller, a realization described in a state-space domain is proposed. The advantages of the designed controller are studied by numerical experiments to control a nonlinear drying plant.

2 NEO-Fuzzy Neural Networks

2.1 NEO-Fuzzy Concept

The idea for a NEO-Fuzzy neuron was proposed by Yamakawa et al. [6] and its simple structure is shown on Fig. 1.

The NEO-Fuzzy neuron has a similar structure to a zero order Sugeno fuzzy inference system where only one input is included in each fuzzy rule, and to a radial basis function network (RBFN) with scalar arguments of the basis functions [12]. The NEO-Fuzzy neuron has a nonlinear synaptic transfer characteristic. The nonlinear synapse is realized by a set of fuzzy implication rules. The output of the NEO-Fuzzy neuron is obtained by the following equation:

$$f(x) = \sum_{j=1}^{m} \mu_{ij}(x(k))w_{ij} \tag{1}$$

where $x(k)$ is the input, w_j is a weight coefficient and μ_j for $j = 1{:}m$ is a defined set of Gaussian membership functions:

$$\mu_{ij}(x_i) = \exp - \left(\frac{x_i - c_{ij}}{2\sigma_{ij}} \right)^2 \tag{2}$$

with corresponding center—c and width (standard deviation)—σ.

Each nonlinear synapse is expressed by a fuzzy rule matching to singleton rule consequents:

Fig. 1 Structure of a single
NEO-Fuzzy neuron

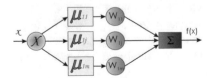

$$if\ x_i\ is\ \tilde{A}_{ij}\ then\ w_{ij} \tag{3}$$

where A is a corresponding fuzzy set.

Using the basic concept for a NEO-Fuzzy neuron it can be easily designed networks of such neurons able to capture the dynamics of a multiple inputs and outputs. It should be mentioned that compared with the classical neuro-fuzzy systems, the neo-fuzzy neurons do not impose adjustment of the fuzzy rule premise parameters, only the rule consequents are under adaptation during each discrete time learning instant.

2.2 Design of Multi-Input–Multi-Output NEO-Fuzzy Network

The five-layer network structure of the proposed MIMO NEO-Fuzzy network is shown on Fig. 2. The nodes in the first layer accept the input variables and transmit them to the next layer directly. The second layer performs fuzzification using a set of a three membership functions described by (2). In the assumed typical case of two outputs, in the third layer the obtained membership degrees are multiplied by two different vectors of weighting coefficients, depending on the cross-relationships

Fig. 2 Structure of a MIMO NEO-Fuzzy network

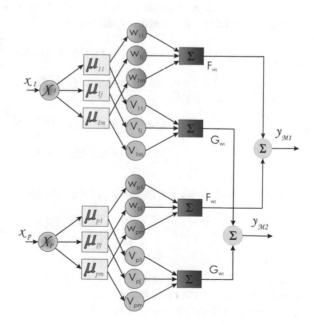

between the inputs and outputs. Thus, in the fourth layer two groups of functionals are computed:

$$F_w(x) = \sum_{j=1}^{m} \mu_{ij}(x(k))w_{ij} \quad G_w(x) = \sum_{j=1}^{m} \mu_{ij}(x(k))v_{ij} \tag{4}$$

where m is the number of the fuzzy rules along a neuron and $w(k)$ and $v(k)$ are the corresponding vectors of synaptic weights. The outputs of the MIMO NEO-Fuzzy Network are obtained in the last fifth layer by simple sum of the outputs of each neuron along the defined output relation:

$$y_{m1} = \sum_{i=1}^{p} F_{wi}(x) \quad y_{m2} = \sum_{i=1}^{p} G_{wi}(x) \tag{5}$$

To train the proposed MIMO NEO-Fuzzy Network an unsupervised learning scheme has been used. For that purpose, a defined error cost terms are being minimized at each sampling period in order to update the two group weights in the consequent part of the fuzzy rules:

$$E_1 = \varepsilon_1^2/2 = (y_{m1}(k) - \hat{y}_{m1}(k))^2/2 \quad E_2 = \varepsilon_2^2/2 = (y_{m2}(k) - \hat{y}_{m2}(k))^2/2 \tag{6}$$

where y_{m1} and y_{m2} are the reference outputs and $\hat{y}_{m1}$ and $\hat{y}_{m2}$ are the outputs being estimated by the model.

To learn the model parameters, the conventional gradient descent procedure results to the following weights update rule:

$$w(k+1) = w(k) + \Delta w(k) = w(k) + \eta\left(\frac{\partial E_1(k)}{\partial w(k)}\right) \tag{7}$$
$$= w(k) + \eta\varepsilon_1(k)\mu_{ij}(x_i(k))$$

$$v(k+1) = v(k) + \Delta v(k) = v(k) + \eta\left(\frac{\partial E_2(k)}{\partial v(k)}\right) \tag{8}$$
$$= v(k) + \eta\varepsilon_2(k)\mu_{ij}(x_i(k))$$

by using in both cases the following chain rule notation:

$$\Delta\alpha = -\eta\frac{\partial E}{\partial\alpha} = -\eta\frac{\partial E}{\partial\hat{y}_{mi}}\frac{\partial\hat{y}_{mi}}{\partial\alpha} = -\eta\varepsilon\frac{\partial\hat{y}_{mi}}{\partial\alpha} \tag{9}$$

where $\alpha = [w, v]$ is a vector of the trained parameters: the synaptic links in the consequent part of the rules and η is an adaptive learning rate whose optimal value is defined by following expression [39]:

$$\eta(k) = 0.1 * \left\| \mu_{ij}(x_i(k)) \right\|^2 \tag{10}$$

2.3 Type-2 NEO-Fuzzy Network

Using Type-2 Interval Fuzzy sets, each element of the model's input vector is being fuzzified in the following way:

$$\mu_{ij}(x_i) = - \exp\left(\frac{x_i - c_{ij}}{2\sigma_{ij}}\right)^2 = \begin{cases} \bar{\mu}_{ij} \text{ as } \sigma_{ij} = \bar{\sigma}_{ij} \\ \underline{\mu}_{ij} \text{ as } \sigma_{ij} = \underline{\sigma}_{ij} \end{cases} \tag{11}$$

where μ is the membership degree defined by a Gaussian membership function with uncertain variance and c and σ represent the center (mean) and the width (standard deviation) depending on the defined FOotprint of Uncertainty (FOU). Graphically the fuzzification procedure is demonstrated on Fig. 3.

The fuzzy inference should match the output of the fuzzifier with fuzzy logic rules performing fuzzy implication and approximation reasoning in the following way:

$$\mu_{ij}* = \begin{cases} \bar{\mu}_{ij}* = \prod_{i=1}^{n} \bar{\mu}_{ij} \\ \underline{\mu}_{ij}{}^* = \prod_{i=1}^{n} \underline{\mu}_{ij} \end{cases} \tag{12}$$

The output of the network is produced by implementing consequence matching and linear combination as follows:

$$\begin{aligned} \hat{y}(k) &= \frac{1}{2} \sum_{j=1}^{l} \left(\bar{\mu}_{ij}*(x_i(k)) + \underline{\mu}_{ij}*x_i(k) \right) f_i(x_i) \\ &= \frac{1}{2} \sum_{i=1}^{l} \left(\bar{\mu}_{ij}*x_i(k) + \underline{\mu}_{ij}*x_i(k) \right) w_{ij} \end{aligned} \tag{13}$$

which in fact represents a weighted product composition of the ith input to jth synaptic weight, as presented on Fig. 4.

Fig. 3 Gaussian Membership Function with uncertain upper and lower membership functions

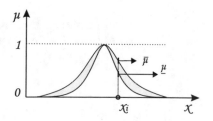

Fig. 4 Structure of the
proposed Type-2 Neo-Fuzzy
Neural Network

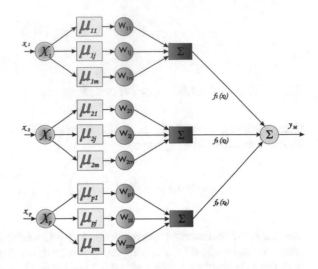

To train the proposed modeling structure, a defined error cost term is being minimized at each sampling period in order to update the weights in the consequent part of the fuzzy rules:

$$E = \varepsilon^2/2 \text{ and } \varepsilon(k) = y_m(k) - \hat{y}_m(k) \tag{14}$$

where y_m is the reference output measured form the process and $\hat{y}_m$ is the output being estimated by the model. As learning approach the gradient descent algorithm with adaptive heuristic feature is proposed. Thus, form (9) and (14) the vector w of the adjusted parameters is calculated using the *signum* of the gradient as:

$$w(k+1) = w(k) + \Delta w(k) = w(k) + \eta sign\left(\frac{\partial E(k)}{\partial w(k)}\right) = w(k) + \eta sign\left(\frac{\partial E(k)}{\partial \hat{y}(k)}\frac{\partial \hat{y}(k)}{\partial w(k)}\right)$$

$$= w(k) + \eta sign\left(\varepsilon(k)(\bar{\mu}_{ij}^*(x_i(k)) + \underline{\mu}_{ij}^*(x_i(k))/2\right) \tag{15}$$

where η is the learning rate which is local to each synaptic weight and it is adjusted by taking into account the extent of the gradient in the current and the past sample period as:

$$\eta_{ij}(k)\begin{cases} \min\left(a\eta_{ij}(k-1), \eta_{\max}\right) & if \ \Delta E_{ij}(k)\Delta E_{ij}(k-1) > 0 \\ \max\left(b\eta_{ij}(k-1), \eta_{\min}\right) & if \ \Delta E_{ij}(k)\Delta E_{ij}(k-1) < 0 \\ \eta_{ij}(k-1) & if \ \Delta E_{ij}(k)\Delta E_{ij}(k-1) = 0 \end{cases} \tag{16}$$

where the constants are: a = 1.2, b = 0.5 and $\eta_{min} = 10^{-3}$, $\eta_{max} = 5$ [40]. The main advantage of the proposed approach is that the information about the gradient is *neglected, which accelerates significantly the learning process.*

2.4   Intuitionistic NEO-Fuzzy Network

Intuitionistic Fuzzy Set

Atanassov [32] defines an Intuitionistic Fuzzy Set (IFS) *A* in over a finite universal set *E* as an object with the following form:

$$A = \{(x, \mu_A(x), \nu_A(x)) | x \in X\} \tag{17}$$

where $\mu_A: X \rightarrow [0, 1]$ and $\nu_A: X \rightarrow [0, 1]$ are such that $0 \leq \mu_A + \nu_A \leq 1$, $\mu_A(x)$ denote a degree of membership of x $\in$ A, $\nu_A(x)$ denote a degree of non-membership of x $\in$ A. For each intuitionistic fuzzy set in X, we call $\pi_A(x) = 1 - \mu_A - \nu_A$ the degree on non-determinacy (uncertainty) or hesitation of x $\in$ A. This parameter expresses a hesitation degree of whether x belongs to A or not and it is obviously $0 \leq \pi_A \leq 1$ for each x $\in$ X.

Intuitionistic NEO-Fuzzy Network

The structure of the proposed Intuitionistic NEO-Fuzzy network (INFN) is presented on Fig. 5.

Fig. 5 Schematic diagram of an Intuitionistic NEO-Fuzzy network

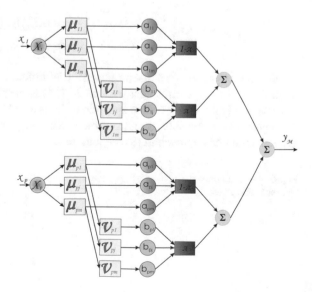

INFN is five-layer multiple input single output (MISO) structure which consists of a number of Neo-Fuzzy neurons. The first layer of the INFN is the input layer. The nodes in this layer only accept the input variables and then transmit them to the next layer directly. The second layer is so-called fuzzification layer, where the degrees of membership μ and non-membership ν, using Gaussian functions are being determined as:

$$\mu_{ij}(x_i) = \exp\left(\frac{-(x_i - c_{ij})^2}{2\sigma_{ij}^2}\right) \quad \nu_{ij}(x_i) = \left(1 - \exp\left(\frac{-(x_i - c_{ij})^2}{2\sigma_{ij}^2}\right)\right)^k, k \geq 1 \quad (18)$$

where x_i is the input value, $i = 1{:}p$ is the number of the inputs of the IFNF, c_{ij} and σ_{ij} are the center and the standard deviation of a Gaussian membership function, $j = 1{:}m$ and m is the number of used membership functions, k is parameter that must be designed. If the $k = 0$, then obviously $\mu + \nu = 1$ and the hesitation degree π (which is also computed on this layer) equals to zero. The schematic representation of an IFS is given in Fig. 6.

The neurons of the third layer, the synaptic weights and the defined hesitation degrees are multiplied with the corresponding membership degrees. In the fourth layer the zero order Sugeno functions are calculated by the following expression:

$$f_i(x_i(k)) = (1 - \pi_{ij}(x_i(k)))\mu_{ij}(x_i(k))a_{ij}(x_i(k)) \\ + \pi_{ij}(x_i(k))\nu_{ij}(x_i(k))b_{ij}(x_i(k)) \quad (19)$$

Thus, in this layer the output of the every single NFN is obtained. Afterwards, the output of the proposed INFN structure is calculated in the fifth layer as a sum of the outputs of the individual NFN's:

$$y_M(k) = \sum_{i=1}^{p} y_i(k) = \sum_{i=1}^{p} \sum_{j=1}^{m} \frac{(1 - \pi_{ij}(x_i(k)))\mu_{ij}(x_i(k))a_{ij}(x_i(k))}{+ \pi_{ij}(x_i(k))\nu_{ij}(x_i(k))b_{ij}(x_i(k))} \quad (20)$$

As a learning approach of the proposed INFN structure a simple gradient algorithm minimizing the same error cost term as in (14) is employed. During the learning procedure only the vectors of the synaptic weights a and b are under adaptation while the degrees of membership μ_A and non-membership ν_A are not trained. Thus, using the above mentioned approach in (9) it could be easily obtained the update rules:

Fig. 6 An Intuitionistic Gaussian fuzzy set

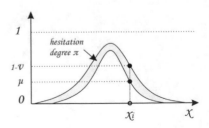

$$a(k+1) = a(k) + \Delta a(k) = a(k) + \eta \left(\frac{\partial E(k)}{\partial a(k)} \right)$$
$$= a(k) + \eta \varepsilon(k)(1 - \pi(k))\mu_{ij}(x_i(k)) \tag{21}$$

$$b(k+1) = b(k) + \Delta b(k) = a(k) + \eta \left(\frac{\partial E(k)}{\partial b(k)} \right)$$
$$= b(k) + \eta \varepsilon(k)\pi(k)\nu_{ij}(x_i(k)) \tag{22}$$

2.5 Neo-Fuzzy State-Space Network

A great challenge when modeling dynamical systems using the state-space approach is to estimate the system of equations:

$$\left| \begin{array}{l} \hat{\mathbf{x}}(k+1) = \mathbf{A}\hat{\mathbf{x}}(k) + \mathbf{B}u(k) \\ \hat{y}(k) = \mathbf{C}\hat{\mathbf{x}}(k) \end{array} \right. \tag{23}$$

where $x(k)$ is an input vector of the states, $u(k)$ is the input/control signal and $y(k)$ is the system output. In order to assume the occurring variances of the parameters and the uncertainties into the data space, a natural choice is to design a fuzzy-neural structure. Such an approach enables the possibility to incorporate knowledge, expressed as simple fuzzy rules while estimating nonlinear kernels to model the complex dynamics. An available approach is to build a network of fuzzy neurons. Thus, using the methodology proposed by Yamakawa, we can rewrite the last equation as:

$$\left| \begin{array}{ll} \hat{x}_1(k+1) & = \left(\sum_{i=1}^{N} f_1(z_i)z_i \right) \\ \hat{x}_2(k+1) & = \left(\sum_{i=1}^{N} f_2(z_i)z_i \right) \\ \vdots & \quad \vdots \\ \hat{x}_n(k+1) & = \left(\sum_{i=1}^{N} f_n(z_i)z_i \right) \end{array} \right. \tag{24}$$

where the input space vector is represented by $z(k) = [x(k), u(k)]$ and n is number of system states and $N = n + 1$ is the number of the neo-fuzzy neurons needed to represent the system. Each neuron comprises a simple fuzzy inference which produces reasoning to singleton weighting consequents:

$$R^{(i)}: \text{if } z_i \text{ is } \tilde{A}_{li} \text{ then } w_{li} \tag{25}$$

Each element of $z(k)$ is being fuzzified using the available Gaussian fuzzy sets relevant to each neo-fuzzy neuron. The fuzzy inference should match the output of the fuzzifier with fuzzy logic rules performing fuzzy implication and approximation reasoning in the following way:

$$\bar{\mu}_{li}(z_i) = \prod_{l=1}^{h} \mu_{li}(z_i) \tag{26}$$

where h is the number of the fuzzy rules in each neuron and l is the relevant synaptic link. The output of the network is produced by implementing consequence matching and linear combination as follows:

$$\left| \begin{array}{ll} \hat{x}_1(k+1) & = \sum_{i=1}^{N} \left(\sum_{l=1}^{h} \bar{\mu}_{li}(z_i) w_{li_1} \right) z_i \\ \hat{x}_2(k+1) & = \sum_{i=1}^{N} \left(\sum_{l=1}^{h} \bar{\mu}_{li}(z_i) w_{li_2} \right) z_i \\ \vdots & \qquad\qquad \vdots \\ \hat{x}_n(k+1) & = \sum_{i=1}^{N} \left(\sum_{l=1}^{h} \bar{\mu}_{li}(z_i) w_{li_n} \right) z_i \end{array} \right. \tag{27}$$

which in fact represents a weighted product composition of the ith input to lth synaptic weight to the corresponding nth rule base.

Thus, the neo-fuzzy implementation of a state-space system for a typical second order system can be represented as:

$$\left| \begin{array}{l} \hat{x}_1(k+1) = (\bar{\mu}_{li}(x_1)a_{11}\hat{x}_1(k) + \bar{\mu}_{li}(x_2)a_{12}\hat{x}_2(k) + \bar{\mu}_{li}(u)b_1 u(k)) \\ \hat{x}_2(k+1) = (\bar{\mu}_{li}(x_1)a_{21}\hat{x}_1(k) + \bar{\mu}_{li}(x_2)a_{22}\hat{x}_2(k) + \bar{\mu}_{li}(u)b_2 u(k)) \\ y(k) = c_{11}\hat{x}_1(k) + c_{12}\hat{x}_2(k) \end{array} \right. \tag{28}$$

where the corresponding outputs of the neo-fuzzy neurons represent the elements of the matrices A and B. The schematic representation of the designed modeling structure is given in Fig. 7.

Fig. 7 Schematic diagram of the NEO-Fuzzy State-Space network

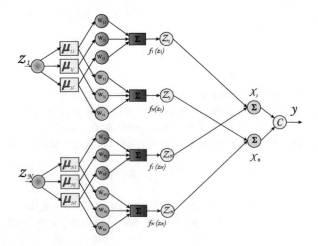

To train the proposed state-space model an unsupervised learning scheme has been used. For that purpose, a defined error cost term is being minimized at each sampling period. Thus, in order to update the weights in the consequent part of the fuzzy rules is defined:

$$E = \varepsilon^2/2 \text{ and } \varepsilon(k) = x_i(k) - \hat{x}_i(k) \tag{29}$$

where $x(k)$ is the reference input (state) measured form the process and $\hat{x}(k)$ is the state being estimated by the model. As learning approach is used the well-known back propagation approach where the synaptic weights are being adjusted using (9) as:

$$\beta(k+1) = \beta(k) + \Delta\beta(k) = \beta(k) + \eta\left(\frac{\partial E(k)}{\partial \beta(k)}\right) =$$
$$= \beta(k) + \eta\varepsilon(k)\mu_{ij}(x_i(k)) \tag{30}$$

where η is the learning rate and $\boldsymbol{\beta} = [\mathbf{a}, \mathbf{b}]$ is a vector of the trained parameters: the synaptic links in the consequent part of the rules. The parameters in the output matrix $\mathbf{C}$ are easily adjusted by using the same approach.

In order to overcome the deficiencies of the Gradient Descent approach, a simple adaptive solution to define at each iteration step the learning rate η, has been employed. The idea lies on the estimation of the *Root Squared Error*:

$$\mathrm{E} = \sqrt{\sum_{k=1}^{M}(x(k) - \hat{x}(k))^2} \tag{31}$$

Afterwards, the following condition is applied:

$$\begin{aligned} & if \text{ E}_i > \text{E}_{i-1}k_w \\ & \eta_{i+1} = \eta_i\tau_d \\ & \quad if \text{ E}_i \leq \text{E}_{i-1}k_w \\ & \eta_{i+1} = \eta_i\tau_i \end{aligned} \tag{32}$$

where $\tau_d = 0.7$ and $\tau_i = 1.05$ are scaling factors and $k_w = 1.41$ is the coefficient of admissible error accumulation [40].

3 NEO-Fuzzy Model Predictive Control

Model predictive control (MPC) is an appropriately descriptive name for a class of model based control schemes that utilize a process model for two central tasks: explicit prediction of future process behavior and computation of appropriate corrective control action required to drive the predicted output as close as possible to the desired target values.

Generally, the term predictive control does not designate a specific control strategy but a wide range of control algorithms which make explicit use of a predictive process model in a cost function minimization to obtain the control signal. This makes the MPC an open framework to implement different modeling and optimization strategies taking into account the specific requirements of the controlled process [41]. Usually, in the MPC the objective of the modeling procedure is to determine a model that can be numerically evaluated quickly and that adequately describes the process dynamics. Therefore, in this study the possibilities of the proposed State-Space Neo-Fuzzy model to be used as a predictor in the MPC scheme is evaluated.

3.1 Explicit NEO-Fuzzy Predictive Control

Using the designed NEO-Fuzzy state-space model the *optimization algorithm* computes the future control actions at each sampling period, by minimizing the following cost under system of constraints:

$$J(k) = \sum_{i=N_1}^{N_2} \|\hat{y}(k+i) - r(k+i)\|^2 Q + \sum_{i=N_1}^{N_u} \|\Delta u(k+i)\|^2 R \tag{33}$$

$$\text{subject to } \Omega \Delta \mathrm{U} \leq \gamma$$

which can be expressed in vector form as:

$$J(k) = \|Y(k) - \mathrm{T}(k)\|^2 \mathbf{Q} + \|\Delta U(k)\|^2 \mathbf{R} \tag{34}$$

$$Y(k) = \begin{bmatrix} y(k+N_1) \\ \vdots \\ y(k+N_2) \end{bmatrix} \mathrm{T}(k) = \begin{bmatrix} r(k+N_1) \\ \vdots \\ r(k+N_2) \end{bmatrix} \Delta U(k) = \begin{bmatrix} \Delta u(k+N_1) \\ \vdots \\ \Delta u(k+N_u) \end{bmatrix} \tag{35}$$

where, Y is the matrix of the estimated plant output, T is the reference matrix, ΔU is the matrix of the predicted controls and Q and R are the matrices, penalizing the changes in error and control term of the cost function:

$$Q = \begin{bmatrix} Q(N_1) & 0 & \cdots & 0 \\ 0 & Q(N_1+1) & \cdots & 0 \\ \vdots & \vdots & \ddots & \vdots \\ 0 & 0 & \cdots & Q(N_2) \end{bmatrix} R = \begin{bmatrix} R(N_1) & 0 & \cdots & 0 \\ 0 & R(N_1+1) & \cdots & 0 \\ \vdots & \vdots & \ddots & \vdots \\ 0 & 0 & \cdots & R(N_u) \end{bmatrix} \tag{36}$$

Taking into account the general prediction form of a linear state-space model [40] we can derive:

$$Y(k) = \Psi X(k) + \Gamma u(k-1) + \Theta \Delta U(k) \tag{37}$$

$$\Psi = \begin{bmatrix} CA \\ CA^2 \\ CA^3 \\ \vdots \\ CA^{N_2} \end{bmatrix} \Gamma = \begin{bmatrix} CB \\ CAB + CB \\ CA^2B + CAB + CB \\ \vdots \\ C \sum_{i=0}^{N_2-1} A^i B \end{bmatrix} \tag{38}$$

$$\Theta = \begin{bmatrix} CB & \cdots & 0 \\ CAB + CB & \cdots & 0 \\ \vdots & \ddots & CB \\ \vdots & \ddots & CAB + CB \\ C \sum_{i=1}^{N_u} A^i B & \cdots & \vdots \\ \vdots & \ddots & \vdots \\ C \sum_{i=1}^{N_2-1} A^i B & \cdots & C \sum_{i=1}^{N_2-N_u} A^i B \end{bmatrix} \tag{39}$$

where the corresponding matrix summations depend on the different predictions instants along the defined prediction horizons within one discrete sampling period.

Then we can define the system error as: $E(k) = T(k) - \Psi X(k) - \Gamma u(k-1)$. This expression is assumed as tracking error in sense of that it is the difference between the future target trajectory and the free response of the system, that occurs over the prediction horizon if no input changes were made, if $\Delta U = 0$ is set. Using the last notation, we can write:

$$J(k) = \Delta U^T H \Delta U + \Delta U^T \Phi + E^T Q E, \\ \Phi = -2\Theta^T Q E(k), H = \Theta^T Q \Theta + R \tag{40}$$

Differentiating the gradient of J with respect to ΔU, gives the Hessian matrix: $\partial^2 J(k)/\partial \Delta U^2(k) = 2H = 2(\Theta^T Q \Theta + R)$. If $Q(i) \geq 0$ for each i (ensures that $\Theta^T Q \Theta$ 0) and if $R \geq 0$ then the Hessian is certainly positive-definite, which is enough to guarantee the reach of minimum.

Since, $U(k)$ and $Y(k)$ are not explicitly included in the optimization problem, the constraints can be expressed in terms of ΔU signal:

$$\begin{bmatrix} F_1 \\ G\Theta \\ W \end{bmatrix} \Delta U \leq \begin{bmatrix} -F_2 u(k-1) + f \\ -G(\Psi X(k) + Y u(k-1)) + g \\ w \end{bmatrix} \tag{41}$$

The first row represents the constraints on the amplitude of the control signal, the second one the constraints on the output changes and the last the constraints on the rate change of the control.

4 Numerical Experiments

Modeling of Chaotic time series

Chaos is a common dynamical phenomenon in various fields [42] and different definitions as series representations exist. Chaotic time series are inherently non-linear, sensitive to initial conditions and difficult to be predicted. For that purpose, the chaotic time series prediction based on measurement is a practical technique for studying characteristics of complicated dynamics [43] and evaluation of the accuracy of different types of nonlinear models as Neural Networks. In this study, a two chaotic time series, Mackey-Glass [3] and Rossler [44] are used to assess the performances of the proposed NEO-Fuzzy Neural Networks.

4.1 Multi-Input–Multi-Output NEO-Fuzzy Network

In this section, numerical experiments with a specific example to demonstrate the performance of the proposed MIMO neo-fuzzy network are provided. The following MIMO nonlinear system is considered as reference dynamical system [45]:

$$
\begin{aligned}
y_1(k) &= \frac{y_1^2(k-1)}{y_1^2(k-1)+1} + 0.5y_2(k-1) \\
y_2(k) &= \frac{y_1^2(k-1)}{y_2^2(k-1)+y_3^2(k-1)+y_4^2(k-1)} + u_1(k-1) \\
y_3(k) &= \frac{y_3^2(k-1)}{y_3^2(k-1)+1} + 0.3y_4(k-1) \\
y_4(k) &= \frac{y_3^2(k-1)}{y_1^2(k-1)+y_2^2(k-1)+y_4^2(k-1)} + 0.5u_2(k-1)
\end{aligned}
\tag{42}
$$

where $u_1(k)$ and $u_2(k)$ are the system inputs, $y_1(k)$ and $y_3(k)$ are the outputs. The above described nonlinear system (42) has four inputs and two outputs. Their values for the current and previous time instants k, $k-1$ are used as inputs for the proposed MIMO neo-fuzzy model, i.e. the model has 12 inputs and 2 outputs. As a test signal the Mackey-Glass chaotic time series is used:

$$
u_1(k) = u_2(k) = \frac{x(i) + ax(i-s)}{(1x^c(i-s)) - bx(i)}
\tag{43}
$$

where $a = 0.2$; $b = 0.1$; $C = 10$; initial conditions, $x(0) = 0.1$ and $s = 17$ s. The learning rate η has a fixed value of 0.05.

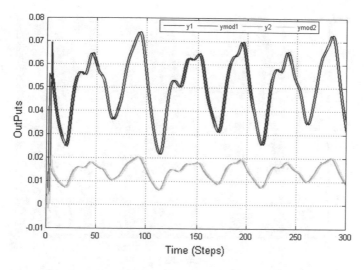

Fig. 8 MIMO NEO-Fuzzy model validation with Mackey-Glass chaotic time series as input

For greater clarity, the results are given in linear and logarithmic scales. Results on model validation by using Mackey-Glass chaotic time series are shown on Fig. 8 where it can be seen that the proposed model structure predicts accurately the generated time series. On Fig. 9 the model errors for both outputs are depicted. It can be observed that they quickly reach values closer to zero. As well, it can be seen that the model error for the second output y_2 is smaller than the error of the first output y_1, which is due to the structure of the considered nonlinear system (42). The instant values of Mean Squared Errors (MSE) and Root Mean Squared Errors

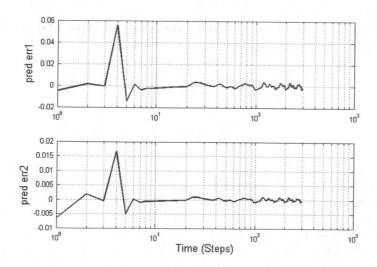

Fig. 9 Model predicted errors in a case of Mackey-Glass chaotic time series

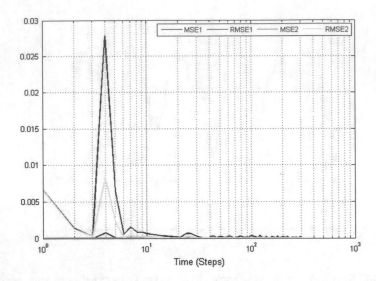

Fig. 10 Plots of MSE and RMSE for both system outputs

Table 1 MSE and RMSE values in a case of Mackey-Glass chaotic time series as input

Steps	RMSE1	MSE1	RMSE2	MSE2
50	$2.0e^{-4}$	$5.5e^{-8}$	$6.1e^{-5}$	$3.7e^{-9}$
100	$1.8e^{-4}$	$4.8e^{-8}$	$4.6e^{-5}$	$3.4e^{-9}$
150	$5.5e^{-5}$	$3.7e^{-8}$	$3.4e^{-5}$	$2.8e^{-9}$
200	$4.2e^{-5}$	$3.1e^{-8}$	$3.1e^{-5}$	$2.2e^{-9}$
250	$3.8e^{-5}$	$2.4e^{-8}$	$2.7e^{-5}$	$1.8e^{-9}$
300	$3.5e^{-5}$	$1.9e^{-8}$	$2.1e^{-5}$	$1.3e^{-9}$

(RMSE) are presented on Fig. 10. The observed values are in the order of e^{-5} to e^{-9} which guarantees the accurate prediction of the time series (Table 1).

Another test of the proposed MIMO NEO-Fuzzy model with Rossler chaotic time series is performed. The Rossler series are described by three coupled first-order differential equations:

$$\frac{dx}{dt} = -y - z \quad \frac{dy}{dt} = x + ay \quad \frac{dz}{dt} = b + z(x - c) \tag{44}$$

where a = 0.2; b = 0.4; c = 5.7; initial conditions x0 = 0.1; y0 = 0.1; z0 = 0.1. The results are given respectively on Figs. 11, 12 and 13 and Table 2.

The obtained results again show that the proposed MIMO NEO-Fuzzy network model predicts accurately the generated time series, with minimum error and fast transient response of the MSE and RMSE, reaching values closer to zero. The main advantages of the proposed MIMO network lie on its simplicity to represent non-linear cross-relationships without the need for statement of many parameters.

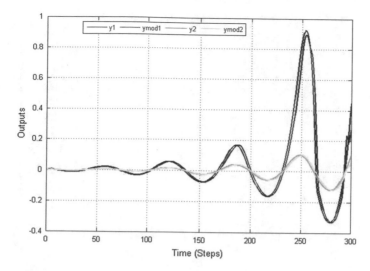

Fig. 11 MIMO NEO-Fuzzy model validation with Rossler chaotic time series as inputs

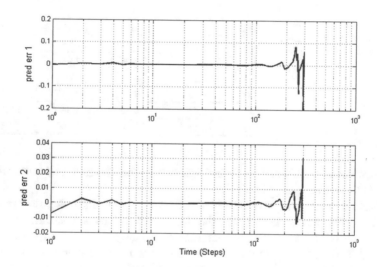

Fig. 12 Model predicted errors in a case of Rossler chaotic time series

4.2 Type-2 NEO-Fuzzy Neural Network

In this section the performance of the proposed interval Type-2 NEO-Fuzzy Network to model the same chaotic time series under the same initial conditions is evaluated. On Figs. 14, 15 and 16 the abilities of the proposed network to model nonlinear dynamical systems under occurring uncertain variations of the input signals are demonstrated. For that purpose three cases are considered: without/with

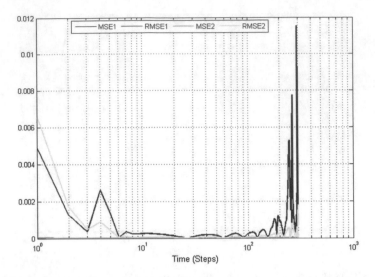

Fig. 13 Plots of MSE and RMSE for both system outputs

Table 2 Table with MSE and RMSE values in a case of Rossler chaotic time series as input

Steps	RMSE1	MSE1	RMSE2	MSE2
50	$1.6\,e^{-4}$	$2.8e^{-8}$	$4.8e^{-5}$	$2.3e^{-9}$
100	$1.6e^{-4}$	$2.6e^{-8}$	$4.9e^{-5}$	$3.1e^{-9}$
150	$1.5e^{-4}$	$2.5e^{-8}$	$5.6e^{-5}$	$6.2e^{-9}$
200	$9.8e^{-4}$	$9.7e^{-7}$	$2.7e^{-4}$	$7.5e^{-8}$
250	$6.4e^{-3}$	$4.1e^{-5}$	$1.5e^{-4}$	$2.1e^{-7}$
300	$3.8e^{-3}$	$1.5e^{-5}$	$1.2e^{-4}$	$4.4e^{-6}$

additive noise of 5% (from the nominal value of the input signal) and two variations of the FOU in each fuzzy set.

Therefore, for each input of a NFN neuron, Gaussian membership functions with 5 and 10% FOU are considered. The simulation experiments are performed on equal initial conditions for the scheduled parameters and the predicted error variations are estimated. As can be seen in the noiseless case, the model performs well following the nonlinear behavior of the chaotic series with relatively small error tolerances. The simulation experiments in the case of 5% additive noise and 5 and 10% of FOU, show a good model performance, as well.

As it can be seen again, the proposed model structure estimate accurately the generated time series, with minimum MSE and RMSE and fast transient response of the RMSE, reaching values closer to zero.

In Tables 3 and 4 are given the estimations of the MSE and RMSE in different steps instants for the considered above three cases. As can be observed, increasing the width of FOU in notion to the input signal variation gives smaller MSE due to the capabilities of the Type-2 fuzzy sets to absorb the noise trough fuzzification. On

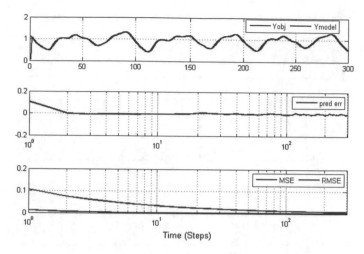

Fig. 14 Modeling of Mackey-Glass chaotic time series and the estimated errors in the noiseless case

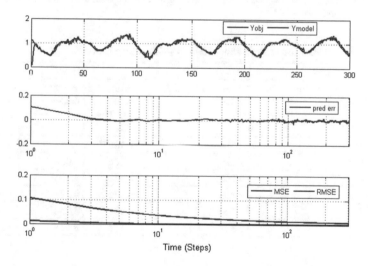

Fig. 15 Modeling of Mackey-Glass chaotic time series and the estimated error in the case of 5% additive noise and 5% FOU

the other hand, to achieve a better model performance with small error variations it is needed carefully to define an appropriate width of FOU in order to filter the occurring uncertain input variations. The estimated values of the of the RMSE show again the same tendency of absorbing the uncertain input variations maintaining lower values when increasing the width of FOU.

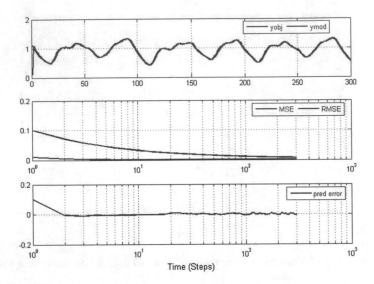

Fig. 16 Modeling of Mackey-Glass chaotic time series and the estimated error in the case of 5% additive noise and 10% FOU

Table 3 Estimation of the MSE in case of Mackey-Glass time series

Steps	Type-2 FNN			IFNN	
	Without noise and 5% FOU	With noise and 5% FOU	With noise and 10% FOU	Without noise	With noise
50	$2.38e^{-4}$	$2.51e^{-4}$	$2.18e^{-4}$	$1.26e^{-6}$	$3.42e^{-6}$
100	$1.26e^{-4}$	$1.32e^{-4}$	$1.15e^{-4}$	$7.05e^{-7}$	$2.25e^{-6}$
150	$8.85e^{-5}$	$9.26e^{-5}$	$8.15e^{-5}$	$5.19e^{-7}$	$1.81e^{-6}$
200	$7.01e^{-5}$	$7.31e^{-5}$	$6.47e^{-5}$	$4.32e^{-7}$	$1.61e^{-6}$
250	$5.89e^{-5}$	$6.10e^{-5}$	$5.43e^{-5}$	$3.75e^{-7}$	$1.48\ e^{-6}$
300	$5.19e^{-5}$	$5.33e^{-5}$	$4.78e^{-5}$	$3.41e^{-7}$	$1.37\ e^{-6}$

4.3 Intuitionistic NEO-Fuzzy Network

The INFN gives a generalization of the Type-2 Neo-Fuzzy concept by introducing in a more flexible manner the way of uncertainty management. In contrast to typical Type-2 fuzzy sets the hesitation bound who is similar to FOU is being defined indirectly by powering the degrees of non-membership of a grade greater than 1. Thus, at each sampling period of algorithm execution depending on the input signal variations the hesitation bound is adaptively varied.

To investigate the potentials of the proposed INFN in comparison to Type-2 fuzzy concept the same experiments as in the previous case are performed. For

Table 4 Estimation of the RMSE in case of Mackey-Glass time series

Steps	Type-2 FNN			IFNN	
	Without noise and 5% FOU	With noise and 5% FOU	With noise and 10% FOU	Without noise	With noise
50	$1.51e^{-2}$	$1.58e^{-2}$	$1.48e^{-2}$	$1.10e^{-3}$	$1.93e^{-3}$
100	$1.09e^{-2}$	$1.15e^{-2}$	$1.08e^{-2}$	$0.83e^{-3}$	$1.52e^{-3}$
150	$0.92e^{-2}$	$0.96e^{-2}$	$0.90e^{-2}$	$0.72e^{-3}$	$1.32e^{-3}$
200	$0.81e^{-2}$	$0.85e^{-2}$	$0.80e^{-2}$	$0.65e^{-3}$	$1.26e^{-3}$
250	$0.74e^{-2}$	$0.78e^{-2}$	$0.74e^{-2}$	$0.61e^{-3}$	$1.14e^{-3}$
300	$0.70e^{-2}$	$0.73e^{-2}$	$0.69e^{-2}$	$0.58e^{-3}$	$1.11e^{-3}$

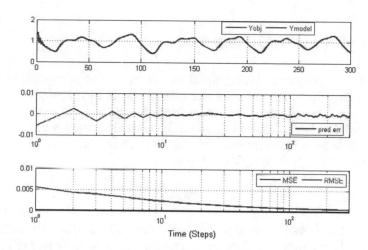

Fig. 17 Modeling of Mackey-Glass chaotic time series and the estimated errors in the noiseless case

simplicity the degrees of non-memberships are set in powers of 2. On Figs. 17 and 18 the results with the considered noiseless case and 5% additive noise are presented.

In Tables 3 and 4 are given the compared results in notion to achieved values of MSE and RMSE.

The results show that the proposed INFN ensures even smaller order of the estimated errors without the need to have a priory information about the possible uncertain input variations in order to set an appropriate value of FOU as in Type-2 fuzzy set case. On the other hand, there is no need to perform the fuzzification procedure twice along the upper and lower membership functions which decrease the number of performed computational operations.

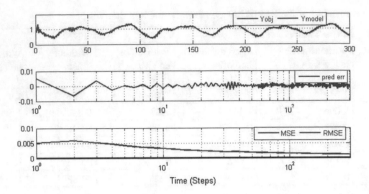

Fig. 18 Modeling of Mackey-Glass chaotic time series and the estimated errors in the case of 5% additive noise

4.4 State-Space NEO-Fuzzy Model

An oscillation pendulum system model to test the modeling capabilities of the proposed State-Space Fuzzy-Neural Network is used. A rigid zero-mass pole with length L connects a pendulum ball and a frictionless pivot at the ceiling. The mass of the pendulum ball is M, and its size can be omitted with respect to L. The pole (together with the ball) can rotate around the pivot, against the friction f from the air to the ball, which can be simply quantified as:

$$f = -sign(v)Kv^2 \tag{45}$$

where $v = L\theta$ r is the line velocity of the pendulum ball, and θ is the angle between the pole and the vertical direction. The item—sgn(v) (sgn(v) is the sign of v) in (45) shows that f always counteracts the movement of the ball, and its direction is perpendicular with the moving pole.

If we exert a horizontal force to the ball, or give the pendulum system a non-zero initial position ($\theta \neq 0$) or velocity ($\theta \neq 0$), the ball will rotate around the pivot. Below is its kinetic equation:

$$\dot{\theta} = \frac{F}{ML}\cos\theta - \frac{g}{L}\sin\theta - sign(\theta)\frac{KL}{M}\theta^2 \tag{46}$$

where $g = 9.8$ m/s^2 is the acceleration of gravity. Using two state variables x_1, x_2 to represent θ and θ respectively, the state-space equation of the system is: (for simplicity, let K, L, M all be 1)

$$\mathbf{X} = (x_1, x_2)^T = (\theta, \theta)^T, \mathbf{U} = (F)$$

$$\dot{\mathbf{X}} = \begin{pmatrix} 0 & 1 \\ -g \sin x_1 / x_1 & -|x_2| \end{pmatrix} \mathbf{X} + \begin{pmatrix} 0 \\ \cos x_1 \end{pmatrix} \mathbf{U} \qquad (47)$$

Applying Runge-Kutta method to (47), we can get the states of the testing system. The input (U) and states (X) are sampled every 0.25 s and for total 70 s, using the following conditions (F = 0, $x_1(0) = 5\pi/12$, $x_2(0) = 0$). To simplify the solution the output is being defined by the states. At first, the model parameters have been randomly generated in order to ensure the directions of the optimization approach adjusting the rule premise and consequent parameters. The amplitude of the signals is gradually changing, to persistent oscillation.

The assumed example generates a fast changing oscillating set of high frequency signals, which in most cases is a challenging task for estimation by a fuzzy-neural network. On Fig. 19 are shown the obtained results of dynamical estimation of the states and the output of the pendulum system. As can be observed the states and the output are correctly predicted with minimal error. The error residuals have been evaluated on Figs. 20 and 21 in a logarithmic scale. As can be seen the errors are converging fast to a minimum.

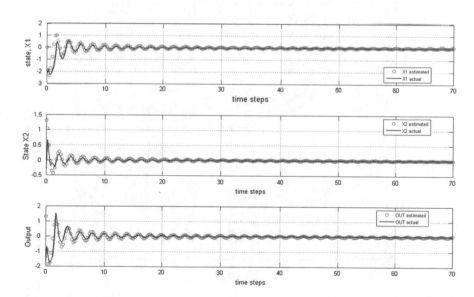

Fig. 19 Estimation of the states and the output of the pendulum oscillating system

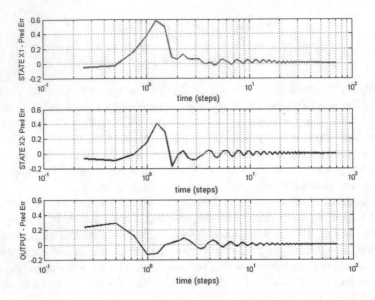

Fig. 20 Estimation of the model predicted errors (in logarithmic scale)

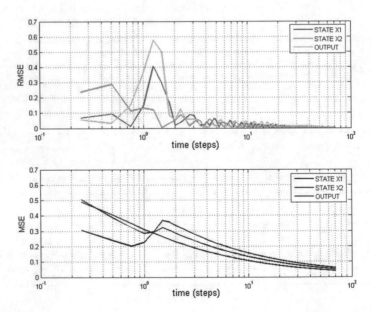

Fig. 21 Estimation of MSE and RMSE for the states and the output (in logarithmic scale)

4.5 NEO-Fuzzy State-Space Predictive Control

To prove the efficiency of the proposed MPC strategy, simulation experiments to control a nonlinear lyophilization plant for drying of 50 vials filled with glycine in water adjusted to pH 3, with hydrochloric acid are performed. The plant is highly nonlinear, since it handles processes of heat and mass transfer over a long period of time. Referring to Fig. 22 a simplified diagram of the main components of the lyophilization plant is shown. The plant consists particularly of a drying chamber (1); temperature controlled shelves (2), a condenser (3) and a vacuum pump (4). The major purposes of the shelves are to cool and freeze or to supply heat to the product. This is supported by the shelves heater and refrigeration system (5). On those shelves the product is placed (6). The chamber is isolated from the condenser by the valve (7). The vacuum system is placed after the condenser. A detailed description of the process is given in [46].

The following initial conditions for simulation experiments are assumed: $N_1 = 1$, $N_2 = 5$, $N_u = 3$; System reference $r = 255$ K; Initial shelf temperature, before the start of the primary drying $Ts_{in} = 228$ K; Initial thickness of the interface front $x = 0.0023$ m; Thickness of the product $L = 0.003$ m. The following constraints on the optimization problem are imposed: constraints on the amplitude of the control signal- the heating shelves temperature 228 K $< T_s < 298$ K; constraints on the output changes—product temperature 238 K $< T_2 < 256$ K; constraints on the rate change of the control 0.5 K $< \Delta U < 3$ K.

On Figs. 23 and 26 are presented the obtained results in control of the drying plant for selected values of $Q_i = 0.7 \times 10^{-4}$ and $R_i = 0.19 \times 10^{-3}$, considering two cases for additive noise (amplitude ± 2 K and sampling period of 1.5 s), added solely to control signal that manipulates the state x_2, in order to assess the controller capabilities in presence of disturbances. On both figures are presented the transient responses of the control and output signals, as well as the response of the control error in logarithmic units. The provided simulation experiments are performed on equal initial conditions for the model (Figs. 24 and 25).

Fig. 22 Schematic diagram of a Lyophilization plant

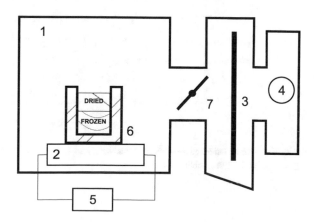

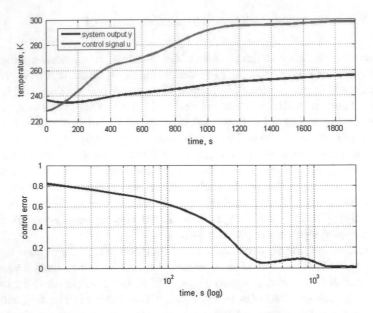

Fig. 23 Simulation results in control and estimated control, output and control error signals (with noise on the control action)

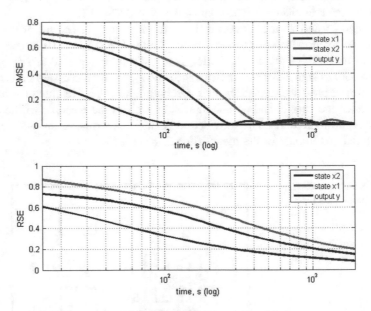

Fig. 24 The transient responses of the RMSE and RSE of the modeling errors in logarithmic units (with noise on the control action)

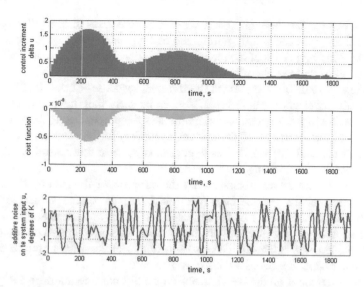

Fig. 25 Bar plots of the instant valuses of the control increment Δu and the cost function J (with noise on the control action)

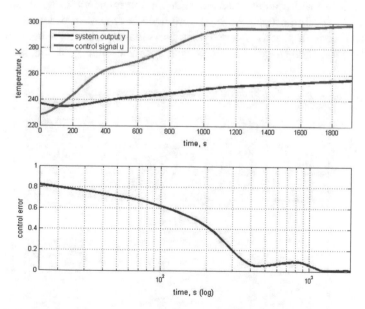

Fig. 26 Simulation results in control and estimated control, output and control error signals (with noise on the control action and the state)

As can be seen from the transient responces in both cases of controller operation, the control error is successfully minimized and all stated constraints are satisfied.

Due to fuzzy properties of the model to generalize the input signals, a significant deterioration of the model/controller operation cannot be observed.

On Figs. 24 and 27 are presented the transient responses of the RMSE and RSE of the modeling errors for both cases. As can be seen, they have a smooth nature reaching values closer to zero, which guarantees the proper operation of the model providing correct estimates of the system dynamics. In the second case the modeling errors are slightly increased as expected. In spite of that reason the dynamics of the model is still unchanged.

On Figs. 25 and 28 are demonstrated the bar plots of the instant values of the control increment Δu and the cost function J. The calculated values for Δu are within the admissible bound of the imposed constraints, which guarantees that we move smoothly on the surface of the defined dynamic optimization problem without violating them. The defined cost function is successfully minimized at each sampling period as it reaches values closer to zero.

When designing a predictive controller a careful consideration of the type and the structure of the model should be made in order to address many issues. Usually, it is required to be designed a modeling structure that allows to capture the process dynamics in fast and accurate manner, rejecting the influence of uncertain factors. On the other hand, the computational simplicity is crucial in order to save computational time in respect to system dynamics. A major problem for implementation of fuzzy-neural state-space controllers is the dimensionality of the input space.

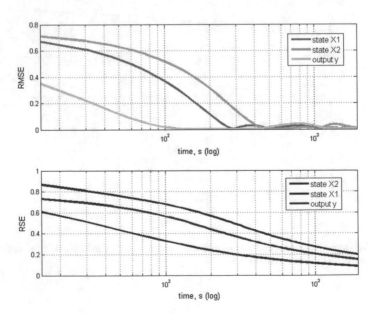

Fig. 27 The transient responses of the RMSE and RSE of the modeling errors in logarithmic units (with noise on the control action and the state)

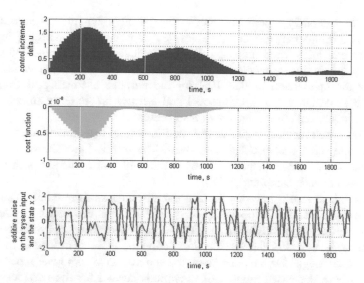

Fig. 28 Bar plots of the instant values of the control increment Δu and the cost function J (with noise on the control action and the state)

Table 5 Average processor time for algorithm execution

Algorithm	Model (s)	Optimization (s)
Neo-Fuzzy model	$1.4e^{-3}$	$6.94e^{-2}$
Dist. TS model	$2.6e^{-3}$	$8.85e^{-2}$

When it is large, the number of associated training parameters increases sharply and often requires an extensive computational effort.

In Table 5, a comparative study between the proposed NEO-fuzzy state-space controller and another previously proposed distributed state-space algorithm on the basis of the classical Takagi-Sugeno approach is shown [47]. As can be observed, the computational time for the proposed algorithm is twice less than the case of using the classical neuro-fuzzy approach. It should be mentioned also that, the chosen model structure affects not only the computational simplicity of the model but the computational abilities of the constructed predictor along the defined controller horizons. Thus, the chosen structure of the model impacts the operation of the controller as a whole.

5 Conclusions

In this chapter were presented several approaches for knowledge based modeling and control of complex dynamical systems based on the NEO-Fuzzy neuron concept. An extension to it was proposed to deal with multidimensional systems that are able to accurately capture and predict the system's dynamics regardless of the cross relationships between the variables. To improve the robust qualities of the NFNs several ideas were proposed on how to include Type-2 and Intuitionistic fuzzy logic in order to handle the uncertain dynamical input data variations. To simplify the application of the NFN for the purpose of Intelligent Control, an approach to design a state-space model and a fuzzy predictor as an extension to explicit model predictive control was also presented.

The proposed methodologies were tested in several benchmark cases to predict chaotic time series and oscillation systems. The results have shown an accurate operation of the models that are following the nonlinear dynamics with relatively small errors at each sampling period.

The obtained results have shown also the advantages of the Intuitionistic Fuzzy Logic over the classical Type-2 in cases where uncertainties occur. The hesitation bound is defined adaptively depending on the input data variations without the need to presume the extent of the variations to define the FOU as needed in the case of Type-2 Fuzzy Logic. On the other hand, the application of IFN gives a better model operation with decreased number of operations needed for the fuzzification procedure and even smaller error terms when compared to Type-2 fuzzy logic.

The proposed state-space approach gives a simplified solution to model complex systems with fewer parameters and to construct a reliable fuzzy predictor needed to provide predictions of the main system variables several steps ahead in order to calculate a horizon of reliable control signals. The provided simulation experiments prove the efficiency of the proposed approach in notion to achieved small modeling errors and decreased number of parameters affecting the computational time of the control algorithm compared to a classical approach.

References

1. Jang, J.-S.R., Sun, C.-T., Mizutani, E.: Neuro-Fuzzy and Soft Computing—A Computational Approach to Learning and Machine Intelligence. Prentice Hall (1997)
2. Jin, Y.: Advanced fuzzy systems design and applications. Physica (2003)
3. Diaconescu, E.: The use of NARX neural networks to predict chaotic time series. WSEAS Trans. Comput. Res. 3(3), 182–191 (2008)
4. Esmaili, A., Shahbazian, M., Moslemi, B.: Nonlinear process identification using fuzzy wavelet neural network based on particle swarm optimization algorithm. J. Basic Appl. Sci. Res. 3(5) (2013)
5. Jang, J.S.R.: ANFIS: adaptive network based fuzzy inference systems. IEEE Trans. Syst. Man Cybern. 23(3), 665–685 (1993)
6. Uchino, E., Yamakawa, T.: Neo-fuzzy-neuron based new approach to system modeling, with application to actual system. In: Proceedings of 6th International IEEE Conference on Tools with Artificial Intelligence, pp. 564–570 (1994)
7. Uchino, E., Yamakawa, T.: High speed fuzzy learning machine with guarantee of global minimum and its applications to chaotic system identification and medical image processing. In: Proceedings of 7th International IEEE Conference on Tools with Artificial Intelligence, pp. 242–249 (1995)

8. Baceiar, A., De Souza Filho, E., Neves, F., Landim, R.: On-line linear system parameter estimation using the neo-fuzzy-neuron algorithm. In: Proceedings of the 2nd IEEE International Workshop on Intelligent Data Acquisition and Advanced Computing Systems: Technology and Applications, pp. 115–118 (2003)
9. Bodyanskiy, Y., Viktorov, Y.: The cascade neo-fuzzy architecture and its online learning algorithm. Int. Book Ser. Inf. Sci. Comput. **17**(1), 110–116 (2010)
10. Bodyanskiy, Y., Tyshchenko, O., Kopaliani, D.: An extended neo-fuzzy neuron and its adaptive learning algorithm. Int. J. Intell. Syst. Appl. **2**, 21–26 (2015)
11. Bodyanskiy, Y., Pliss, I., Vynokurova, O.: Flexible neo-fuzzy neuron and neuro-fuzzy network for monitoring time series properties. Inf. Technol. Manag. Sci. **16**, 47–52 (2013)
12. Bodyanskiy, Y., Kokshenev, I., Kolodyazhniy, V.: An adaptive learning algorithm for a neo-fuzzy neuron. In: Proceedings of the 3rd Conference of the European Society for Fuzzy Logic and Technology, pp. 375–379 (2005)
13. Landim, R., Rodriguez, B., Silva, S., Caminhas, W.: A neo-fuzzy-neuron with real time training applied to flux observer for an induction motor. In: Proceedings of 5th IEEE Brazilian Symposium of Neural Networks, pp. 67–72 (1998)
14. Soualhi, A., Clerc, G., Razik, H., Rivas, F.: Long-term prediction of bearing condition by the neo-fuzzy neuron. In: Proceedings of 9th International IEEE Symposium of Diagnostic of Electric Machines, Power Electronics and Drives, pp. 586–591 (2013)
15. Kim, H.D.: Optimal learning of neo-fuzzy structure using bacteria foraging optimization. In: Proceedings of the ICCA 2005 (2005)
16. Zaychenko, Y., Gasanov, A.: Investigations of cascade neo-fuzzy neural networks in the problem of forecasting at the stock exchange. In: Proceedings of the IVth IEEE International Conference "Problems of Cybernetics and Informatics" (PCI 2012), pp. 227–229 (2012)
17. Pandit, M., Srivastava, L., Singh, V.: On-line voltage security assessment using modified neo fuzzy neuron based classifier. In: IEEE International Conference on Industrial Technology, pp. 899–904 (2006)
18. Silva, A.M., Caminhas, W., Lemos, A., Gomide, F.: A fast learning algorithm for evolving neo-fuzzy neuron. Appl. Soft Comput. **14**, 194–209 (2014)
19. Silva, A.M., Caminhas, W., Lemos, A., Gomide, F.: Evolving neo-fuzzy neural network with adaptive feature selection. In: Proceedings of 2013 BRICS IEEE Congress on Computational Intelligence and 11th Brazilian Congress on Computational Intelligence, vol. 209, pp. 341–349 (2014)
20. Zadeh, L.: The concept of a linguistic variable and its applications to approximate reasoning-1. Inf. Sci. **8**, 199–249 (1975)
21. Karnik, N., Mendel, J.: Introduction to type-2 fuzzy logic systems. In: Proceedings of the IEEE FUZZ Conference, Anchorage, pp. 915–920 (1998)
22. Karnik, N., Mendel, J.: Operations on type-2 fuzzy sets. Fuzzy Sets Syst. **122**, 327–348 (2000)
23. Castillio, O., Melin, P.: Type-2 Fuzzy Logic: Theory and Applications. Studies in Fuzziness and Soft Computing, vol. 223. Springer, Berlin, Heidelberg (2008)
24. Melin, P., Castillo, O.: A review on type-2 fuzzy logic applications in clustering, classification and pattern recognition. Appl. Soft Comput. **21**, 568–577 (2014)
25. Fayek, H.M., Elamvazuthi, I., Perumal, N., Venkatesh, B.: A controller based on optimal type-2 fuzzy logic: systematic design, optimization and real-time implementation. ISA Trans. **53**(5), 1583–1591 (2014)
26. Miccio, M., Cosenza, B.: Control of a distillation column by type-2 and type-1 fuzzy logic PID controllers. J. Process Control **24**(5), 475–484 (2014)
27. Terziyska, M., Todorov, Y.: Modeling of chaotic time series by interval type-2 NEO-fuzzy neural network. In: International Conference on Artificial Neural Networks (ICANN'2014). Lecture Notes on Computer Science, vol. 8681, pp. 643–650 (2014)
28. Juang, C.-F., Jang, W.-S.: A type-2 neural fuzzy system learned through type-1 fuzzy rules and its FPGA-based hardware implementation. Appl. Soft Comput. **18**(2014), 302–313 (2014)

29. Tung, S.W., Quek, C., Guan, C.: eT2FIS: an evolving type-2 neural fuzzy inference system. Inf. Sci. **220**, 124–148 (2013)
30. Aliev, R.A., Guirimov, B.G.: Type-2 Fuzzy Neural Networks and Their Applications. Springer International Publishing (2014)
31. Abiyev, R.H., Kaynak, O.: Type 2 fuzzy neural structure for identification and control of time-varying plants. IEEE Trans. Ind. Electron. **57**(12), (2010)
32. Atanassov, K.: Intuitionistic Fuzzy Sets. Springer, Heidelberg (1999)
33. Castillo, O., Melin, P., Tsvetkov, R., Atanassov, K.: Short remark on fuzzy sets, interval type-2 fuzzy sets, general type-2 fuzzy sets and intuitionistic fuzzy sets. In: IEEE Intelligent Systems'2014, Advances in Intelligent Systems and Computing, vol. 322, pp. 183–190 (2015)
34. Kuncheva, L., Atanassov, K.: An intuitionistic fuzzy RBF network. In Proceedings of EUFIT'96, pp. 777–781 (1996)
35. Zhou, X., Zhao, R., Zhang, L.: An intuitionistic fuzzy neural network with triangular membership function. In: Proceedings of 2013 Chinese Intelligent Automation Conference. Lecture Notes in Electrical Engineering, vol. 254, pp. 813–820 (2013)
36. Zhou, X., Zhao, R., Shang, X., Zhang, L.: Intuitionistic fuzzy neural networks based on extended Kalman filter training algorithm. In: Proceedings of the 1st International Workshop on Cloud Computing and Information Security (2013)
37. Hájek, P., Olej, V.: Adaptive intuitionistic fuzzy inference systems of Takagi-Sugeno type for regression problems. In: Artificial Intelligence Applications and Innovations, IFIP Advances in Information and Communication Technology, vol. 381, pp. 206–216 (2012)
38. Li, L., Yang, J., Wu, W.: Intuitionistic fuzzy Hopfield neural network and its stability. Expert Syst. Appl. **29**, 589–597 (2005)
39. Bodyanskiy, Y., Kokshenev, I., Kolodyazhniy, V.: An adaptive learning algorithm for a neo fuzzy neuron. In: Proceedings of the 3rd International Conference of European Union Society for Fuzzy Logic and Technology (EUSFLAT '2003), Zittau, Germany, pp. 375–379 (2003)
40. Osowski, S.: Sieci neuronowe do przetwarzania infromacji. Oficyna Wydawnycza Policehniki Warzawsikej (2000)
41. Huang, B., Kadali, R.: Dynamic Modeling, Predictive Control and Performance Monitoring. LNCIS, vol. 374. Springer (2008)
42. Yao, J., Mao, J., Zhang, W.: Application of fuzzy tree on chaotic time series prediction. In: IEEE Proceedings of International Conference on Automation and Logistics, pp. 326–330 (2008)
43. Lai, Y.: Recent developments in chaotic time series analysis. Int. J. Bifurc. Chaos **13**(6), 1383–1422 (2003)
44. Archana, R., Unnikrishnan, A., Gopikakumari, R.: Bifurcation analysis of chaotic systems using a model built on artificial neural networks. In: Proceedings of International Conference on Computer Technology and Artificial Intelligence, pp. 198–202 (2013)
45. Pan, Y., Wang, J.: Model predictive control of unknown nonlinear dynamical systems based on recurrent neural networks. IEEE Trans. Ind. Electron. **59**(8), 3089–3101 (2012)
46. Schoen, M.: A simulation model for primary drying phase of Freeze-drying. Int. J. Pharm. **114**, 159–170 (1995)
47. Todorov, Y., Terziyska, M., Doukovska, L.: Distributed state-space predictive control. In: Proceedings of the International IEEE Conference "Process Control'15", Stribske Pleso, Slovakia, pp. 31–36 (2015)

Sign-Based Representation and World Model of Actor

Gennady Osipov

Abstract In accordance with modern ideas about relations of the neurophysiological processes and subjective sensations, an emergence of psychological function is associated with the existence of specific information structures or their synthesis during the communication process. These structures are generated by return of excitation by diffuse projections and enable the integration of different stimulus features into a single image. Such information structures contain three main components that connect three types of information, different in their origins: information coming from external environment, information retrieved from memory and information coming from motivation centers. Binding of these components into a single unit is realized by their subsequent naming, which also provides a steadiness of emerged structures. Such information structures were named by us signs because of their resemblance to similar structures, studied in semiotics. A set of signs forming by subject during his actions and communication composes his sign-based model of the World. In this paper a sign-based model of the Word is considered, procedures of sing component's forming, a set of signs operations and a set of relationships in model of the World are explored, and it is also shown that a model of the World of a subject reflects his view of an external environment, of him-self and of other subjects. Sign-based model of the World allows to raise and solve a number of tasks in modeling of intelligent agents and their coalitions behavior, for example, goal-setting, synthesis of goal-directed behavior, distribution of roles and interactions of agents in coalition.

G. Osipov (✉)
Federal Research Center, "Computer Science and Control" of the Russian
Academy of Sciences, Moscow, Russia
e-mail: gos@isa.ru; gosipov.osipov@gmail.com

© Springer International Publishing AG, part of Springer Nature 2018   215
V. Sgurev et al. (eds.), *Practical Issues of Intelligent Innovations*,
Studies in Systems, Decision and Control 140,
https://doi.org/10.1007/978-3-319-78437-3_9

1 Introduction

In cognitive sciences the problem of goal-oriented behavior includes, in particular, setting the goal of behavior. However, in artificial intelligence research, this problem is not even set. I think, this limitations related to symbol-based formalism.

Here we consider a fragment of the formalism that may be called a sign-based or a semiotic one. that originated in the semiotics and that is used, in an informal way, in cognitive psychology.

It is based on the concept similar to concept of the sign [1, 2]. Concept of sign, in an informal way, is used in cognitive psychology [3].

According [3], the representation of each object in consciousness includes three components—image of the object, its cultural significance and personal meanings. For brevity, we will below use the term *consciousness element instead of the term representation of each object in consciousness.*

The image of a potential element of consciousness, its significance, and meanings are not always connected in a whole; in this case, the sign is not formed (in the phylogenesis) or actualized (in the microgenesis), and the psychic reflection fixes for the actor the *biological significance* of the object rather than its personal meaning, not the *consciousness image* but rather the *perception image* and the *functional significance* of the object in a specific task instead of the *significance* developed in practical activities. Below we will use the term *percept* as a synonym of *perception image* and *image* as a synonym of *consciousness image.* Such a nonsign reflection of reality makes it possible to perform only "paired" transitions between two components of the knowledge about the object: from the percept to the functional significance (selection of a method of using a concrete object), from the functional significance to the biological significance (selection of a "goal" for a specific action), and from the biological significance to the percept (selection of a specific object satisfying the given requirements). Since the three aspects of knowledge about the object are at best connected by paired relationships, an "external observer" is required to see that these three components reflect the same real object [2].

Before describing mechanisms of sign formation, we consider relationships between the elements of consciousness and elements of the sign structure in semiotics [1, 2]. It is easy to see that

(a) the concept of *image* in psychology is identical to the concept of *representation* in semiotics [2]: according to the concept of image developed in cognitive psychology, perception is interpreted as the process of categorization [4], which exactly corresponds to the concept of *representation* in semiotics, where the representation is used to differentiate the objects corresponding to the sign under examination from other objects;

(b) the concept of *significance* in psychology corresponds to *meaning* in semiotics and semantics, that is, to the semantic component of the sign;

(c) *personal meanings* are interpreted by the sets of actions that are applied by the actor to an object [3]. In applied semiotics [5], this corresponds to the *pragmatic* component of the sign, that is, to the set of actions associated with this sign.

For the components of sign listed above, we retain the names adopted in psychology—image, significances, and personal meaning. Up to the time when these components are linked into a sign, they are called perception, biological significance, and functional significance, respectively. Such linking becomes possible due to naming the emerging structure, which leads to the construct called *sign*.

The sign and its components become elements of the language structure; that is, the sign is incorporated into the world model of the actor (which does not happen without naming). Then, the object acquires a stable and conventional significance, personal experience in dealing with this object is reflected in the personal meaning as a sign component; and the event of object perception, which is a reflection in the simultaneous "picture" of the procedure of reproducing the object's properties in the motor functions of the perceiving organ is fixed as the image or representation of the object.

Also a number of neurophysiologic researches [6, 7] a indicate the possibility of existence of sign structures in a world model of the subject of activity.

In [8] the transition from the neurophysiological level on the psychological one, i.e. appearing of consciousness is connected with formation of signs.

In [9] neurophysiological mechanisms of some cognitive functions and their relations with formation of a world model are considered.

Paper [10] is devoted to the appearing of mechanisms of communication on the basis of semiotics approach.

In [11] offers a sign-based world model as a basis for an operational component of the robot manipulator.

2 Sign Formation

According with the above reasoning, we assume that the formation (actualization) of a sign includes the following phases (see also [12, 13]).

2.1 Phases of Sign Formation

0. *Object localization.* This occurs in the space in which, in addition to the four dimensions of the physical space-time, there is the fifth quasidimension—the *dimension* of significances (because each person as a carrier of consciousness

lives in two realities—physical and language ones). The actor estimates the position of an object relative to itself. This means that he must realize the selfconsciousness function (reflection), know his "coordinates" in this space, that is, reside in the clear consciousness state as is said in psychiatry (know how to determine not only physical but also social parameters of himself in the situation where the person finds himself).

1. Percept *formation* is based on the procedure of reproducing the object's properties by the motor functions of the perception organ (for living organisms) or on processing the data obtained from sensors using pattern recognition methods (for artificial intelligence systems).
2. *Generation* of the set of pairs "percept–functional significance" of the functional significance of the object based on earlier experience or precedents.
3. *Evaluation* of the degree of closeness of the functional significance obtained in phase 2 to the functional significance obtained in phase 0 using a special procedure. If these significances are not close enough, then the percept formation is continued by returning to phase 1 (in psychology of sensor perception processes, this mechanism is called sensory confidence).
4. Phases 1–3 are executed until a degree of closeness is reached that is sufficient from the viewpoint of special procedure mentioned in the description of phase 3.
5. Using a special procedure, the actor obtains from the cultural environment accumulated in a natural language system the pair *sign name–significance* and evaluates the degree of closeness of the functional significance obtained in phase 4 to the significance obtained from the cultural environment. If these significances are not close enough, then the percept formation is continued by returning to phase 1.
6. Linking the name from the pair *sign name–significance* to the percept constructed after the completion of phases 1–5. At this time, the percept turns into an image.
7. Formation of personal meanings of the sign based on precedents of actions with the object.
8. Linking the name from the pair *sign name–significance* to each personal meaning. From this time on, the functional significance turns into the significance and the biological significance turns into the personal meaning.
9. Continuing the mapping *biological significance–percept* by including the personal meaning (formed in the preceding phase) in the domain and by including the image formed in phase 6 in the set of values.

As a result, a sign corresponding to the object is formed.

Remark It is implied by phase 2 that the sign cannot be formed outside of the cultural environment.

It is clear that phases 0–9 are described only schematically. This scheme will be elaborated in the further presentation.

2.2 Refinement of a Concept of a Sign

Thus a signs is an ordered tuple $<n, p, m, a>$ where $n \in N, p \subseteq P, m \subseteq M, a \subseteq A; N$ is a set of words of finite length in an some alphabet or a set of names—set of names a name of sign; P is a set of closed atomic formulas for first order predicates calculus—a set of properties. M is referred to as a set of significances, and A—as a set of personal meanings. The overlap of the sets M and A is, generally speaking, non-empty; S is a set of sign: $s \in S$.

Note here, that the terms 'significance' and 'personal meaning' denote the assignment of the item (or phenomenon) to reality and the way of using (applying) the item that is preferred by the actor, respectively. Thus, the M and A can be considered as a set of actions. More precisely, the M-set of actions with an object, admissible from the cultural and historical point of view, A—a set of the actions preferred by the subject of activity.

Let us remember that the actions in Artificial Intelligence can be describing by rules. Each rule generally means an ordered triad of sets: $r = <C, B, D>$ where C is a condition of the rule r, B is a set of the facts added by the rule r, and D is the set of the facts removed by the rule r. Generally, each of those sets is a set of the atomic formulas for first order predicates calculus.

Then, we introduced next *linking* operators:

$\Psi_p^m : 2^P \rightarrow 2^M$ is the operator of linking images p to significances m: $\Psi_p^m (p^{(i)}) = m^{(i)}$. In other words, significance m is formed by such set of rules which conditions are satisfied on a set of atomic formulas of an image p (Fig. 1).

On Fig. 1 $P(p_i)$—set of atomic formulas (properties) of image p_i, $P(r_i)$—set of atomic formulas of condition of rule r_i.

Second operator $\Psi_p^m : 2^M \rightarrow 2^A$ links significances to personal meanings. Personal meaning is formed by such set of rules r* for which intersection of condition C(r*) with C(r) is not empty (Fig. 2).

Fig. 1 Schema of linking images to significances

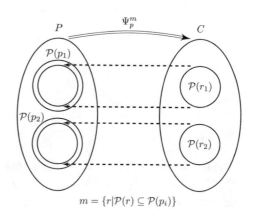

$$m = \{r | \mathcal{P}(r) \subseteq \mathcal{P}(p_i)\}$$

Fig. 2 Schema of linking significances to meanings

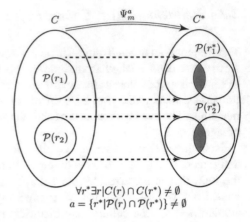

$$\forall r^* \exists r | C(r) \cap C(r^*) \neq \emptyset$$
$$a = \{r^* | \mathcal{P}(r) \cap \mathcal{P}(r^*)\} \neq \emptyset$$

Fig. 3 Schema of linking meanings to images

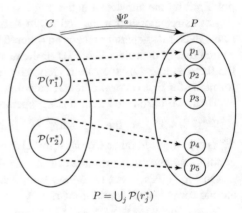

$$P = \bigcup_j \mathcal{P}(r_j^*)$$

Fig. 4 Four components of the sign

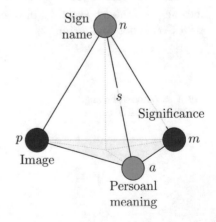

Third operator $\Psi_a^p : 2^A \to 2^P$ links personal meaning to images so that $\Psi_a^p(a^{(i)}) = p^{(i+1)}$ where $p^{(i+1)} = \cup_j \mathcal{P}(r*J)$, $a^{(i)} \in 2^A$, $p^{(i+1)} \in 2^P$, and J is the number of the rule in the set $a^{(i)}$ (Fig. 3).

Generally, $p^{(i+1)} \neq p^{(i)}$. One can show that for some initial approximation this iterative process converges to some p.

We have then $\|\mathcal{P}(\mathbf{r}) \cap \mathcal{P}(\mathbf{r}^*)\| \geq 2$. The sufficient condition of convergence is $\mathcal{P}(\mathbf{r}) \subseteq \mathcal{P}(\mathbf{r}^*)$.

We introduced an operator $\Psi_m^p = \Psi_a^p \Psi_m^a$, and showed that the sign is a fixed point of the Galois closure of the operators Ψ_p^m and Ψ_m^p.

Thus, arise structure, shown in Fig. 4.

3 Relations Over a Set of Signs

Consider the structures that can emerge on the set of signs as a result of self-organization. Simulation of self-organization in the world model makes it possible to operationalize the idea of "knowledge activity", which was formed in artificial intelligence under the influence of the concept of the stimulating role of knowledge in human behavior proposed by Festinger in 1956. According to Festinger, knowledge is not only accumulated and used by a actor; the knowledge live their own life, enter into relations, form harmonized consistent systems of notions or are involved into conflicts and are opposed to each other. In the latter case, the knowledge dissonance shows itself as a behavior stimulating force. Views and attitudes have the property of combining into a system in which the elements are consistent; the existence of contradictory relations between certain elements in the system of knowledge is a motivating factor. First, processes of self-organization lead to emergence of some structures on a set of signs. Such structures are generated by family of the relationships.

The existence of internal structure of signs components allows define three class of relationships over a set of signs (see also [13]).

First class: relationships, generated by images of signs;
Second class: relationships, generated by significances of signs;
Third class: relationships, generated by personal meaning of signs.
Let $S = \{s_1, s_2, ..., s_k\}$ be a set of signs, $p = (\pi_1, \pi_2, ..., \pi_g)$ and $q = (\theta_1, \theta_2, ..., \theta_h)$ are images of signs s_p and s_q, respectively.
The ordered sets $\tau_p = <i_1, i_2, ..., i_g>$ and $\tau_q = <j_1, j_2, ..., j_h>$ will be referred to as types of the images of signs s_p and s_q, respectively.
First class of relationships.

Definition 1 If, for signs s_p and s_q, $\tau_p = \tau_q$ and $\forall i \ \pi_i = \theta_i$, $(i \in 1, 2, ..., g)$, then $R_1 := R_1 \cup \{(s_p, s_q)\}$.

One can readily see that relation R_1 is the relation of equivalence over the set of images of signs form S. The relations R_2, R_3, and R_4 are the relations of inclusion, similarity, and opposition, respectively.

Definition 2 If, for signs s_p and s_q, $\tau_p \subset \tau_q$ and $\forall i \in \tau_p$, we have $\pi_i = \theta_i$, then $R_2 := R_2 \cup \{(s_p, s_q)\}$ is the relation of inclusion (Fig. 5).

Definition 3 If, for signs s_p and s_q, $\tau_p \cap \tau_q \neq \varnothing$ and $\forall i \in (\tau_p \cap \tau_q)$, we have $\pi_i = \theta_i$, then $R_3 := R_3 \cup \{(s_p, s_q)\}$ is the relation of similarity (Fig. 6).

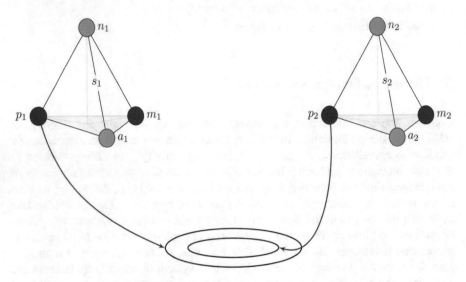

Fig. 5 Schema of the relation of image inclusion

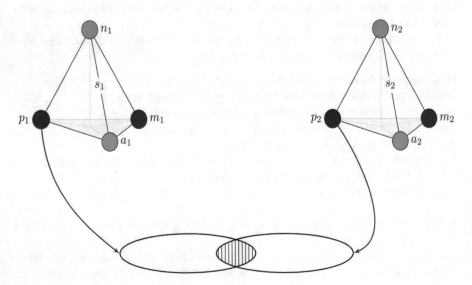

Fig. 6 Schema of the relation of image similarity

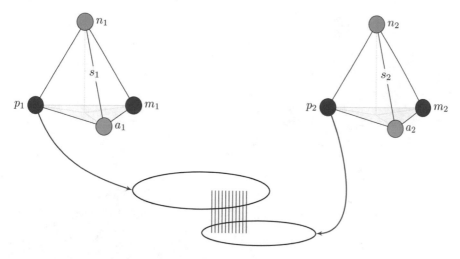

Fig. 7 Schema of the relation of image opposition

Definition 4 If, for sign s_p and some sign s_q, $\tau_p \cap \tau_q \neq \varnothing$ and $\forall i \in (\tau_p \cap \tau_q)$, we have $\pi_i \neq \theta_i$, then $R_4 := R_4 \cup \{(s_p, s_q)\}$ is the relation of opposition (Fig. 7).

Second class of relationships.

The significance is a set of actions which the actor can perform with the item.

Each action corresponds to a set of some roles that are substituted by participants of the action in, for example, the way described by Fillmore.

Therefore, we will associate the significance of each sign with some ordered set of its roles.

So, if $I = \{i_1, i_2, ..., i_q\}$ is a set of all possible roles, the significance of each sign is a subset of that set. (For simplicity it is assumed that each significance includes one action.). Let now s_p be a sign with the significance $m(s_p) = <i_1, i_2, ..., i_k>$ where $i_1, i_2, ..., i_k \in I$, and s_q is a sign.

Definition 5 If, for signs s_p and s_q, we have s_q/i_j, (the sign s_q substitutes role i_j), $i_j \in m(s_p)$, then $R_5 := R_5 \cup \{(s_p, s_q)\}$. As it was stated above, it is reasonable to refer this relation as the script-based one. It is clear that the relation R_5 enables one to generate complex constructions, i.e. that are essentially the networks of signs related by the significances and names of those signs (Fig. 8).

Third class of relationships.

Over the sets of signs there is a natural way for generating relationships of subsumed and opposition on the basis of their personal meanings.

It should be reminded that each meaning is associated with a set of actions. As before, here for simplicity we assume that each meaning is associated with one action described by the rule r = <C, A, D>.

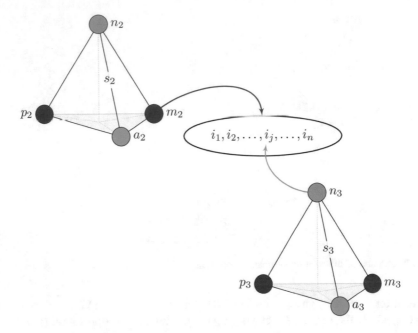

Fig. 8 Significances of the signs

Let, as before, $S = \{s_1, s_2, ..., s_k\}$ be a set of signs and a1 and a2 be the meanings of signs s_1 and s_2, respectively.

We define the following relations over the set of personal meanings:

1. $\subseteq$(a1, a2) or (a1 $\subseteq$ a2) (read 'meaning a1 is subsumed by meaning a_2'), if $A(r_1)$ $\subseteq A(r_2)$ or $D(r_1) \subseteq D(r_2)$ where $A(r_1)$, $A(r_2)$, and $D(r_1)$, $D(r_2)$ are sets of the facts added or removed by the rules r_1 and r_2, respectively; then $R_6 := R_6 \cup \{(s_1, s_2)\}$ (Fig. 9).
2. $\perp(a_1, a_2)$ or $a_1 \perp a_2$ (read 'meaning a_1 is opposed to meaning a_2'), if $\forall P(x_1, x_2, ..., x_n) \in A(r_1) \exists P(x_1, x_2, ..., x_n) \in D(r_2)$; then $R_7 := R_7 \cup \{(s_1, s_2)\}$ (Fig. 10).

Operations over a set of signs and screens.

Operations of the set signs used for creating of new signs and scripts.

For example we will describe first the operation of *generalization*.

The operation of generalization Θ is defined over the set of pairs of signs that belong to the relation R_3; applying the operation Θ create a new image that includes all *common* features of the initial images. Namely, if π is a set of images, $p_1, p_2 \in \pi$, $p_1 = (x_1, x_2, ..., x_g)$ and $p_2 = (y_1, y_2, ..., y_h)$, then $\Theta : \pi \times \pi \to \pi$ so that for each $p_1, p_2 \in \pi$ such that $(p_1, p_2) \in R_3 \Theta(p_1, p_2) = p_3$, where $p_3 = (z_1, z_2, ..., z_l)$ so that for $\forall i \exists j, k$ such that $z_i = x_j = y_k$.

One can show that R_3 is a lower semi-lattice with respect to operation Θ.

Second example is the operation of closure over significances.

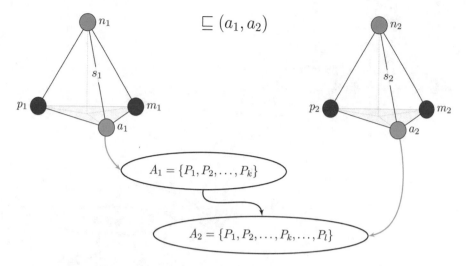

Fig. 9 Schema of the relation of meaning subsumption

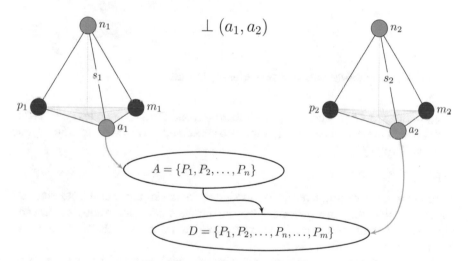

Fig. 10 Schema of the relation of meaning opposition

Operation of closure over significances of $\prod (s_1, i_j, s_2)$. If s_1 is a sign with a significance $m(s_1)$ and $i_j \in m(s_1)$ is a role of that significance, the operation creates a screen s_1^* where role i_j is substituted with the sign s_2 (s_2/i_j). In this case, the meanings and significances of the initial signs are combined (Fig. 11).

Next is operation of agglutination.

Operation of agglutination $\nabla(s_1, s_2) = s_3$. If s_1, s_2 are signs and a_1 and a_2 are their meanings, the operation of agglutination creates a new sign s_3 with meaning a_3 where $A(r_3) = A(r_1) \cup A(r_2)$ or $D(r_3) = D(r_1) \cup D(r_2)$. It is clear that in both cases

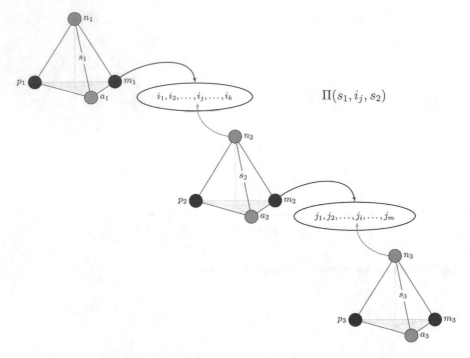

Fig. 11 Schema of the operation of closure over significances

$C(r_3) = C(r_1) \cup C(r_2)$ where $C(r_3)$ are the set of conditions of the rules r_3, r_1, and r_2 and A and D are the sets of the facts that are added and removed by the rules r_1 and r_2, respectively (Fig. 12).

Models of the World

It was shown above that, over the set of signs, three main types of the structure are formed so that they are generated by the families of relationships on images, significances, and meanings. It is reasonable to refer to each of them as an heterogeneous semantic network.

Therefore, we have a semantic network H_P over a set of images, semantic network H_A over a set of personal meanings, and semantic network H_M over a set of the significances of the signs.

We will refer to the triad of objects $H = <H_P, H_A, H_M>$ as a *semiotic* network.

As it follows from the above, the transitions between the networks H_P, H_A, H_M are implemented using the procedures Ψ_m^a, Ψ_a^p, and Ψ_a^p described earlier.

In psychology, there exist a concept of the *Model of the World* of the actor.

Nine main types of the Model of the Worlds may be distinguished: Three of them the rational one, the common life, and the mythological—non degenerate. Six types—degenerate models.

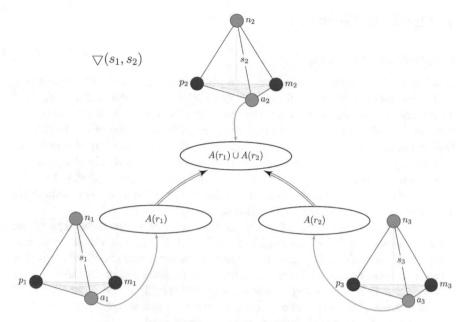

Fig. 12 Schema of the operation of meaning agglutination

All types of the World Models use networks over images, meanings, and significances.

However, there is some 'controlling' network that is used for defining a goal, searching for adequate actions, calling up scripts, and changing personal meanings.

For example, in the rational Model of the World, in the network over images, first, a goal is set. Then, on the network over significances, appropriate roles are looked for in a script as a condition for performing actions to attain the goal.

Next, the meanings of the objects are taken into account. They may perform motivations or obstacles or means for attaining the goals.

The characteristic feature of the common-life Model of the World consists in following some stereotypes or scenarios of behavior.

In the mythological Model of the World, each role has an invariable meaning and an image associated to it, i.e. in this case the network over meanings is the leading one. In other words, inheritance of the network H_A to the level of names of signs results in forming a *mythological* Model of the World.

Degenerated World Models in which two networks instead of three are used, are known as well.

It should be added here that it is the actors' Models of the World are the environments in which different cognitive functions are implemented such as introspection, reflection, and goal setting. We describe below one of the most important functions, the goal setting function as an example.

4   Sign-Based Synthesis of Behavior

The problem of goal setting

Goal setting is a complicated process that also involves, in addition to finding a goal, determination of the conditions and specific way for attaining that goal. As it was told above, the character of the goal setting process is determined by the type of the actors' Model of the World. In case of a common-life Model of the World, the leading component is *significance*. In this case actor us the existing structure and existing signs to construct a suitable situation -screen, which will be the goal.

Because the process of goal-setting is carried out within an activity, then the motive of activity included in the Model of the World of the actor. In common-life Model of the World it is the significance of the sign.

The basic idea underlying of the process is as follows: since the sign of the matter of need included in the Model of the World of Actor, it should be to find such a sign (or build script), personal meaning (personal meanings) which would provide the achievement of the matter of need. On the semantic or operational level this means search or building of such a sign (or script), in the structure of the personal meaning of which there is action (or there are actions), the result of which to the current situation is the image of the matter of need.

Further, s_4—a sign, significance of which—m_4 is a motive.

In the algorithm below we will use the syntax and semantic considerations, especially not stressing the circumstances.

Step 1: Transition $m_4 \rightarrow m_1$. The search for the subnet of significances such significance m_i, which sign has personal meaning a_i contains such a rule r1 so that the action in a set of added facts $A(r1)$ contains of image p_4 of the sign s_4 i.e.

$p_4 \subseteq A(r1)$;

Else, go to step 2.

Step 2: Transition $m_1 \rightarrow m_2$. Search for the subnet of significances such significance m_2, that the corresponding sign s_2 in set of personal meanings a_2 contains such a rule r2, which in the set of the added facts $A(r2)$ contains the set $p_4 \backslash A(r1)$. If $p_4 \subseteq A(r1) \cup A(r2)$, the process is completed, built a script is a goal.

Else the search process on the network of the significances is repeated until the will be built the script from the signs $s_1, s_2, ..., s_n$, such that the $p_4 \subseteq A(r1) \cup A(r2) \cup ... \cup A(r n)$, where the r1, r2, ..., rn—actions corresponding signs $s_1, s_2, ... s_n$, respectively.

A few words on the synthesis of the plan of behaviour.

It is well known that the task of AI-planning in general belong to class EXPSPACE complete. In other words, it complexity is $O(2^{p(N)})$ space, where $p(N)$ is a polynomial function of N. The power exponent is the polynomial function of total number of rules in the system—N.

In case of planning in a sign Model of the World, the task is reduced to creation of the plan of achievement of the script constructed at a goal-setting stage. It means that for creation of the plan only signs from the constructed scripts will be used. Thus, in case of planning in a sign-based Model of the World, complexity will also be exponential, however, exponent will be $p(N^*)$, where N^*—number of rules in structure of personal meaning of such sign from the constructed script in which this number is maximal.

It is clear, that the less signs in model of the world of the subject of activities, i.e. the more simply, the world model is less differentiated, the it is more than N^*. In a degenerate case when the model of the world contains one sign, $N = N^*$ (in more details see [14, 15]).

5 Conclusion

In this paper, the basis of a new formalism that correlates with the available neurophysiological and psychological data on the structure and function of the actor's Model of the World, is considered. The proposed formalism enables describing a number of cognitive functions, such as introspection, reflection, goal setting, and some others, and enhancing the extent of understanding cognitive processes. A description of a cognitive function, the goal setting function, is provided. Further development of the formalism is related to handling multi-agent and robotic systems, in which Models of the World may be generated in automatic (or semiautomatic) way as a result of communication, interaction of visual and audio systems. On this basis it becomes possible to distribute tasks and coordinate goals in cognitive robots coalitions. Implementation of such functions will enable one to boost the extent of autonomy of such systems and their coalitions.

Acknowledgements This work was supported by the Russian Science Foundation (Project No. 16-11-00048).

References

1. Peirce, C.S.: Collected Papers, vols. 1–8. Harvard University Press, Cambridge, MA (1931–1958); Alteiya, St. Petersburg (2000)
2. Frege, G.: Logic and Logical Semantics. Aspekt, Moscow (2000) (in Russian)
3. Leontiev, A.N.: Activity, Consciousness, Personality. Politizdat, Moscow (1975) (in Russian)
4. Bruner, J.: Psychology of Knowledge. Progress, Moscow (1977) (in Russian)

5. Pospelov, D.A., Osipov, G.S.: Introduction to applied semiotics. Novosti iskusstv. intellekta, No. 6, 28–35 (2002)

6. Edelman, G., Mountcastle, V.: The Mindful Brain. Cortical Organization and the Group-Selective Theory of Higher Brain Function. The MIT Press, Cambridge, MA (1978)

7. Desmedt, J., Tomberg, C.: Neurophysiology of preconscious and conscious mechanisms of the human brain. In: Abstracts of the Xth International Congress of Electromyography and Clinical Neurophysiology, Kyoto, Japan. Electroenceph. Clin. Neurophysiol. **97**(4), S4 (1995)

8. Ivanitsky, A.M.: Information synthesis in key parts of the cerebral cortex as the basis of subjective experience. Neurosci. Behav. Physiol. **27**(4), 414–426 (1997)

9. Friederici, A.D., Singer, W.: Grounding language processing on basic neurophysiological principles. Trends Cogn. Sci. **19**(6), 329–338 (2015). https://doi.org/10.1016/j.tics.2015.03.012

10. Loula, A., Queiroz, J.: Synthetic Semiotics: on modelling and simulating the emergence of sign processes. In: AISB/IACAP World Congress 2012: Computational Philosophy, Part of Alan Turing Year 2012, p. 102129, Birmingham (2012)

11. Roy, D.: Semiotic schemas: a framework for grounding language in action and perception. Artif. Intell. **167**(1–2), 170–205 (2005). https://doi.org/10.1016/j.artint.2005.04.007

12. Osipov, G.S.: Signs-based vs. symbolic models. In: Sidorov, G., Galicia-Haro, S.N. (eds.) Advances in Artificial Intelligence and Soft Computing, pp. 3–11. Springer International Publishing (2015)

13. Osipov, G.S., Panov, A.I., Chudova, N.V.: Behavior control as a function of consciousness. I. World model and goal setting. J. Comput. Syst. Sci. Int. **53**(4), 517–529 (2014)

14. Osipov, G.S., Panov, A.I., Chudova, N.V.: Behavior control as a function of consciousness. II. Synthesis of a behavior plan. J. Comput. Syst. Sci. Int. **54**(6), 882–896 (2015)

15. Panov, A.I.: Behavior planning of intelligent agent with sign world model. Biol. Inspired Cogn. Archit. (2017). https://doi.org/10.1016/j.bica.2016.12.001

Responsive Production in Manufacturing: A Modular Architecture

Maria Marques, Carlos Agostinho, Gregory Zacharewicz, Raul Poler and Ricardo Jardim-Goncalves

Abstract This paper proposes an architecture aiming at promoting the convergence of the physical and digital worlds, through CPS and IoT technologies, to accommodate more customized and higher quality products following Industry 4.0 concepts. The architecture combines concepts such as cyber-physical systems, decentralization, modularity and scalability aiming at responsive production. Combining these aspects with virtualization, contextualization, modeling and simulation capabilities it will enable self-adaptation, situational awareness and decentralized decision-making to answer dynamic market demands and support the design and reconfiguration of the manufacturing enterprise.

Keywords Modular production architectures · Cyber-physical systems Internet of Things · Virtualization · Contextualization

M. Marques (✉) · C. Agostinho
Center of Technology and Systems, CTS, UNINOVA 2829-516 Caparica, Portugal
e-mail: mcm@uninova.pt

C. Agostinho
e-mail: ca@uninova.pt

G. Zacharewicz
University Bordeaux, CNRS, UMR 5218, IMS, F 33405 Talence Cedex, France
e-mail: gregory.zacharewicz@u-bordeaux.fr

R. Poler
Research Centre on Production Management and Engineering (CIGIP),
Universitat Politècnica de València, Valencia, Spain
e-mail: rpoler@cigip.upv.es

R. Jardim-Goncalves
DEE/FCT, Universidade Nova de Lisboa, 2829-516 Caparica, Portugal
e-mail: rg@fct.unl.pt

© Springer International Publishing AG, part of Springer Nature 2018
V. Sgurev et al. (eds.), *Practical Issues of Intelligent Innovations*,
Studies in Systems, Decision and Control 140,
https://doi.org/10.1007/978-3-319-78437-3_10

231

1 Introduction

Manufacturing represents 16% of the EU GDP employing more than 30 million persons and the objective of revitalization of the EU economy calls for the endorsement of the reindustrialization efforts to raise the contribution of industry to GDP to as much as 20% by 2020 [1]. Yet, nowadays, manufacturing companies are facing one of the most challenging moments in their existence, i.e. to cope with the increasingly stricter requirements in terms of flexibility while maintaining their production capacity. The advances in digital automation can contribute significantly to help unleashing their potential to respond to these challenges [2]. Indeed, the explosion in the Internet of Things (IoT), and the shift towards a new industrial paradigm based on cyber physical systems (CPS) are paving the way. If well applied at manufacturing processes, they will potentiate a new industrial revolution with enormous value, changing manufacturing nature forever. European industry needs to meet the increasing global consumer demand for greener, more customized and higher quality products through the necessary transition to a more efficient, flexible, responsive, digitalized and demand-driven industry with lower waste generation [3]. Manufacturing enterprises are pushed to take 'glocal' actions, i.e. thinking globally but acting and staying economically compatible with the local context [4]. The same can happen inside the factory, where enterprise level strategy needs to be accompanied by local action at the resources and devices. As an example, integrated systems for machining (e.g. CNC machines) are used world-wide, and organizational strategies need to be flexible to accommodate highly variable domains of application and consumer policy restrictions, configuring and allocating resources in-house depending of the product variant. Similar situation happens in more traditional SME-based environments such as furniture manufac-turing, where product variability implies a great number of changes locally and is only possible with automation and reconfigurable production lines.

To answer these challenges the European Commission has identified the need for the advances in production architectures so that they become more responsive to dynamic market demands. This requires radical change of production topologies to achieve dynamic production re-configurability, scaling and resource optimization, fully exploiting the digital models of processes and products and to synchronization of the digital and physical world [5]. Based on these motivational factors, this paper proposes an architecture to support responsive production in manufacturing com-panies aiming to contribute for reindustrialization of manufacturing. Promoting the convergence of the physical and the digital worlds through CPS and IoT tech-nologies, to accommodate more customized and higher quality products, the approach will also contribute to build trust in European companies and reinforce their position worldwide.

Recently, the term "Industry 4.0" has invaded all conversations about the future of industrial production. What started as a national initiative in Germany, rapidly evolved to a much more extended concept that is being used to identify what it is seen as the next industrial revolution [6]. In an initiative launched in April 2016, the

European Commission recognizes the importance of promoting measures to support the development of a digitized European industry aiming to ensure that Europe is ready for the emerging challenges of digital products and services. Moreover, similar approaches are being followed outside EU (e.g. China has launched "Made in China 2025", United States created the "Smart Manufacturing Leadership Coalition" (SMLC) initiative, etc.).

The main goal of all these initiatives is to achieve the intelligent factory characterized by adaptability, resource efficiency, and ergonomics, as well as the integration of customers and business partners in business and value processes [7]. To achieve this challenging objective companies must cope with the increasingly stricter requirements in terms of flexibility, while maintaining their production capacity [8].

Currently, industrial companies are pushed to take 'glocal' actions, i.e. thinking globally but acting and staying economically compatible with the local context [9]. The same can happen inside the factory, where enterprise level strategy needs to be accompanied by local action at the resources and devices level. As an example, integrated systems for machining (e.g. CNC machines) are used worldwide, and organizational strategies need to be flexible to accommodate highly variable domains of application and consumer policy restrictions, configuring and allocating resources in-house depending of the product variant.

Manufacturing is typically associated to the transformation of raw material and assemblage of components into final products that fit the needs of many and can be sold worldwide. As a consequence, research and development (R&D) in this domain has targeted the acceleration and mass-replication of more or less static production processes, construction of production machinery and the development of software to control such systems. However, the last decade has demonstrated clear signs that industry cannot proceed with 'business as usual' practices. As identified by the European Commission's Future Internet Enterprise Systems cluster (FInES) roadmap in 2012, a change of paradigm is required to maintain and improve the current standard of life [10].

Mass replication, although still an important part of production, tends to lose space for customised products tailored to fit consumer needs and demands [9]. Also, as production stages and technologies have become more mobile, a single final manufactured good is nowadays often processed in different companies and countries, crossing several information systems (IS) with sequential tasks in the value chain. Therefore, the survival of enterprises in the near and long term future will depend on their ability to see their own role within the physical and social environment and to become flexible to changes in paradigm that can give them a competitive advantage.

The advances in digital automation can contribute significantly to help unleashing their potential to respond to these challenges [2]. Indeed, the explosion in the Internet of Things (IoT), and the shift towards a new industrial paradigm based on cyber physical systems (CPS) are paving the way.

In a Smart Factory (SF), everything is connected [11]. Production machines, humans, products, transport options and IT tools communicate with each other and are organized with the objective of improving overall production, not only within the physical boundaries of the company but also beyond them. One of its key characteristics is the ability to decentralize control and decision as it facilitates modifications in the production process contributing to meet the increasing demand for mass customization.

This paper starts by identifying opportunities, challenges and main characteristics of Industry 4.0 followed by an analysis of the main barriers to its implementation. Then, related work is presented together with the highlights and the challenges that are to be addressed by the proposed architecture. The objectives to be achieved and the concept to support their achievement are also presented as well as the proposed architecture detailing the different modules that are being proposed. Finally, some conclusions are drawn and some future developments are discussed.

2 Industry 4.0: Opportunities, Challenges and Main Characteristics

When discussing how companies should be addressing the opportunities provided by Industry 4.0 it becomes obvious that some of them are still waiting for more clear advantages in joining the new paradigm. On the other hand, the examples of companies that are modernizing themselves keeps growing. This difference in attitude can be critical for the success, in the middle to long term, of companies that are not accompanying the trend.

Despite their position towards Industry 4.0 companies identify a set of opportunities regarding its adoption [6], namely:

- Increased competiveness: digitalization of industry will increase their competitiveness as it will impact both local and global value chains
- Easier adaptation to market changes—from "push into the market" to "pull from the consumer": the immense potential of industry 4.0 will facilitate the integration of customers needs and preferences into the development of new products and adaptation of production processes
- Risk and fault reduction: data integration and data analysis contribute for improved monitoring and thus reduce down times and faults. Cyber security will also be reinforced reducing *hackerism* risks
- Skilled workers and IT: investments on the education of workers as well as in appropriate infrastructure, although mandatory in most cases, are also seen as an opportunity for improving performance
- Use of currently growing technologies: smart sensors, 3D printing, etc., are seen as the key for accelerating the transformation for industry 4.0 by enabling rapid testing, prototyping and production adaptation.

It is possible to analyze the identified set of opportunities and to correlate them with some of the most promising technological developments that are considered the biggest contributors for making Industry 4.0 a reality. Concepts such as CPPS (Cyber-Physical Production Systems), and IIoT (Industrial Internet of Things) have emerged and represent challenges that need to be tackled to answer the requirements of a continuously changing environment.

With roots on CPS, CPPS are systems of collaborating and autonomous across all levels of production, from processes through machines up to production and logistics networks [12]. They are capable of accessing, providing and using production data from real world at real-time. Aspects such as sovereignty, collaboration, optimization and responsiveness need to be especially addressed by CPPS. In addition to that, simulation, sensor networks, big data and security issues represent an important part to deal with challenge of CPPS implementation.

The recent advances on smart sensors, wireless networks and embedded systems, together with the consistent decrease of technology costs have contributed for the rapid development of industrially oriented IoT. Thus, IIoT appears as the main driver for the implementation of Industry 4.0 and is directly related with the successful development of CPPS. Making use of technologies such as data acquisition and data integration, capabilities to capture and fuse information from various sensors/objects and cloud-based data centers, IIoT facilitates the adjustment of production parameters, opening new perspectives in easy reconfiguration of production lines, effective detection of failures, autonomous maintenance triggering and prompt reaction to unexpected changes in production. Although the possibilities for their application are immense there are still some technological issues that need to be addressed to ensure the full implementation of IIoT (e.g. semantic integration and analysis for which additional developments on standards and protocols are needed).

CPPS together with IIoT enable the creation of a smart network of machines, ICT systems, smart products and people across the entire value chain and the full product life cycle. Interfacing with other infrastructures is also a reality enabling access to information coming from other smart platforms (e.g. smart logistics).

One key challenge is the introduction of the consumer perspective in the production process with the trend on customization growing every day. Manufacturing, especially in western countries, is no longer based on mass production but on mass customization in a "lot size 1" approach. Thus, links with social media networks are fundamental to access consumer expectations and desires.

To answer to the identified challenges and needs Industry 4.0 is composed by four main approaches that demonstrate its enormous potential for change (see Fig. 1):

- Vertical integration: responsible by the vision about high flexibility on production towards full customization ("lot size 1") by fully integration of the internal value chain of the company.
- Horizontal integration: enables the inclusion of worldwide value network, and allows to work both on the processes as well on the systems till delivery of product to the customer.

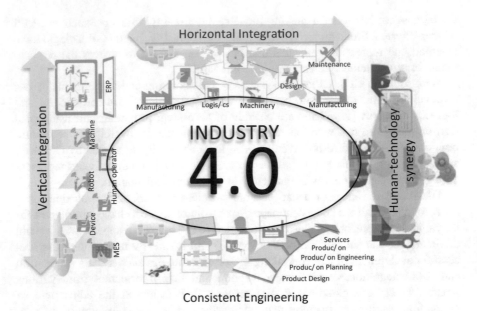

Fig. 1 Main characteristics of Industry 4.0 (adapted from [13])

- Consistent engineering: complete systems engineering, through production digitalization, from product design to product distribution, disposal and after-sales services.
- Human-technology synergy: promoting new skills and competences from the workforce, adapting working conditions (more attractive and productive) and safeguarding jobs.

3   Barriers for Industry Digitalization

All the aforementioned aspects can only be succeeded if supported by the developments on the areas previously mentioned. But, on the other hand, technological developments must be combined with social sciences and humanities in order to achieve human-technology synergy.

In fact, companies recognize that having the right people in place is critical for leveraging technological gain, and to accomplish the goals of smart manufacturing. The current perception about the so-called "skills gap" is that it will continuing growing as the percentage of new jobs needing highly skilled workers keeps increasing [11].

In addition to this, currently there is still a notion of manufacturing jobs as being less important in society, with lower incomes and poorly recognized. Many companies are facing huge difficulties in attracting talented resources to work with them as they are still very influenced by these views and companies need to work on their message to attract talents for industry [14].

Despite advanced digitization, horizontal integration, with suppliers, customers and other value chain partners, is progressing a slower than the vertical one [15]. Although the potential of greater horizontal integration is broadly recognized (e.g. offers the prospect of coordination of orders, materials flow and production data, with all companies along the value chain being able to add their own value-adding steps) there are still barriers that need to me removed to achieve it. These barriers are not only technological ones but also related with confidence and trust.

The World Economic Forum published the results of a 2014 [16] survey on Industrial Internet in which a set of barriers were identified by companies (see Fig. 2). The results demonstrated that almost two-thirds of the two major issues are related with security and interoperability. Other significant barriers cited include the lack of clearly defined return on investment (ROI), legacy equipment and technology immaturity.

Thus, in the latest years, interoperability and standards has been one of the areas capturing major attention from public and private institutions. In 2008, European Union launched Public-Private Partnership (PPP) for Factories of the Future (FoF). The FoF work program for 2018–2020 expects to mobilize more than €50bn of public and private investment with great focus on the development of standards for: 5G, Cloud Computing, Internet of Things, Data technologies and Cybersecurity.

Interoperability is an essential problem of sharing information and exchanging services. It goes far beyond the simple technical problems of computer hardware and software, but encompasses the broad but precise identification of barriers not

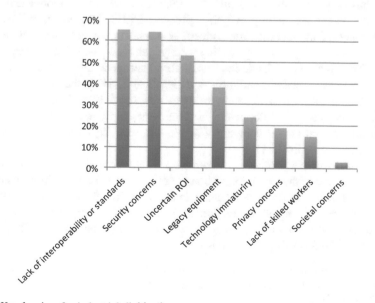

Fig. 2 Key barriers for industrial digitization

only concerning data and service but also process and business as well [17]. In fact, difficulties are observed when actors from various fields of expertise or with different types of resources are forced to exchange information [18, 19]. These difficulties arise at two levels:

- The difficulties to exchange data: observed when an actor does not have the proper tools to send or receive data. This is also observed when an actor is facing problems in accessing the content of the data files after receiving them.
- Difficulties in understanding the information exchanged: once the data exchanged, it must be ensured that the actors have a mutual understanding of the data. In fact, the diversity of actors and their areas of expertise, as well as differences in language and models complicate the uniqueness of meaning and understanding.

In what regards interoperability, traditional integration and interoperability services are often inflexible and difficult to adapt to meet dynamic requirements. Most development is either relying on international accepted standards for data exchange, e.g. STEP, EDI/EDIFACT, ebXML, UBL, or is implemented on a peer-to-peer basis [20–22]. Architectures on integration and interoperability [23], modeling frameworks and tools, as well as methodological [24–26] are available but the real challenge resides in applying them to streamline data integration and interoperability while sustaining collaboration throughout market adaptation and innovation.

This sustainability convenes the needs of the present without compromising the ability of future changes, meeting new system requirements [27]. Integration and interoperability is acknowledged with many researchers working in related domains, such as the digital and sensing enterprise [28], smart networks [29], or digital business innovation and big data [30]. In many cases, model driven and knowledge-based technology is being promoted [31], however they are rarely applied together and there is little concern on the company network sustainability, currently addressing interoperability only at the network design time. Jardim-Goncalves et al. [32] define sustainable interoperability as the development of novel strategies, methods and tools to maintain and sustain the interoperability of enterprise systems in networked environments as they inevitably evolve with their environments. These developments are fundamental to establish and maintain interoperability within the company (intra-interoperability) and also in the value chain (extra-interoperability).

Additional work is also needed to develop widely accepted methods for privacy and security. One of the technologies that is positioning as being a strong possibility to deal with these issues is blockchain technology. It offers a way of recording transactions or any digital interaction in a way that is designed to be secure, transparent, highly resistant to outages, auditable, and efficient [33]. Its potential results from the following characteristics:

- Reliability and availability: in a network of participants, in case of node failure, the others will continue to operate, maintaining the information's availability and reliability.

- Transparency: transactions are visible to network participants, increasing auditability and trust.
- Immutability: changes are almost impossible to be made without being detected, increasing confidence in the information and reducing fraud opportunities.
- Irrevocability: transactions can be made irrevocable, increasing the accuracy of records and simplifying back-office processes.
- Digitalization: as almost any document or asset can be expressed in code, applications are endless.

Also, the development of mechanisms to support collaboration and information flow are also a key aspect for full horizontal integration. Nonetheless, to get there, companies have to get their vertical integration done first, starting at the heart of their production processes.

4 Related Work and Proposed Novelties

4.1 CPS Production Architectures for Manufacturing

Higher availability and affordability of sensors, data acquisition systems and computer networks as well as the competitive nature of the industrial sector have lead more companies to implement these technologies in their manufacturing processes and products. Today, an increasing number of industrial companies are introducing Internet of Things (IoT) and Cyber-Physical Systems (CPS) concepts starting with embedding sensors in manufacturing equipment or tagging products with RFID tags. Consequently, the growing use of sensors and networked machines has resulted in the continuous generation of high volume data that is known as Big Data [34]. In such an environment, CPS can be further developed for managing Big Data and leveraging the interconnectivity of machines to reach the goal of intelligent, resilient and self-adaptable machines [35]. Furthermore by integrating CPS with production, logistics and services in the current industrial practices, it would transform today's factories with significant economic potential [36]. An example of the potential benefits of transforming data coming from these devices into decisions is using embedded sensors in manufacturing equipment to predict equipment wear or diagnose possible faults with a reduction of maintenance costs by nearly 40% [37]. Cyber-Physical Systems (CPS) is a transformative technology for managing interconnected systems between its physical assets and computational capabilities [38]. Since CPS is in the initial stage of development, it is essential to clearly define the structure and methodology of CPS as guidelines for its implementation in industry. In this sense, new CPS architectures are demanded so that they provide advanced connectivity that ensures real-time data acquisition from the physical world and information feedback from the cyber space; and intelligent data management, analytics and computational capability that constructs the cyber space [38]. An example is the proposed CPS 5C-level architecture [39] where Smart connection, Data-To-information conversion, Cyber, Cognition and Configuration levels are associated to applications and techniques.

CPS architectures must take into account modularity and scalability issues, e.g. in case of a machine manufacturer a modular approach consist of component level, machine level, fleet level and enterprise level [40]. The architecture proposed by the authors will support the virtualization and synchronization of data coming from real world assets belonging to the whole supply chain where an industrial company is involved. De-centralization, modularity and scalability are key characteristics associated to this architecture in order to enhance the efficiency of production processes inside and outside of factories allowing the introduction of a more personalized and diversified product portfolio at competitive costs.

4.2 Smart Factory and Factory Virtualization

Smart Factory (SF) is a Factory that context-aware assists people and machines in execution of their tasks [41]. Mark Weiser [42] has coined the term ubiquitous computing for this new world. His vision as regards smart environments involves a physical world, closely and invisibly interwoven with sensors, actuators, displays and computer elements, which are seamlessly embedded into daily life objects and connected with each other by a network. Mark Weiser's approach of smart environments is transferred to manufacturing issues [41]. After the development of digital and virtual factories, the next step is the fusion of physical and digital/virtual world [43] under a so-called SF. The SF concept enables the real-time collection, distribution and access of manufacturing relevant information anytime and anywhere. Systems working in background accomplish their tasks based on information coming from physical and virtual world. SF represents a real-time, context-sensitive manufacturing environment that can handle turbulences in production using decentralized information and communication structures for an optimum management of production processes [41]. SF products, resources and processes are characterized by cyber-physical systems where materials are moved efficiently across the factory floor. This provides significant real-time quality, time, resource, and cost advantages in comparison with the traditional production systems. Sensing components such as actuators and sensors within the industrial set-up are expected to become "smart" as they are, increasingly, becoming self-sufficient with integrated computing abilities and low power consumption. SF will involve consolidation of existing solutions based on a holistic integration of field devices and technologies, including context-aware applications, federation platform, sensor fusion, status recognition, embedded systems, calm-systems (hardware), communication technologies (wireless), auto ID technologies, positioning technologies, and assistance of people and machines [41]. This integration is being driven by the need for seamless exchange of business intelligence to enhance the efficiency by the optimization of resource planning, scheduling, and controlling in real time [44]. Industrial automation platforms are experiencing a paradigm shift. New technologies and production strategies are enabling a synchronization of the digital and real world, providing real-time access to sensorial information, as well as giving technological infrastructures advanced networking and processing capabilities to actively cooperate and form a sort of 'nervous system' within the factory [45, 46]. Enterprise resources (e.g. assets in the form of materials, devices, people, etc.) can be transformed or consumed to produce such benefit. Yet, the development of applications that exploit knowledge from such heterogeneous resources will require a clear understating of all relations and inter-dependencies. Factory resources virtualization exposes an abstraction layer that removes inherent complexity and softens the inner-company operations, creating the conditions to improve agility, responsiveness, and decentralized decision-making [47]. Either by applying simple resource virtualization or mashup, factory virtualization allows to abstract, model and simulate the full automation pyramid, uniquely identifying and virtually representing the real physical entities (e.g. specific sensor) or some aggregation of them (e.g. combined knowledge) [48].

The architecture proposed handles major smart factory aspect through a total virtualization of the manufacturing production pyramid and by offering the capability of designing flexible production processes. Moreover, it will contribute to achieve the full virtualization process (i.e. total virtualization of the traditional automation pyramid from sensor-control to enterprise-level and/or methods and models for the synchronization of the digital and real world) and optimize the knowledge extraction for a comprehensive reasoning, visualization of factory reconfiguration and decentralized decision.

4.3 Situational Awareness and Contextualization in Manufacturing

Endsley [49] defines Situational Awareness (SA) as the perception of the elements in the environment within a volume of time and space, the comprehension of their meaning and a projection of their status in the near future. Three main theoretical approaches dominate the research of SA [50]: the information processing approach, the activity approach, and the ecological approach. The first one has been best represented by Endsley's [51] information processing-based three-level model, describing SA as a product comprising three hierarchical levels: Level 1, the perception of task relevant elements in the environment; Level 2, the comprehension of their meaning in relation to task goals; and Level 3, the projection of their future states. The activity theoretic approach presents SA as one of many components of reflective-orientational activity [52]. The model of the perceptual cycle presents SA as a dynamic interaction between humans and their environment. Proponents of this approach suggest that it is the context of the interaction that defines the SA [53]. Context simplifies and enriches human-human interaction. However, enhancing human-computer interaction through the use of contextualization remains a difficult task. Applications in pervasive and mobile environments need to be context-aware so that they can adapt themselves to rapidly changing situations [54]. Also, contextualization can represent a radical improvement in knowledge enrichment [55]. Industrial applications can be found in [56].

The architecture is based on the optimization of the knowledge extraction for the factory virtualization taking into account the physical and social context. Evolving from a context-specific and objective-oriented Situational Awareness (SA) to a shared SA, it enables the knowledge and understanding common to all the components involved in a situation to support an effective, collective response. SA will allow the generation of novel services, resource sharing and service quality development. By enhancing and promoting the SA capabilities, it will contribute to the factory virtualization process achieving much greater agility to more quickly meet changing business needs and an advanced and improved decentralized decision process.

4.4 Decentralized Decision-Making

Spatial Integration can be achieved in a centralized or decentralized way. In a centralized decision-making (CDM) process, a single, decisional center (DC) is acquainted with all the system information. The central node is in charge of the system planning and owns the power to manage the operations performed by all the network nodes. The central node performs the decision-making in terms of optimizing the objectives of the entire network. In the decentralized decision-making (DDM) models each individual independent network entity makes its own decisions, trying to optimize its own objectives. More than one decision-maker is identified. Depending on the collaboration degree, the nodes will take into account (to a lesser or larger extent) the decisions of other nodes. In a DDM Collaborative mechanisms are needed to coordinate node's decisions and exchange the information [57, 58]. In a DDM model each independent entity, or DC, has its own objective function, which is subject to its constraints. In addition, the decision variables for each entity are often influenced by other entities' decisions, and the flows between levels [59].

To manage interdependent relationships of DDM, it is necessary to define mechanisms that are capable of coordinating the decisions made by the different nodes, as well as the information they exchange. These coordination mechanisms can be found in pre-agreed business rules, and assessment and comparison of alternatives using performance measurement techniques [60–63]. Information is a key aspect for decision-making. The structure of the information systems (IS) is usually supported by legacy software such as Enterprise Resource Planning (ERP) or other centralized software. First challenge is to synchronize the information stored in the ERP and the information managed in the Manufacturing System and second is to achieve a real-time updating of the information regarding unforeseen events.

System Theory [64] states that "an organization reacts to conflict by using analytical processes or bargaining processes". In this way, the recommendation [65] is that the problem solving process of conflict resolution is to identify a solution that satisfies the shared criteria building a decisional structure in order to decentralize the decision-making, taking in account the coherence of the objectives between the levels of decision.

This approach enables the classification of decision considering the time frame associated to the decision process (strategic, tactical, operational, real time) and will provide the criteria to decentralize the decision and facilitate the reactivity facing unpredicted events.

In the context of multi-stage Supply Chains (SC), when focusing on a certain decision-making temporal level it is usual to connect the decisions of a specific SC part with the decisions of the rest of the SC parts, especially with those that are immediately upstream or downstream. This is similar when the focus is in the factory at different levels of the automation pyramid.

Value chains are distributed and dependent on complex information and material flows requiring new approaches to reduce the complexity of manufacturing management systems. They need ubiquitous tools supporting collaboration among value chain partners and providing advanced algorithms to achieve holistic global and local optimization of manufacturing assets and to respond faster and more efficiently to unforeseen changes.

The way manufacturing and service industries manage their businesses is changing due to the emerging new competitive environments. According to [66] The enterprises' success in the new dynamic environments is associated to the improved competencies in terms of new business models, strategies, governance principles, processes and technological capabilities of manufacturing enterprises of 2020. Moreover, especially for SMEs, the participation in collaborative networks is also a key issue for any enterprise that is willing to achieve differentiated and competitive strengths. In the light of this, establishing collaborative relationships becomes an important issue to deal with customer needs, through sharing competencies and resources.

Collaborative Networks consist of a variety of heterogeneous autonomous entities, geographically distributed, in which participants collaborate to achieve a common goal and base their interactions through computer networks. SMEs are characterized by limited capabilities and resources therefore, in order to overcome possible barriers that can appear when establishing collaboration, joint efforts must be performed to achieve the desired collaborative scenarios. When establishing collaboration, networked partners share information, resources and responsibilities to jointly plan, implement, and evaluate a program of activities to reach a common goal and therefore jointly generate value. Thus, establish collaborative relationships imply sharing risks, resources, responsibilities, losses, rewards and trust.

The last decades show a clear trend in business: away from big comprehensive trusts which can cover all stages of a value creation chain, and away from long-standing, well-established and stable supply chains [67]. Most of the companies are moving their focus on their core business competencies and enter into flexible alliances for co-value creation and production. This requires flexible business process integration strategy and interoperable models.

Recent works in interoperability have provided promising results and have been partly responsible for initial commercial products and service offerings and operational deployed applications as discussed in [68, 69]. Collaborative systems need to be agile enough to address the changing needs in manufacturing processes. Agile and resilient enterprises have to cope with the complexity of information presented in many interconnected dimensions, and continuously adapt and re-organize themselves [70]. Representing the enterprise as a static system neglects issues raised by the dynamics of today's business [71].

To achieve this level of collaboration companies must be capable of interoperate. Enterprises today face many challenges related to the lack of interoperability, as most of their Information and Communication Technology (ICT) solutions are often inflexible and difficult to adapt to meet the requirements of those changing enterprises [72].

Assuming that the problems regarding interoperability are solved and that all the network participants are able to communicate and understand each other, a new range of collaboration opportunities is open. Among these, the possibility to implement mechanism for decentralized decision-making. In this context we are basically looking for decision that are strategic (i.e. related to network's mission and objectives) and/or tactical (decisions that will contribute to the longevity, profitability, and continued improvement of all areas of operation).

As we are focusing on decisions that are made within a network of stakeholders, procedures that enable decentralized decision-making are sought. In these cases the decision-making methods are not the central part of the problem. Instead, the most important aspects are related with the establishment of the decision process. For this [73] proposes a six-step approach to facilitate the collaborative decision-making:

1. Ensure leadership and commitment: despite the collaborative nature of the process, the existence of facilitator that owns the process is defended to ensure the success;
2. Frame the problem: specifying known policies, givens, and constraints; identifying problem areas and uncertainties; and defining assumptions and details that are follow-on parts of the decision
3. Develop evaluation models and formulate alternatives: achieving consensus about how success will be measured is fundamental. Alternatives must be developed based on the network vision, framing of the problem and understanding of the issues requiring consideration and alternatives that overlap or are not independent must be identified.
4. Collect meaningful, reliable data: all decision processes require colleting the right information (i.e. the one that is critical for the decision to be made) in an appropriate amount (excluding information that can contribute to turn the decision process messy). The use of decision analysis tools (e.g. Nominal Group Technique—NTG [74]) can be helpful for identifying what information is meaningful to the process and how it should be collected.

5. Evaluate alternatives and make decision: Evaluate identified alternatives and select the one that best fits the criteria. Several decision making tools can be used in this step from a simple cost-benefit analysis to multi criteria decision methods (e.g. Analytical Hierarchy Process—AHP [75])
6. Develop an implementation plan: success of the process depends on how decisions are implemented. The definition of an implementation plan allows to consider barriers, performance interventions, and project management issues. To support the development action plans and decision tree diagrams can be used.

Collaborative decision-making is a complex task especially in what regards the human factor that is involved. Thus, the criticality along the process resides in the steps, which involve interaction among participants and to reach a consensus. This is the reason why Step 1: Ensure leadership and commitment is the most important. The selection of the facilitator can also be made following different strategies. When focusing on value chain networks the facilitator this selection can be made using two different strategies:

- Select an impartial and trustworthy entity, external to the network;
- Within the network, select a different facilitator depending on aim of decision process.

In the second case, a set of rules must be defined at network setup to make this selection clear and accepted by all involved stakeholders. Also, if new partners join the network the set of rules must be communicated and accepted.

The architecture here presented establishes the relation between CPS and the 'glocal' Decision/Information Structure allowing modular virtualization and modeling of the factory automation pyramid. This approach will enable the classification of decision considering the time frame associated to the decision process (strategic, tactical, operational, real-time) and will provide the criteria to decentralize the decision and facilitate the reactivity facing unpredicted events.

5 Vision for Smart Factory Responsive Production

5.1 Objectives

The main objective of the proposed approach is to support the design and reconfiguration of manufacturing plants so that they can easily respond to new demands. These demands can be external (e.g. market demands) or internal (e.g. energy efficiency demands). To achieve this objective the approach proposes a decentralized modular production architecture to support the design and reconfiguration of the manufacturing plant.

The proposed architecture targets at providing support for the following functionalities: (i) total virtualization of the factory automation pyramid enabling remote control and (re)programing of production lines contributing for a reduction on the

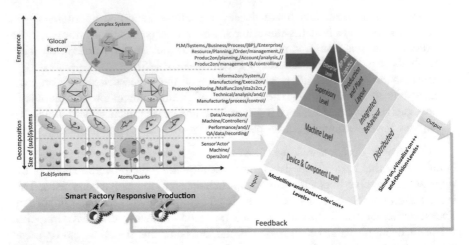

Fig. 3 Vision for smart factory responsive production

number of on-site changes and time to start-up, (ii) reconfiguration/adaptation/ evolution of factories to facilitate reaction to uncertainties/disruptions, (iii) modularization of the production process and product life-cycle to reach mass-personalized product port-folio, and (iv) decentralization of decision-making, allowing responsiveness improvement.

The architecture is used to develop the smart factory responsive production platform (Fig. 3). It acts as the operating system of the smart factory, and will support the entire production process and product lifecycle, contributing for a more personalized, diversified and mass-produced product portfolio and for rapid, flexible and responsive reaction to market changes. The baseline idea is that the virtualization of the entire automation pyramid will enable synchronization of the digital and real world.

Thus, starting from the virtualization of the automation pyramid, and using a modular approach to decompose all the production processes, reconfiguration and/ or readjustments will be facilitated taking advantage of a complex systems perspective, enabling factory 'glocality'. The development of modular blocks, modeling the production steps, which can be automatically reconfigured and/or reorganized at the different levels, enables decentralized process co-simulation and optimization (e.g. testing several combinations) while keeping costs and risks at a low level.

5.2 Concept

Manufacturing is typically associated to the transformation of raw material and assemblage of components into final products that fit the needs of many and can be sold worldwide. As a consequence, research and development (R&D) in this domain has targeted the acceleration and mass-replication of more or less static production processes, construction of production machinery and the development of software to control such systems. Mass replication, although still an important part of production, tends to lose space for customized products tailored to fit consumer needs and demands. Also, as production stages and technologies have become more mobile, a single final manufactured good is nowadays often processed in different companies and countries, crossing several information systems (IS) with sequential tasks in the value chain. Therefore, the survival of enterprises in the near and long term future will depend on their ability to see their own role within the physical and social environment and to become flexible to changes in paradigm that can give them a competitive advantage. Together with flexibility and mobility, complexity has also risen, representing an immense opportunity for technologies such as Cyber Physical Systems (CPS), Internet of Things (IoT) and data analytics. These technologies, when correctly used are capable of providing a huge impact on the factories daily operations. Supported by smart components delivering global and physical awareness to the business systems, they will enhance overall context awareness and the opportunity for better decisions. Complexity science theorizes simple causes for complex effects, with rules that determine how a set of agents behave and interact over time within their environment. It does not predict an outcome for every state, and uses feedback and learning algorithms to enable systems to adapt to its environment over time. In the domain of factory systems, the application of these rules to a large population leads to emergent behaviour that may emulate real-world phenomena. Figure 1, can be used to better explain the concept of the Smart Factory Responsive Production. Highlighting the link between emergent complexity and an enterprise organization, it is easy to observe that the automation pyramid can be directly associated to complexity theories and the 'Glocal' factory idea:

- **The bottom-up perspective (emergence)** enables to understand how simpler systems can be aggregated to provide more complex functions in the frame of the decomposition structure. The proposed architecture relies on this idea and seeks to collect information and act on systems as simple as possible, making use of the pyramid to take input data and knowledge directly from the lower level devices and components up-to the enterprise level where production planning can be reconfigured, products redesigned, etc. Depending on the level of decision required, the emergent flow might not reach the top of the pyramid, decentralizing control and increasing automation.
- **The top-down perspective (decomposition)** allows to define the global structure of the pyramid by introducing decentralization mechanisms and modularization that will be responsible for decreasing levels of complexity to a point where simulation, visualization and decision can be distributed and used to facilitate control and actuation on the real world.

The combined bottom-up and top-down perspectives, enables the resolution of problems, offering a structure of decomposition that discovers inputs from localized mechanisms (e.g. self-adaptation triggered by smart objects), instead of always starting the "responsive" process from the higher level of the automation pyramid. This will create a feedback loop into the manufacturing process that allows reconfiguring and reorganizing physical, human and computational resources in a better form to respond to new trends in mass-customization and re-shoring, as well as unforeseen problems in the daily operation. This concept is also tightly connected with the vision of Sensing Enterprise that was created to reconcile traditional non-native "Internet-friendly" organizations with the tremendous possibilities offered by the cyber worlds [37]. It envisions the enterprise as a smart complex entity capable of sensing and reacting to stimuli, by integrating decentralized intelligence, context awareness, dynamic configurability and sensorial technology into its decision-making process. The enterprise uses visualization and simulation techniques to anticipating future behaviour and taking decisions on multi-dimensional information captured through physical and virtual objects.

6 Architecture to Support Responsive Production

The proposed architecture for Smart Factory Responsive Production was developed taking into consideration aspects such as: decentralization, modularity, scalability and responsiveness. Figure 4 presents the developed architecture and the following sections describe the proposed set of building blocks and services.

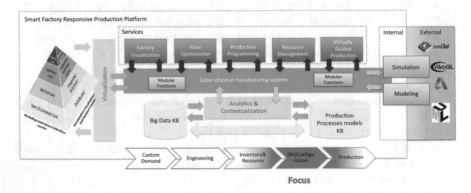

Fig. 4 Architecture for smart factory responsive production

6.1 Modular Building Blocks

- **Virtualization**: Responsible for providing bi-directional access to the different automation pyramid levels, including data collection, reading and writing functionalities. Also supports decentralized decision-making providing systematic decision mechanisms to deal with multi-level hierarchical systems enabling the rapid resolution of local problems.
- **Analytics and Contextualization**: Analyses and contextualizes the collected data enriching it and enabling anomalies detection, future evolution predictions, data interpretation and complex event detection.
- **Cyber-physical manufacturing systems**: Implements the modular functions needed to describe the different manufacturing systems available, together with the appropriate communication, decision and control mechanisms. All information from all related manufacturing systems is closely monitored and synchronized between the physical factory floor and the cyber computational space.
- **Modeling and Simulation**: Provides modeling and simulation functionalities, at both business and physical layers, and across the different levels of the automation pyramid. Whenever possible enables integration with external modeling and simulation tools making use of their capabilities.
- **Knowledge bases**: Namely "Big Data KB", which will gather and contextualize relevant knowledge for the virtualization process, and "Production Process Models KB", responsible for storing the selected production processes that were found to be relevant for the approach.

6.2 Set of Services

The set of Services assemble the intelligent and automated functions at the CPS layer. As the functions at the CPS layer of the architecture are modular, they can be combined in a number of different ways for the elaboration of further services. In addition to this, the approach envisages to that development should open for integration with other (external) systems as well as to allow different combinations of the building blocks to support the development of new services. The envisaged services provide:

- **"Factory Visualization"**: enables users to access and view, in a digital and virtual interface, the factory and its resources, at a certain point in time. This functionality can become quite useful providing support to human decision, providing simple interaction to navigate layouts and correlate past, present and future resource configurations.
- **"Plant Optimization"**: focused on the plant and production line layouts. Based on the needs of different products and customizations, plant layout configuration can be optimized for a certain time-frame.

- **"Resource Management"**: exposes, through a simple interface, factory resources (Human, Physical and Computational). Using this service, and depending on their access rights, users (e.g. production managers) can perform activities such as: resource re-allocation within a certain space or time frame, material management, resource sharing within certain tasks, physical or computation resource reconfiguration, etc.
- **"Production Programming"**: enables users to have an harmonized interface to quickly be able to reprogram machine controllers and smart objects in a way that they can behave in the desired form, and respond to the reconfigurations necessary.
- **"Virtually guided production"**: changes to the planning, provisioning, resource allocation and production tools of the ongoing projects become difficult to manage. This service complements the automatic rearrangements with a guideline for management and workers, providing a step-by-step checklist for readjustments and changes to perform.

7 Conclusions and Future Work

Industry 4.0 represents a huge opportunity, as well as very demanding challenge, for companies. On the other hand, companies that choose to ignore it may be at stake and will for sure struggle in this new production approach. In this paper the authors start by contextualizing the potential that represents industrial digitalization and how technological advances can contribute for a new perspective on manufacturing production.

The architecture here proposed aims at combining virtualization, contextualization and decentralized decision-making to improve production responsiveness and promote plant adaptation. Moreover, the utilization of multiple technologies, such as IoT and CPS, is the key for the convergence of the physical and digital worlds contributing to achieve increased production quality (i.e. by reducing risks through testing combinations) and mass customization. From technological point of view, the access to multiple sources of information together with the processing of that information to generate new and improved knowledge (data analytics) is fundamental for the implementation of the proposed approach.

The developed work answers to specific needs and challenges that must be addressed to solve problems related with dynamic market changes, which are intimately connected with the design and reconfiguration of the manufacturing enterprise. Specifically, the proposed architecture, aims at providing a baseline for further developments in terms of supporting platforms that combine smart factory responsive production, combining virtualization, contextualization, modeling and simulation functionalities to enable self-adaptation to dynamic market demands.

Future work will be directly related with the implementation of the proposed architecture in real industrial scenarios in order to test its appropriateness.

Moreover, and in what regards validation, scenarios will focus in design and reconfiguration of the manufacturing enterprise to test the capacity of answer easily to dynamic market demands.

Acknowledgements The research leading to these results has received funding from the European Union H2020 project C2 NET (FoF-01-2014) nr 636909.

References

1. European Commission: For a European Industrial Renaissance, Communication from the Commission to the European Parliament, the Council, the European Economic and Social Committee and the Committee of the Regions (2014)
2. Hartmann, B., King, W.P., Narayanan, S.: Digital manufacturing: the revolution will be virtualized. McKinsey & Company (2015)
3. European Forum for Manufacture: Driving Innovation and Growth in European Manufacturing (2015)
4. European Factories of the Future Research Association (EFFRA): Factories of the Future: Multi-annual Roadmap for the Contractual PPP under the Horizon 2020 (2013)
5. European Commission: Horizon 2020—Work Programme 2016–2017: 17. Cross-cutting Activities (2016)
6. Schlaepfer, R.C., Koch, M., Merkofer, P.: Industry 4.0 challenges and solutions for the digital transformation and use of exponential technologies. Deloitte AG (2015)
7. 7iD: Industry 4.0. https://www.7id.com/technology/industry-4-0/ (2016)
8. European Commission: Horizon 2020—Work Program 2016-2017—Cross-cutting Activities, 25 July 2016
9. EFFRA: Factories of the Future: Multi-annual Roadmap for the Contractual PPP under the Horizon 2020 (2013)
10. FInES Research Roadmap Task Force (2012)
11. Jacinto, J.: Smart manufacturing? Industry 4.0? What's it all about? Siements Totally Integrated Automation, Automation World & Design World (2014)
12. Monostori, L.: Cyber-physical production systems: roots, expectations and R&D challenges. Procedia CIRP **17**, 9–13 (2014)
13. Adolphs, P.: RAMI 4.0—An architectural Model for Industrie 4.0. Platform Industrie 4.0 (2015)
14. Collins, M.: Why America has a shortage of skilled workers. Industry Week (2015)
15. Forbes, J., Naujok, N., Geissbauer, R., Vedso, J., Schrauf, S.: Industry 4.0: building the digital enterprise. PWC (2016)
16. World Economic Forum Industrial Internet Survey (2014)
17. Chen, D., Vernadat, F.B.: Enterprise interoperability: a standardisation view. Enterprise Inter- and Intra-Organizational Integration, Volume 108 of the series IFIP—The International Federation for Information Processing, pp. 273–282 (2003)
18. Yan, L., Li, Z., Yuan, X.: Study on method-of-robust-multidisciplinary-design-collaborative-decision for product design. Inf. Technol. J. **8**(4), 441–452 (2009)
19. Ruiz Dominguez, G. A.: Caractérisation de l'activité de conception collaborative à distance: étude des effets de synchronisation cognitive (2005)
20. Jung, J.J.: Reusing ontology mappings for query routing in semantic peer-to-peer environment. Inf. Sci. (2010). https://doi.org/10.1016/j.ins.2010.04.018
21. Ranjan, R., Zhao, L., Wu, X., Liu, A., Quiroz, A., Parashar, M.: Peer-to-Peer Cloud Provisioning: Service Discovery and Load-Balancing. https://doi.org/10.1007/978-1-84996-241-4_12

22. Agostinho, C., Pinto, P., Jardim-goncalves, R.: Dynamic adaptors to support model-driven interoperability and enhance sensing enterprise networks. In: 19th World Congress of the International Federation of Automatic Control (IFAC'14), Cape Town, South Africa (2014)

23. Chen, D., Doumeingts, G., Vernadat, F.: Architectures for enterprise integration and interoperability: past, present and future. Comput. Ind. **59**, 647–659 (2008). https://doi.org/10.1016/j.compind.2007.12.016

24. Ducq, Y., Chen, D., Alix, T.: Principles of servitization and definition of an architecture for model driven service system engineering. In: 4th International IFIP Working Conference on Enterprise Interoperability (IWEI 2012), Harbin, China, 2012. https://doi.org/10.1007/978-3-642-33068-17_12

25. Elvesæter, B., Hahn, A., Berre, A., Neple, T.: Towards an interoperability framework for model-driven development of software systems. In: 1st International Conference on Interoperability Enterprise Software and Applications. Springer. http://www.springerlink.com/index/L10NU4306N054T6G.pdf (2005)

26. OMG: MDA Guide Version 1.0.1 (omg/2003-06-01), Object Management Group. http://www.omg.org/cgibin/doc?omg/03-06-01.pdf (2003)

27. Agostinho, C., Ducq, Y., Zacharewicz, G., Sarraipa, J., Lampathaki, F., Poler, R., Jardim-Goncalves, R.: Towards a sustainable interoperability in networked enterprise information systems: trends of knowledge and model-driven technology. Comput. Ind. (2015). https://doi.org/10.1016/j.compind.2015.07.001

28. Santucci, G., Martinez, C., Vlad-câlcic, D.: The sensing enterprise. In: FInES Work. FIA 2012, Aalborg, Denmark. http://www.theinternetofthings.eu/sites/default/files/%5Buser-name%5D/Sensing-enterprise.pdf (2012)

29. Sriram, R.: Smart networked systems and societies: what will the future look like? In: IEEE IT Professional Conference (IT Pro). IEEE Computer Society (2014)

30. Manyika, J., Chui, M., Brown, B., Bughin, J., Dobbs, R., Roxburgh, C., et al.: Big data: the next frontier for innovation, competition, and productivity. http://www.mckinsey.com/insights/business_technology/big_data_the_next_frontier_for_innovation (2011)

31. Zacharewicz, G., Diallo, S., Ducq, Y., Agostinho, C., Jardim-Goncalves, R., Bazoun, H., Wang, Z., Doumeingts, G.: Model-based approaches for interoperability of next generation enterprise information systems: state of the art and future challenges. Inf. Syst. e-Bus. Manag. (2016). https://doi.org/10.1007/s10257-016-0317-8

32. Jardim-Goncalves, R., Agostinho, C., Steiger-Garcao, A.: A reference model for sustainable interoperability in networked enterprises: towards the foundation of EI science base. Int. J. Comput. Integr. Manuf. **25**(10) (2012). (Special Issue on Collaborative Manufacturing and Supply Chains). https://doi.org/10.1080/0951192x.2011.653831

33. Schatsky, D., Muraskin, C.: Blockchain is coming to disrupt your industry. Deloitte (2015)

34. Shi, J., Wan, J., Yan, H., Suo, H.: A survey of cyber-physical systems. In: International Conference on Wireless Communications and Signal Processing, pp. 1–6 (2011)

35. Rajkumar, R.: Workshop report on foundations for innovation in cyber-physical systems. NIST. http://www.nist.gov/el/upload/CPS-WorkshopReport-1-30-13-Final.pdf/ (2013)

36. Lee, J., Lapira, E., Yang, S. Kao, H.-A.: Predictive manufacturing system trends of next generation production systems. In: 11th IFAC Workshop on Intelligent Manufacturing Systems, vol. 11, issue 1, pp. 150–156 (2013)

37. IDC: The digital universe of opportunities: rich data and increasing value of the internet of things. EMC Digital Universe. emc.com/collateral/analyst-reports/idc-digital-universe-2014.pdf. (2014)

38. Baheti, R., Gill, H.: Cyber-physical systems. Impact Control Technol. 1–6 (2011)

39. Lee, J., Bagheri, B., Kao, H.-A.: A cyber physical systems architecture for Industry 4.0-based manufacturing system. Manuf. Lett. 2015, 3, 18–23 (2014). https://doi.org/10.1016/j.mfglet.2014.12.001

40. Bagheri, B., Lee, J.: Big future for cyber-physical manufacturing systems. Design World. http://www.designworldonline.com/big-future-for-cyber-physical-manufacturing-systems/ (2015)

41. Lucke, D., Constantinescu, C., Westkämper, E.: Smart factory-a step towards the next generation of manufacturing. Manufacturing Systems and Technologies for the New Frontier, pp. 115–118. Springer, London (2008)

42. Weiser, M.: The Computer for the 21st Century. Scientific American, Special Issue on Communications. Comput. Netw. (1991)

43. Westkämper, E., Jendoubi, L., Eissele, M., Ertl, T.: Smart factory—bridging the gap between digital planning and reality. Manuf. Syst. **35**(4), 307–314 (2006)

44. Goryachev, A., Kozhevnikov, S., Kolbova, E., Kuznetsov, O., Simonova, E., Skobelev, P., Tsarev, A., Shepilov, Y.: Smart factory: intelligent system for workshop resource allocation, scheduling, optimization and controlling in real time. Adv. Mater. Res. **630**, 508–513 (2012)

45. Agostinho, C., Marques-Lucena, C., Sesana, M., Felic, A., Fischer, K., Rubattino, C., Sarraipa, J.: Osmosis process development for innovative product design and validation. 2015 ASME IMECE, Houston, USA (2015)

46. Ko, J., Lee, B., Lee, K., Hong, S.G., Kim, N., Paek, J.: Sensor virtualization module: virtualizing IoT devices on mobile smartphones for effective sensor data management. Int. J. Distrib. Sens. Netw. (2015). https://doi.org/10.1155/2015/730762

47. Guo, T., Papaioannou, T.G., Aberer, K.: Efficient indexing and query processing of model-view sensor data in the cloud. J. Big Data Res. **1**, 52–65 (2014)

48. Kumra, S., Sharma, L., Khanna, Y., Chattri, A.: Analysing an industrial automation pyramid and providing service oriented architecture. Int. J. Eng. Trends Technol. **3**(5), 586–594 (2012)

49. Endsley, M.: Design and evaluation for situational awareness enhancement. In: Proceedings of the Human Factors Society 32nd Annual Meeting. HFES, Santa Monica, pp. 97–10 (1988)

50. Stanton, N.A., Chambers, P.R., Piggott, J.: Situational awareness and safety. Saf. Sci. **39**(3), 189–204 (2001)

51. Endsley, M.: Toward a theory of situation awareness in dynamic systems. Hum. Factors (The Journal of the Human Factors and Ergonomics Society) **37**, 32–64 (1995)

52. Bedny, G., Meister, D.: Theory of activity and situation awareness. Int. J. Cogn. Ergon. **3**(1), 63–72 (1999)

53. Smith, K., Hancock, P.A.: Situation awareness is adaptive, externally directed consciousness. Hum. Factors (The Journal of the Human Factors and Ergonomics Society) **37**(1), 137–148 (1995)

54. Ranganathan, A., Campbell, R.H.: An infrastructure for context-awareness based on first order logic. Pers. Ubiquit. Comput. **7**(6), 353–364 (2003)

55. Ning, K., Scholze, S., Marques, M., Campos, A, Neves-Silva, R. O'Sullivan, D.: A service oriented platform for context aware knowledge enhancing. In: 5th IFAC Conference on Management and Control of Production and Logistics (2010)

56. Marques, M., Sucic, B., Vuk, T.: Context-based decision support for sustainable optimization of energy consumption. KES Trans. Sustain. Des. Manuf. **1**(1), 899–910 (2014)

57. Schneeweiss, C.: Distributed decision making in supply chain management. Int. J. Product. Econ. **84**, 71–83 (2003)

58. Alemany, M.M.E., Alarcón, F., Lario, F.C., Boj, J.J.: An application to support the temporal and spatial distributed decision-making process in supply chain collaborative planning. Comput. Ind. **62**(5), 519–540 (2011). https://doi.org/10.1016/j.compind.2011.02.002

59. Hong, I.H., Ammons, J.C., Realff, M.J.: Centralized versus decentralized decision-making for recycled material flows. Environ. Sci. Technol. **42**(4), 1172–1177 (2008)

60. Pibernik, R., Sucky, E.: An approach to inter-domain master planning in supply chains. Int. J. Product. Econ. **108**, 200–212 (2007). https://doi.org/10.1016/j.ijpe.2006.12.010

61. Lee, H., Whang, S.: Decentralized multi-echelon supply chains: incentives and information. Manag. Sci. **45**(5), 633–640 (1999)

62. Jung, H., Chen, F., Jeong, B.: Decentralized supply chain planning framework for third party logistics partnership. Comput. Ind. Eng. **55**(2), 348–364 (2008). https://doi.org/10.1016/j.cie.2007.12.017

63. Wang, K.-J., Chen, M.-J.: Cooperative capacity planning and resource allocation by mutual outsourcing using ant algorithm in a decentralized supply chain. Expert Syst. Appl. **36**(2), 2831–2842 (2009)
64. Simon, H.A.: The Science of the Artificial, 1st edn. MIT Press, Cambridge, Mass, (1969). (3rd ed. in 1996, MIT Press)
65. Mesarovic, M.D., Masko, D., Takahara, Y.: Theory of Hierarchical Multilevel Systems. Academic Press, New York and London (1970)
66. Camarinha-Matos, L.M., Afsarmanesh, H.J.: Collaborative networks: a new scientific discipline. J. Intell. Manuf. **16**(4), 439–452 (2005)
67. Popplewell, K., Stojanovic, N., Abecker, A., Apostolou, D., Mentzas, G., Harding, J.: Supporting adaptive enterprise collaboration through semantic knowledge services. In: Enterprise Interoperability Iii: New Challenges and Industrial Approaches, pp. 381–393 (2008). http://doi.org/10.1007/978-1-84800-221-0_30
68. Agostinho, C., Ducq, Y., Zacharewicz, G., Sarraipa, J., Lampathaki, F., Jardim-Goncalves, R., Poler, R.: Towards a sustainable interoperability in networked enterprise information systems: trends of knowledge and model-driven technology. Accepted for Publication at Computers in Industry. http://doi.org/10.1016/j.compind.2015.07.001
69. Agostinho, C., Jardim-Gonçalves, R.: Sustaining interoperability of networked liquid-sensing enterprises: a complex systems perspective. Annu. Rev. Control **39**, 128–143 (2015). https://doi.org/10.1016/j.arcontrol.2015.03.012
70. Weichhart, G., Molina, A., Chen, D., Whitman, L. E., Vernadat, F.: Challenges and current developments for sensing, smart and sustainable enterprise systems. Computers in Industry (2015). http://doi.org/10.1016/j.compind.2015.07.002
71. Weichhart, G.: Supporting Interoperability for Chaotic and Complex Adaptive Enterprise Systems. On the Move to Meaningful Internet Systems: OTM 2013 Workshops. Confederated International Workshops: OTM Academy, OTM Industry Case Studies Program, ACM, EI2N, ISDE, META4eS, ORM, SeDeS, SINCOM, SMS, and SOMOCO 2013. Proceedings: LNCS 8186, 86–92. (2013). http://doi.org/10.1007/978-3-642-41033-8_14
72. Truex, D.P., Baskerville, R., Klein, H.: Growing systems in emergent organizations. Mag. Commun. ACM CACM Homepage Arch. **42**(8), 117–123 (1999)
73. Weiberg, S.: Facilitating collaborative decision-making in six steps. International Association of Facilitators Annual Meeting, pp. 14–15 (1999)
74. Delbecq, A.L., VandeVen, A.H.: A group process model for problem identification and program planning. J. Appl. Behav. Sci. **7**, 466–492 (1971). https://doi.org/10.1177/002188637100700404
75. Saaty, T.L.: The Analytic Hierarchy Process. McGraw-Hill, New York, USA (1980)

Multisensor Data Association
by Using the Polar Hough Transform

Ivan Garvanov

Abstract The data association problem is important in the process of building up multiple sensors system for detection of target and trajectories. In this chapter, we research a multisensor data association approach that uses the polar Hough transform (PHT). After every one radar scan the system associate data in global range-azimuth co-ordinate system and apply polar Hough transform. The advantage of the proposed approach is the data unification in a net of asynchronously working radars with different accuracy characteristics and each radar varied observation-sampling period. The study is performed through Monte-Carlo simulations in MATLAB computing environment.

1 Introduction

The data association problem is of significant importance in the process of building up multiple sensor system for detection of target and its trajectory. Data association problem can be mathematically formulated as a well-studied assignment problem. Conventional approaches for data association of the multi-radar system are: the centralized and the decentralized approach [1–3]. In centralized radar nets, the processing consists of two parts. The first detection is performed on signals for each of the monostatic/bistatic cases, i.e. in a decentralized pre-processing. Next, all the decisions are jointly fused, so the system can provide a final output. Based on the way the networks, systems are divided in two groups—synchronous and asynchronous. Radar nets have better detection efficiency than the single radars.

In the chapter, we researched a multisensor data association approach, which uses the Polar Hough Transform (PHT). There are two approaches for Hough transformation of data—standard and polar Hough transforms (SHT and PHT). The SHT is more suitable for image transformation, while the PHT is very

I. Garvanov (✉)
University of Library Studies and Information Technologies,
119, Tsarigradsko Shose Blvd., 1784 Sofia, Bulgaria
e-mail: i.garvanov@unibit.bg

© Springer International Publishing AG, part of Springer Nature 2018
V. Sgurev et al. (eds.), *Practical Issues of Intelligent Innovations*,
Studies in Systems, Decision and Control 140,
https://doi.org/10.1007/978-3-319-78437-3_11

convenient for the use in radar because the output radar parameters (range and azimuth) are the input parameters of the transform.

The concept of using the Hough transform to improve radar target detection in white Gaussian noise is firstly introduced by Carlson, Evans and Wilson in [4]. An approach for CFAR detection by means of SHT for track and target detection in condition of non-homogeneous background is considered in [5–11]. The Hough detection scheme includes CFAR signal detector in the area of observation, SHT of the target distance measurements from the observation area into the parameter space, binary integration of data in the parameter space and linear trajectory detection. These CFAR Hough detectors have been studied in cases when the target flies in one azimuth and the speed is constant. There are cases when the SHT is used for image processing after the conversion of search radar data from range-azimuth coordinate system to the Cartesian system connected with the radar [12].

As continuation of this research work in [13, 14] is proposed a polar Hough transform detector to be used, which is suitable for search radar. This transform is analogous to the standard Hough transform, where the input parameters are target range and azimuth obtained from the search radar. The technique used combines data from previous search scans into one large multi-dimensional polar data map. This transform is very comfortable for use in radar detection and track determination, when the target changes its speed and flies at different azimuths. The general structure of an adaptive polar Hough detector with binary integration is similar to that of a standard Hough detector. The difference between them is that the polar detector uses (range-azimuth-time) space while the SHT employs (range-time) space. The detection probability of a polar Hough detector can be calculated by Brunner's method as for a standard Hough detector. Finally, the TBD-PHT approach is applied to the design of a multi-channel Polar Hough detector for multi-sensor target detection and trajectory estimation in conditions of randomly arrival impulse interference (RAII) [15–21].

The possibility to minimize time of radar signal detection providing the required values of the probabilities of false alarm and detection has appeared in result of the sequential analysis that has been developed in [22]. The priority of the sequential detector over the conventional detector is in the radar energy reduction at the stage of target detection.

Three different structures of a nonsynchronous multi-sensor Polar Hough detector, decentralized with track association (DTA), decentralized with plot association (DPA) and centralized with signal association (CSA), are considered and analyzed in [23]. The detection probabilities of the three multi-sensor Hough detectors are evaluated using the Monte Carlo approach. The results in [23] show that the detection probability of the centralized detector is higher than that of the decentralized detector.

In the paper we study the basic principles of polar Hough transform and its application for data association in multisensor sistem. The first stage is data association of the N-th radar co-ordinate systems to the Global Co-ordinate System. The second stage is a polar Hough transform, which maps points (targets) from everyone associated local observation space (associated data map) into curves in the

Hough parameter space. If a line trajectory exists in the global (range-azimuth) space, by means of polar Hough transform it is represented as a point of intersection of sinusoids defined by PHT. The paper study the probability characteristics of a decentralized multichannel polar Hough data association detector in asynchronous radar nets with and without measurements errors in conditions of randomly arriving impulse interference. The obtained results are compared with the results from [23].

The chapter includes the following paragraphs—abstract, introduction, Hough transform, Discretization of Hough space, Estimation of line parameters from polar Hough transform, Multisensor data association algorithm by using polar Hough transform, Performance analysis of a multi-sensor polar Hough detector and finally conclusion.

2 Hough Transform

The standard Hough transform and the related Radon transform have received much attention in recent years. Using them makes possible the transformation of two-dimensional images with lines into a domain of possible line parameters, where each image line corresponds to a peak, positioned at the respective line parameters. For these reasons, many line detection applications appeared within the image processing, computer vision, and seismic research areas. The use of the standard Hough transform (SHT) for target detection and track determination in white Gaussian noise is introduced by Carlson, Evans and Wilson in [4]. According to this concept, it is assumed that a target moves in a straight line within in a single azimuth resolution cell (Fig. 1).

The standard Hough transform maps points from trajectory of the observation space termed as range-time $(r - t)$ data space into curves in Hough parameter space. The trajectory from the observation space can be defined by the angle θ of its perpendicular from the origin and the distance ρ from the origin to the line along the perpendicular.

$$\rho = r \cos(\theta) + t \sin(\theta) \qquad (1)$$

where r and t are coordinates measured from the origin of ρ and θ axis in the lower left. The result of transformation is a sinusoid with magnitude and phase depending on the value of the point in range-time $(r - t)$ space.

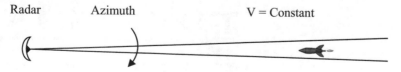

Radar   Azimuth   V = Constant

Fig. 1 Model of moving target

Each point in the Hough parameter space corresponds to one straight line in the $(r-t)$ space with two parameters (ρ, θ). Each of the sinusoids corresponds to a set of possible straight lines through the point. If a straight line exists in the $(r-t)$ space, by means of the Hough transform it can be viewed as a point of intersection of sinusoids defined by the Hough transform. The parameters ρ and θ define the linear trajectory in the Hough parameter space, which could be transformed back to the $(r-t)$ space showing the current distance to the target. Figures 2, 3 and 4 illustrate the $(r-t)$ space, the Hough parameter space and Binary integration of data in Hough parameter space.

In [13] is proposed the Hough transform to be defined by polar coordinates—target range and azimuth (r, a). These parameters are the output of the search radar and they are more suitable for PHT. The proposed polar Hough transform represents each line point in the form:

$$\rho = r \cos(a - \theta), \quad 0 < (a - \theta) \leq \pi \tag{2}$$

where r and a are the target range (distance) and azimuth, θ (theta) is the angle and ρ (rho) is the smallest distance to the origin of polar coordinate system. The mapping can be viewed as stepping through θ from $0°$ to $180°$ and calculating the corresponding ρ.

After N_S radar scans a polar coordinate data map $(r-a)$ is formed containing one trajectory, and it is presented on Fig. 5.

The points' coordinates in $(r-a)$ space form the polar parameter space. A single $(\rho - \theta)$ point in the parameter space corresponds to a single straight line in the $(r-a)$ data space with that ρ and θ values. The result of transformation is a sinusoid with unit magnitudes (Fig. 6).

Each point in the polar Hough parameter space corresponds to one line in the polar data space with parameters ρ and θ. A single ρ and θ point in the parameter

Fig. 2 Range-time
$(r\text{-}t)$ space

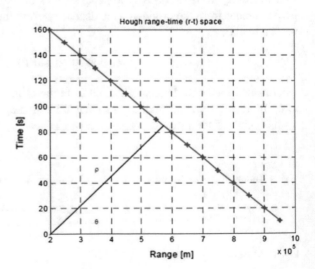

Fig. 3 Hough parameter space (ρ-θ)

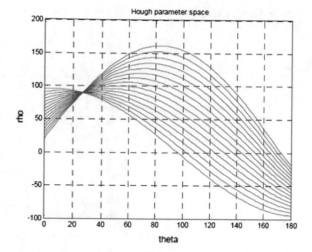

Fig. 4 Binary integration of data in Hough space (ρ-θ)

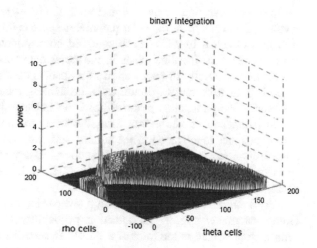

Fig. 5 Cartesian polar coordinate system

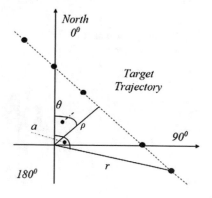

Fig. 6 Hough parameter
space

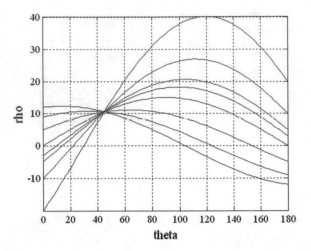

space corresponds to a single straight line in the $(r - a)$ data space with these ρ and θ values. Each cell from the polar parameter space is intersected by a limited set of sinusoids obtained by PHT. Each sinusoid corresponds to a set of possible lines through the point. If a line exists in the polar data space, by means of PHT it is represented as a point of intersection of sinusoids defined by PHT. The polar data space is built from range-azimuth cells, containing the coordinates of targets after NS scans. The parameters ρ and θ have the linear trajectory in the polar Hough parameter space and can be transformed back to the polar data space showing the current distance to the target. If the number of binary integrations (BI) of data in the polar Hough parameter space (of intersections in any of the cells in the parameter space) exceeds the detection threshold TH, both target and linear trajectory detection are indicated (Fig. 7).

Target and linear trajectory detection are carried out for all cells from the polar Hough parameter space. The detection probability of the polar Hough detector cannot be presented in the form of a simple Bernoulli sum as for a standard Hough detector. When a target moves, the SNR of the received signal changes, depending

Fig. 7 Binary integration of
data in Hough space

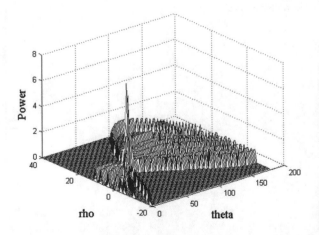

on the distance to the target, and the probability of target detection changes as well. Then the probability Hough PD can be calculated by Brunner's method [4].

The structure of a Polar Hough detector proposed by Garvanov in [13] is shown on Fig. 8.

3 Discretization of Hough Space

The size of discretization of Hough space defines the shape of the obtained platform. If the range and azimuth of the target are measured without errors ($acc(r) = 0$ m and $acc(a) = 0°$), the smaller size of accumulation cells will lead to smaller shape of the platform and the pronounced peak in the Hough space (Fig. 9); the larger accumulator size will lead to the larger platform in the Hough space (Fig. 10).

In a real radar system, it is known that the target coordinates (range and azimuth) are measured with errors [24, 25]. The radars measurement errors (accuracy) $acc(r_n)$ and $acc(a_n)$ can be expressed as measurement oscillations around the considered trajectory. We assume a normal distribution for $acc(r_n) \approx N(0, \sigma_r^n)$ and $acc(a_n) \approx N(0, \sigma_a^n)$ [25], were σ_r^n and σ_a^n are the standard deviations of the measurement errors. The measurement errors (σ_r^n and σ_a^n) of the co-ordinates (r_n, a_n) decrease the binary integration in accumulator cells in the Hough space. In this case, the sinusoids of one linear trajectory will be intersected in different accumulator cells. The correspondence between the Hough parameters (ρ, θ and there discretization $\delta\rho$, $\delta\theta$) and polar range-azimuth space is shown on Fig. 11.

In order to obtain maximal integration (peak), it is necessary to increase the size of the accumulator cells in the Hough space ($\delta\rho$, $\delta\theta$), but this leads to larger errors of ρ and θ (Figs. 12, 13 and 14). Also availability on false alarms will increase the platform. The strip defined by the "maximal" values of these errors has complicated shape depending on the measurement coordinates (Fig. 11).

The polar Hough transform is only efficient if a high number of radar measurements fall in the right range-azimuth area which correspondence of one accumulator cell in Hough space, so that the binary integration can be easily detected amid the background noise. This means that the accumulator cell must not be too small, or else some radar measure will fall in the neighboring Hough cells, thus reducing the visibility of the main cell.

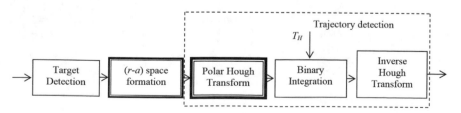

Fig. 8 Structure of a Polar Hough detector

Fig. 9 $acc(r) = 0$ m,
$acc(a) = 0°$, $\delta(\rho) = 100$ m,
$\delta(\theta) = 0.1°$

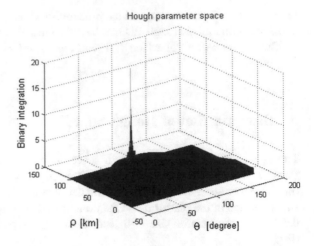

Fig. 10 $acc(r) = 0$ m, acc
$(a) = 0°$, $\delta(\rho) = 500$ m,
$\delta(\theta) = 10°$

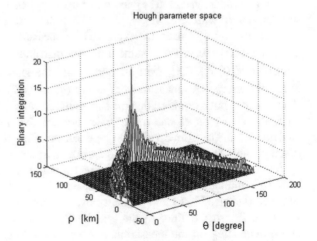

Fig. 11 The Hough
parameters

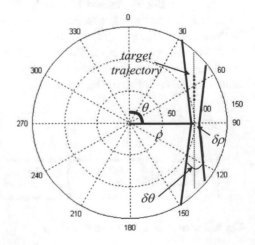

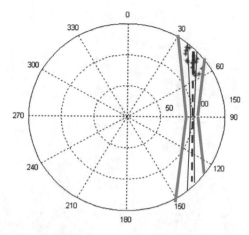

Fig. 12 $acc(r) = 500$ m, $acc(a) = 1°$

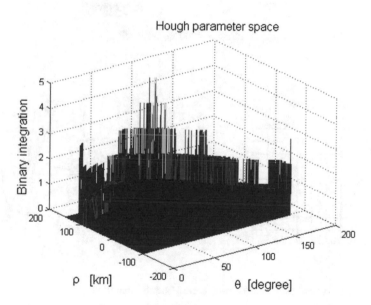

Fig. 13 $acc(r) = 500$ m, $acc(a) = 1°$, $\delta\rho = 500$ m, $\delta\theta = 0.5°$

Also, much of the efficiency of the Polar Hough Transform is dependent on the quality of the input data: the edges must be detected well for the PHT to be efficient. Use of the PHT on noisy background (presence of false alarm) is a very delicate matter and generally, is necessary more radar measurement, since it has the nice effect of attenuating the noise through summation.

The correspondence between the accumulator cells in Hough space and polar range-azimuth space is shown on Fig. 15.

Hough parameter space

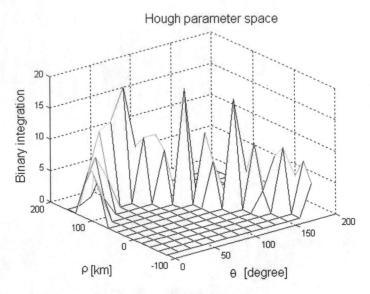

Fig. 14 $acc(r) = 500$ m, $acc(a) = 1°$, $\delta\rho = 2000$ m, $\delta\theta = 15°$

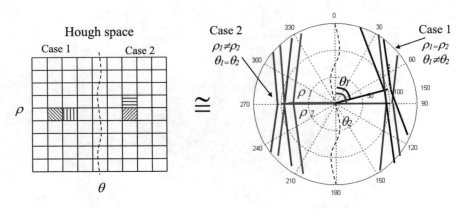

Fig. 15 Every one cell from Hough space correspond the area in range-azimuth space

For example, the binary integration of every one cell in Hough space is obtained from trajectory in the correspond area in range-azimuth space. On the analogy of this, every on cell is obtained from a given area in range-azimuth space. On Fig. 15 are shown two situations, when ρ is constant (case 1) and when θ is constant (case 2). The increasing of value of accumulator cells leads to bigger probability of presence of trajectory in obtained range-azimuth area. Also this will leads to bigger cover area in range-azimuth space and the radar measurements will fall in the neighboring Hough cells. In this case the peak will increase, but platform also will be increase.

Fig. 16 The Hough and
radar parameters

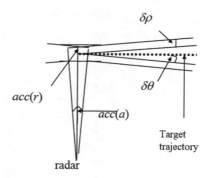

To obtain max binary integration (pick) in the Hough space is necessary to increase size of the accumulator cells but it leads to bigger error for ρ and θ. The strip defined by the "maximal" values of these errors has complicated shape depending on the measurement coordinates (Fig. 16). To improve Hough detector performance by experiments we optimize the size of Hough detector cells and detection threshold value.

4   Estimation of Line Parameters from Polar Hough Transform

The presence of errors when measuring the range and the azimuth causes problems in sinusoid integration in the Hough parameter space (Fig. 17). Increasing the maximal binary integration peak size can be achieved by increasing of the samples in the Hough space.

Thus the platform also gets increased. At the other hand increasing the sampling leads to a bigger error in ρ, θ measuring. In order to improve the estimation

Fig. 17 Binary Hough
parameter space, before
processing with moving
window

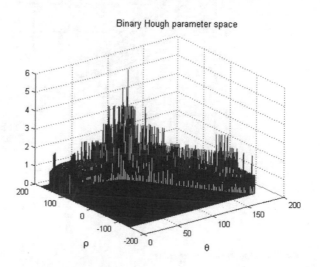

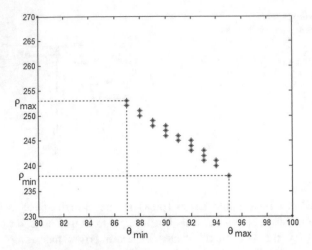

Fig. 18 Estimation of line parameters (ρ, θ)

accuracy of the line parameters, we propose not to increase the sampling in the Hough space, but to use the center of gravity of the area detected (Fig. 18).

$$\rho = \rho_{min} + (\rho_{max} - \rho_{min})/2, \quad \theta = \theta_{min} + (\theta_{max} - \theta_{min})/2 \qquad (3)$$

To make the parameter estimations more accurate we propose the Hough space firstly to be processed in a moving window (Fig. 19) in order to smooth the platform and get more salient area of sinusoids crossing (Fig. 20).

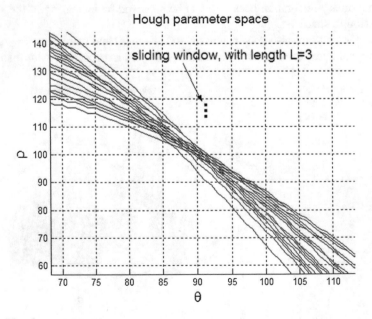

Fig. 19 Hough parameter space

5 Multisensor Data Association Algorithm by Using Polar Hough Transform

The data association problem is of significant importance in the process of building up multiple radar system for detection of target and trajectories. It is known that in radar nets information unification is carried out for signal, detected target or trajectory [25].

The netted radar systems additionally improve the signal processing and especially the target detectability (Fig. 21). The important advantage of this approach is that both the detection probability and the speed of the detection process increase in condition of false alarms. The speed of the detection process and the number of channels are directly proportional quantities. The use of multiple radars, however, complicates the detector structure and requires data association, global time and a processing in a universal coordinate system. All factors, such as technical parameters of radar, coordinate measurement errors, rotation rate of antennas and etc. are taken into account in the sampling the Hough parameter space.

The block scheme of one channel Hough detector is shown on Fig. 22 [25]. The scheme consists of the following blocks: receiver; adaptive detector (CFAR processor); plot extractor; polar Hough transform; binary integration in Hough space; inverse polar Hough transform. The block diagram shows that the data association from different channels is possible to become a: signal level; plot extractor; Hough space and trajectory association.

In [22], three different structures of a nonsynchronous multi-sensor Polar Hough detector, decentralized with track association (DTA), decentralized with plot association (DPA) and centralized with signal association (CSA), are considered and analyzed. The results obtained show that the detection probability of the centralized (CSA) detector is larger than the other.

Fig. 20 Binary Hough parameter space, after processing with moving window

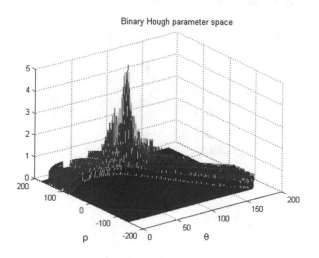

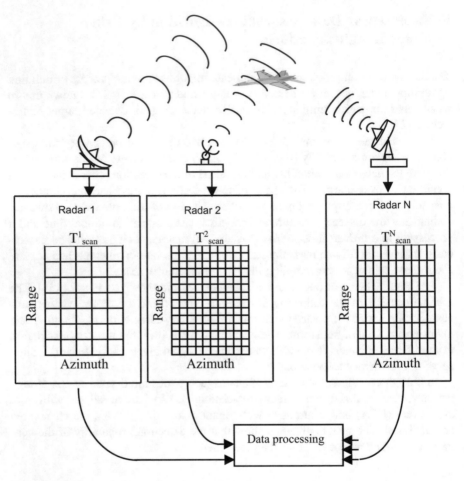

Fig. 21 Multisensor radar system

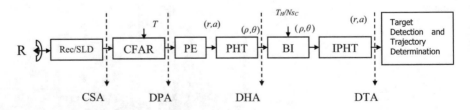

Fig. 22 Block scheme of one channel Hough detector

In CSA multi-sensor detector data association means association of signals processed in all channels of a system. The effectiveness of a centralized Hough detector is conditioned by the minimal information and energy losses in the multi-sensor signal processing. However, a centralized Hough detector requires the usage of fast synchronous data buses for transferring the large amount of data and the large computational resources.

Unlike the centralized structure of a multi-sensor Hough detector, in the decentralized (DPA) Hough detector the process of data association is carried out in the global (r-a-t) space of a Hough detector. The global (r-a-t) space associates coordinates of the all detected target (plots) in radars, i.e. associates all the data at the plot extractor outputs.

The structure of decentralized (DTA) Hough detector shown in [22] consists of three single-channel detectors (Fig. 22). The final detection of a target trajectory is carried out after association of the output data of channels, where a target trajectory was detected or not detected. Local decisions are transferred from each channel to the fusion node where they are combined to yield a global decision.

In many papers, the conventional algorithms for multi-sensor detection do not solve problems of data association in the fusion node, because it is usually assumed that the data are transmitted without loses.

In this chapter, we research a multisensor data association approach that uses the polar Hough transform (PHT). It is so-called decentralized Hough detector with Hough association (DHA), Fig. 23. It can be seen that the decentralized Hough association (DHA) detector has a parallel multi-sensor structure.

The proposed algorithm associates data from sensors with different technical characteristics, operating asynchronously. This algorithm does not associates data into a single coordinate system and a single time. The associations is in Hough space, and it is the same for all channels of the system. The requirement for this system is all Hough transforms in all channels to operate with the same parameter values ρ and θ. The advantage of the proposed approach is the data unification in a net of asynchronously working radars with different accuracy characteristics and each radar varied observation-sampling period.

After radar scan, each of the radars forms the local polar data space (r_n, a_n) where $r_n \in [0, r_n^{max}]$ and $a_n \in [0, 360°]$ are the target range and azimuth, respectively of the N-th radar. All co-ordinate systems are oriented to the "North", and the

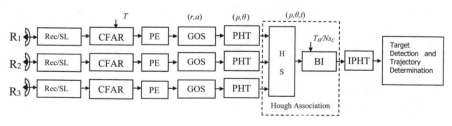

Fig. 23 Decentralized Hough detector, with Hough association (DHA)

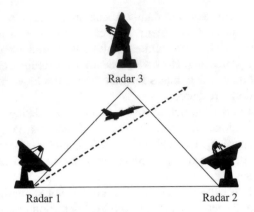

Fig. 24 Multi-sensor target and track detection

earth curvature is neglected. The first stage is transformation of the *N*-th radar co-ordinate systems to the Global Coordinate system resulting into the Global polar observation space (GOS). Each channel has its own specifications on range, azimuth and time of scanning. These parameters do not have translated to a single time and uniform coordinates. Data association is performed in Hough parameter space.

For example, the radar positions form the equilateral triangle, as it is shown on Fig. 24. The received signals from the search radars and 10 radar scans, include moving target, randomly arriving impulse interference (false alarm) and radar receiver noise. The all detected target (plots) from the radars after 10 scans, i.e. plot extractor outputs, as shown on Fig. 25.

The next stage is a polar Hough transform (PHT), which maps points (targets and false alarms) from the observation space (polar data map) into curves in the polar Hough parameter space. All factors, such as technical parameters of radar, coordinate measurement errors, rotation rate of antennas and etc. are taken into account when sampling the Hough parameter space.

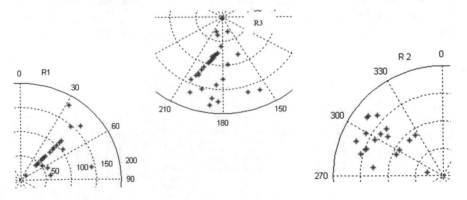

Fig. 25 An observation of a target in the range-azimuth plane after 10 scans

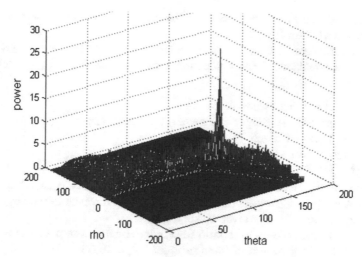

Fig. 26 Binary data integration n Hough parameter space for the Fig. 25

The results of transformation are sinusoids with unit magnitudes. Each point in the polar Hough parameter space corresponds to one line in the polar data space with parameters ρ and θ. A single (ρ, θ) point in the parameter space corresponds to a single straight line in the range-azimuth data space with these ρ and θ values. Each cell from the polar parameter space is intersected by a limited set of sinusoids obtained by PHT.

The sinusoids obtained by the transform are integrated in the Hough parameter space (Fig. 26). If the number of binary integration (BI) of data in the Hough parameter space (of intersections in any of the cells in the parameter space) exceeds the detection threshold, both target and linear trajectory detection are indicated.

If a line trajectory exists in $(r-a)$ space, by means of this transform it is represented as a point of intersection of sinusoids defined by polar Hough transform. The parameters ρ and θ present the linear trajectory in the Hough parameter space and can be transformed back to the data space showing the current distance to the target. The polar coordinates of the detected trajectory are obtained through the inverse polar Hough transform (IPHT) applied to the Hough parameter space, by:

$$r = \rho / \cos(a - \theta) \qquad (4)$$

6 Performance Analysis of a Multi-sensor Polar Hough Detector

The example, given in this section, illustrates the advantages of a four-radar system (CSA, DPA, DHA, DTA) that operates in the presence of randomly arriving impulse interference. The three radars have the same technical parameters as those in [4, 22, 25]. The radar positions form the equilateral triangle, where the lateral length equals 100 km (Fig. 24).

The performance of a multi-sensor polar Hough detector is evaluated using Monte Carlo simulations. The simulation results are obtained for the following parameters:

- Azimuth of the first radar—45°;
- Target trajectory—a straight line toward the first radar;
- Target velocity—1 Mach;
- Target radar cross section (RCS)—1 sq. m;
- Target type—Swerling II case;
- Average SNR is calculated as $S = K/R^4 \cong 15$ dB, where $K = 2.07 * 10^{20}$ is the generalized power parameter of radar and R is the distance to the target;
- Average power of the receiver noise—$\lambda_0 = 1$;
- Average interference-to-noise ratio for random interference noise—$I = 10$ dB;
- Probability of appearance of impulse noise—$P_I = 0.033$;
- Size of a CFAR reference window—$N = 16$;
- Probability of false alarm in the Hough parameter space—$P_{FA} = 10^{-2}$;
- Number of scans—$N_{SC} = 20$;
- Size of an observation area—100×30 (the number of range resolution cells is 100, and the number of azimuth resolution cells is 30);
- Range resolution—1 km;
- Azimuth resolution—2°;
- Size of the Hough parameter space—91×200 (θ cells—91, and ρ cells—200);
- Sampling in θ—2°;
- Sampling in ρ—1 km;
- Binary detection threshold in the Hough parameter space—$T_H = 2 \div 20$.

The performance of the four multi-sensor polar Hough detectors, centralized with signal association (CSA), decentralized with plot association (DPA), decentralized with Hough association (DHA) and decentralized with track association (DTA) are compared against each other. The detection performance is evaluated in terms of the detection probability calculated for several binary decision rules applied to the Hough parameter space. The simulation results are show in Table 1 and Fig. 29. They show that the detection probability of a centralized detector is better than that of a distributed detector. It can be seen that the detection probability of the two types of detectors, centralized and decentralized, decreases with increase of binary decision rules (T_H/N_{SC}). The maximum detection probability is obtained when the binary decision rule is 7/20.

The results have shown that the detection probability of the Polar Hough Data Association detector is between the curves of detector with binary rules in distributed Hough detector 1/3—3/3.

It is apparent from Fig. 29 that the potential characteristics of a decentralized with plot association (DPA) and decentralized with Hough association (DHA) detectors (Fig. 29) are close to the potential curve of the most effective multi-sensor

Table 1 Detection probability of the four multi-sensor Hough detectors—centralized (CSA), decentralized (DPA), decentralized (DTA) and decentralized (DHA) detectors for different binary rules in Hough parameter space

Radar systems	acc(r) acc(a)	(ρ,θ)	Binary threshold—T_H									
			2	4	6	8	10	12	14	16	18	20
CSA	acc(r) = 0 m, acc(a) = 0°	ρ = 500 m, θ = 1°	0.96	0.97	0.98	1	0.97	0.9	0.85	0.83	0.81	0.78
DTA (1/3)			0.8	0.81	0.85	0.84	0.6	0.5	0.5	0.35	0.3	0.15
DTA (3/3)			0.5	0.51	0.55	0.5	0.3	0.27	0.25	0.1	0.5	0.1
DPA DHA			0.95	0.955	0.96	0.98	0.95	0.88	0.8	0.75	0.5	0.1
DPA DHA	100 m 1°		0.9	0.88	0.85	0.78	0.6	0.15	0.1	0	0	0
DPA DHA	100 m 1°	1 km, 1°	0.95	0.954	0.96	0.98	0.94	0.45	0.15	0.05	0	0

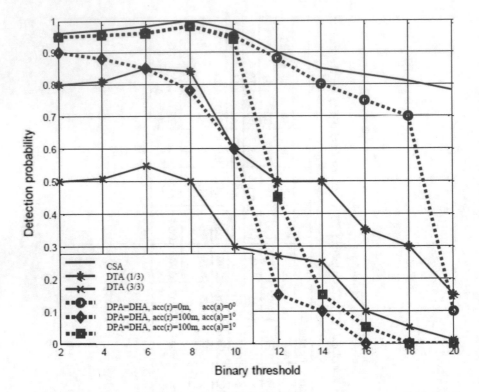

Fig. 29 Detection probability of the four multi-sensor Hough detectors—centralized (CSA), decentralized (DPA), decentralized (DTA) and decentralized (DHA) detectors for different binary rules in Hough parameter space

centralized Hough detector (Fig. 29). It follows that the effective results can be achieved by using the communication structures with low-rate-data channels. The target coordinate measurement errors in the $(r - a)$ space mitigate the operational efficiency of multi-sensor Hough detectors. The needed operational efficiency requires the appropriate sampling of the Hough parameter space.

It is obvious from the results obtained that in conditions of RAII a multi-sensor Hough detector is more effective than a single-channel one. The higher effectiveness is achieved at the cost of complication of a detector structure.

7 Conclusions

In this study we apply a data association in a radar network by using a Polar Hough Transform. The proposed algorithm is applied in Multiple Input Multiple Output (MIMO) radar system. The polar Hough transform allows us to employ a conventional Hough detector in such real situations when targets move with variable speed along arbitrary linear trajectories and clutter and randomly arriving impulse interference are present at the detector input. The polar Hough transform is very comfortable for the use in search radar because it can be directly applied to the output search radar data. Therefore, the polar Hough detectors can be attractive in different radar applications.

Four different structures of a multi-sensor polar Hough detector, centralized (CSA) and decentralized (DPA, DHA, DTA), are studied for target/trajectory detection in the presence of randomly arriving impulse interference. The detection probabilities of the multi-sensor Hough detectors, centralized and decentralized, are evaluated using the Monte Carlo approach. The results obtained show that the detection probability of the centralized detector is higher than that of the decentralized detectors.

The proposed DHA algorithm associates data from sensors with different technical characteristics, operating asynchronously. All factors, such as, coordinate measurement errors, rotation rate of antennas and etc. are taken into account when sampling the Hough parameter space.

Acknowledgements This work is partly financially supported by the project DN 07/1 from 14.12.2016 with title: "Investigation of parameters, properties and phenomena of radio signals from pulsars and their interaction with objects".

References

1. Chernyak, V.: Fundamentals of Multisite Radar System. Gordan and Breach Science Publishers (1998)
2. Griffiths, H.: From a Different Perspective: Principles, Practice and Potential of Biostatics Radar. Radar'03, France (2003)
3. Derham, T., Woodbridge, K., Griffiths, H., Baker, C.: The design and development of an experimental netted radar system. In: Proceedings of Radar'03, France (2003)
4. Carlson, B., Evans, E., Wilson, S.: Search radar detection and track with the Hough transform. IEEE Trans., vol. AES—30.1.1994, Part I, pp. 102–108; Part II, pp. 109–115; Part III, pp. 116–124
5. Behar, V., Kabakchiev, C., Doukovska, L.: Target trajectory detection in monopulse radar by Hough transform. Compt. Rend. Acad. Bulg. Sci. 53(8), 45–48 (2000)
6. Behar, V., Vassileva, B., Kabakchiev, C.: Adaptive Hough detector with binary integration in pulse jamming. In: Proceedings of ECCTD'97, Budapest, pp. 885–889 (1997)
7. Behar, V., Kabakchiev, C.: Hough detector with adaptive non-coherent integration for target detection in pulse jamming. In: Proceedings of ISSSTA'98, South Africa, pp. 1003–1008 (1998)
8. Kabakchiev, C., Doukovska, L., Garvanov, I.: Hough radar detectors in conditions of intensive pulse jamming. Sensors & Transducers Magazine (S&T e-Digest), Special Issue, August 2005, pp. 381–389
9. Kabakchiev, C., Doukovska, L., Garvanov, I.: Adaptive censoring CFAR PI detector with Hough transform in randomly arriving impulse interference. Cybern. Inf. Technol. 5(1), 115–125 (2005)

10. Kabakchiev, C., Doukovska, L., Garvanov, I.: Cell averaging constant false alarm rate detector with Hough transform in randomly arriving impulse interference. Cybern. Inf. Technol. **6**(1), 83–89 (2006)
11. Kabakchiev, C., Garvanov, I., Doukovska, L.: Excision CFAR BI detector with Hough transform in the presence of randomly arriving impulse interference. In: Proceedings of IRS'05, Berlin, Germany, pp. 259–264 (2005)
12. Grishin, Y., Swiercz, E., Janczak, D.: Using the Hough transform as a track initiator in A TBD system. In: Proceedings of IRS 2004, Warszawa, Poland, pp. 291–296 (2004)
13. Garvanov, I., Kabakchiev, C.: Radar detection and track determination with a transform analogous to the Hough transform. In: Proceedings of IRS'06, Krakow, Poland, pp. 121–124 (2006)
14. Garvanov, I., Kabakchiev, C.: Radar detection and track in presence of impulse interference by using the polar Hough transform. J. Eur. Microw. Assoc. **3**, 170–175 (2007)
15. Garvanov, I., Kabakchiev, C.: Sensitivity of track before detect multiradar system toward the error measurements of target parameters. Cybern. Inf. Technol. **8**(2), 85–92 (2007)
16. Kabakchiev C., Garvanov, I., Rohling, H.: Netted radar Hough detector in randomly arriving impulse interference. In: Proceedings of Radar'07, UK, CD ROM 7a.1, pp. 5 (2007)
17. Kabakchiev, C., Garvanov, I., Doukovska, L., Kyovtorov, V., Rohling, H.: Data association algorithm in TBD multiradar system. In: Proceedings of IRS'07, Cologne, Germany, pp. 521–525 (2007)
18. Kabakchiev, C., Garvanov, I., Doukovska, L., Kyovtorov, V., Rohling, H.: Data association algorithm in multiradar system. In: Proceedings of the 2008 IEEE Radar Conference, Rome, Italy, pp. 1771–1774 (2008)
19. Kabakchiev, C., Garvanov, I., Kyovtorov, V., Rohling, H.: Multi-cannel polar TBD Hough detector with a CFAR processor for multi-antennae track elevation measurement. In: Proceedings of IRS'09, Hamburg, Germany, pp. 175–179 (2009)
20. Kabakchiev, C., Behar, V., Rohling, H., Garvanov, I., Kyovtorov, V., Kabakchieva, D.: Analysis of multi-sensor radar detection based on the TBD-HT approach in ECM environment. In: Proceedings of the IEEE Radar Conference—RADAR'10, Washington DC, USA, pp. 651–656 (2010)
21. Garvanov, I., Kabakchiev, C., Vladimirova, M.: Multisensor data fusion and track in presence of randomly arriving impulse interference by using the polar Hough transform. In: Proceedings of ESAVS'10, Berlin, Germany (2010)
22. Kabakchiev, C., Rohling, H., Garvanov, I., Behar, V., Kyovtorov, V.: Multisensor detection in randomly arriving impulse interference using the Hough transform. In: Chapter of the Book "Radar Technology", pp. 179–204. INTECH (2010)
23. Garvanov, I., Kabakchiev, C., Rohling, H.: Detection acceleration in Hough parameter space by K-stage detector. In: 6th International Conference on Numerical Methods and Applications. NM&A-2006. Lecture Notes in Computer Science, LNCS 4310, pp. 558–565. Springer, Berlin, Heidelberg (2007)
24. Garvanov, I., Kabakchiev, C., Doukovska, L., Kyovtorov, V., Rohling, H.: Improvement in radar detection through window processing in the Hough Space. In: Proceedings of IRS'08, Wroclaw, Poland, pp. 139–144 (2008)
25. Garvanov, I.: Detection of moving targets and their trajectories. Za bukvite – O pismeneh, Sofia, Bulgaria (2013) г (in Bulgarian). ISBN 978-954-2946-90-8

Scientific Research Funding Criteria: An Empirical Study of Peer Review and Scientometrics

D. Devyatkin, R. Suvorov, I. Tikhomirov and O. Grigoriev

Abstract In this paper we investigated the problem of scientific research funding from the perspective of data-mining. The object was to conduct versatile retrospective analysis of decisions made by the Russian Foundation for Basic Research regarding scientific research funding. The central task of the analysis was to compare the impact of various items of information on final decision making. In other words, we tried to answer two questions: (a) what does an evaluation committee mainly look at when it selects projects for funding; (b) are scientometric indicators (or science metrics) useful in decision analysis? To achieve this, we built predictive models (classifiers), performed introspection (extracted feature importance) and compared them. The input data was a set of review forms (questionnaires) from the Russian Foundation for Basic Research completed in by peer reviewers. Final decision is made by the foundation board (an evaluation committee). Finally, we concluded that the available input (project proposals, expert assessments and scientometric data) was not enough to explain all the decisions. We showed that scientometric data does not have any significant influence on project proposals assessment. It also means that h-index, mean impact factor, publication and citation number cannot supersede the peer review procedure.

Keywords Decision making · Decision analysis · Research funding
R&D support · Peer review · Feature selection · Criteria importance
Machine learning · Random forest · ReliefF · Gini importance · Linear SVM

D. Devyatkin (✉) · R. Suvorov · I. Tikhomirov · O. Grigoriev
Federal Research Center "Computer Science and Control" of the Russian Academy
of Sciences, Moscow, Russia
e-mail: devyatkin@isa.ru

R. Suvorov
e-mail: rsuvorov@isa.ru

I. Tikhomirov
e-mail: tih@isa.ru

O. Grigoriev
e-mail: oleggpolikvart@yandex.ru

© Springer International Publishing AG, part of Springer Nature 2018
V. Sgurev et al. (eds.), *Practical Issues of Intelligent Innovations*,
Studies in Systems, Decision and Control 140,
https://doi.org/10.1007/978-3-319-78437-3_12

1   Introduction

Grant funding is a very important instrument for science development [2]. There is a
number of scientific foundations, which use peer review to make decisions regarding
research funding. Each proposal is assessed by selected experts (we will refer to them
as *peers* or *peer reviewers*). During review, a peer reviewer completes a form (questionnaire), which we will refer to as a review form. Elements of such questionnaires
(questions to be answered by peers) are called "*criteria*". After all the peer reviewers have submitted their forms, panel review follows and finally the foundation board
summarizes the opinions and makes the decision.

Review form structure depends on research area, funding program etc. Various
elements of the review form have different impact on the final decision. To establish a
new review form structure and the corresponding methodological recommendations,
a group of experts is engaged and complex and often obscure heuristics are used [19].

There are some drawbacks of peer review approach, such as:

1. Peer reviewers often have strong opinion about important problems and promising methods in their own area. This may prevent alternative point of view from
   being funded [16].
2. Conflicts of interest are probable.
3. Various non-scientific attributes may highly affect peer reviewers opinion: personal acquaintanceship, affiliated organizations, nationality, gender, presentation bias [17].
4. Previous achievements may influence the decision, instead of the current scientific level of the group (so called Matthew Effect [25]).

Moreover, there is a trend towards an excessive regularization of science, with
indices as the most important criteria: some researchers [9] promote the idea of
replacing the peer review procedure with scientometric indicators. Some owners of
citation databases also carried out studies related to that topic [7].

We do not agree with this for the next reasons:

1. Scientometric indicators could be artificially inflated by unscrupulous researchers (bogus citations, etc.).
2. One can not guarantee that these indicators are estimated using quality and representative collections of scientific literature.
3. Indicators are domain-dependent, e.g. h-index of chemists cannot be compared
   to that of mathematicians. They also cannot be used to assess interdisciplinary
   projects.

Our point is that peer review process cannot be superseded by automated scientometric based review, but the latter may be useful to reduce bureaucracy, make peer
review process more transparent. Scientific foundations already have accumulated
large amounts of completed review forms over the years. Therefore, we suggest that
machine learning methods may be useful to estimate and compare importance of

various criteria. Such technique makes possible to double-check peer review process, detect outliers, discover how additional criteria might affect decisions, without including such "experimental" criteria into real-life review forms, etc [8, 33]. Another suggestion is to show empirically, that it is not possible to supersede peer review with scientometric indicators of applicants.

One of the main problems with studies of scientific projects evaluation is that most foundations do not disclose their data, either providing a very limited access or requiring to sign NDA. The data we use in this work describe projects, which were funded by the Russian Foundation for Basic Research (RFBR).[1] This is the oldest and the largest scientific foundation in the Russian Federation. Thus, we depend on the data provider terms of usage and we cannot put the dataset in public. However, we believe that this paper will help to make a step towards popularization of open-data approach.

The main contribution is a versatile retrospective analysis of funding decisions made by Russian Foundation for Basic Research.

The rest of the paper is organized as follows: in Sect. 2 we overview the state of the art in quality management of science, in Sect. 3 we briefly describe the dataset and methods we use, in Sect. 4 we present empirical evaluation results and discuss them. Finally, in Sect. 5 we summarize the work done and discuss possible directions of future research.

2   Related Work

It is well known that purely quantitative approaches do not work well for evaluation of scientific projects [15, 31].

Certain researchers [31, 32] stated that mixed set of quantitative and qualitative (or categorical) criteria improves review quality. Thus, we will use only those machine learning methods, which can deal with mixed feature sets.

A number of studies have been conducted in the field of scientific projects evaluation. Schilling et al. [27] applied classification and regression trees (CART) to reveal dependency between research funding and subsequent change of scientometric indicators of project participants. Authors referred to this dependency as cost-effectiveness. They noted that the hierarchical structure of the trees is quite appropriate to deal with complex relationships. They were able to reveal conditions under which funding leads to increase of h-index etc. Also Shilling et al. compared their approach with earlier studies and showed that CART complies with baseline regression approaches while providing more information about the impact of the funding environment and the structure of decision-maker preferences. Zhu et al. [33] described another technique for grant proposals review, which is based on missing value imputation. A special similarity measure [10] over experts is used to reconstruct (impute) missing elements of review forms.

---

[1] Russian Foundation for Basic Research, http://www.rfbr.ru/.

Some papers describe methods for establishing an integrated score over scientific projects and grants. Authors of [24] proposed a special verbal decision analysis-based technique to reduce feature space dimensionality. In [22] authors proposed to use interval valued intuitionistic fuzzy sets [1] to implement a multi-criteria approach for project proposal review. The idea is to build an interval valued intuitionistic fuzzy matrix of expert preferences and then use it to impute missing values and evaluate proposals. This allowed reducing original 24 criteria to 6 top-level ones.

However, these studies do not address the criteria importance estimation problem. On the other hand, researchers from the field of machine learning do. Relief [13] is one of the pioneer methods for feature selection. The idea is to iteratively raise importance (weight) of features that are common for samples from the same category and lower for others. Kononenko et al. [14] (ReliefF) suggested using Manhattan distance instead of Euclidean and to retrieve multiple nearest neighbors instead of just one (to improve robustness). Also, they extended the technique to multi-class problems. Another approach to feature selection can be referred to as a classifier-based. They train a classifier or regression model and then compare internal feature weights [18]. Random forest is another well-known model [12]. Saeys et al. [26] empirically showed that random forest is more robust than SVM and ReliefF. Random forest [6] is a method to construct classifier ensembles using independent decision trees. Each tree is trained on a randomly drawn subset of features and samples (bagging approach). The algorithm is embarrassingly parallel and suits big data very well. The inference employs voting procedure (the most popular outcome among trees becomes the outcome of the entire ensemble). However, Strobl et al. [28] noticed that random forest produces biased estimations for categorical features. To sum up, all of the aforementioned techniques have their pros and cons: convergence difficulties, bias, over- or under-fitting etc.

There are plenty of papers describing investigation of relationship between scientometric indicators and peer review results. However, the obtained findings are rather controversial. For example, a considerable empirical study [29] of 147 university chemistry research groups (covering about 700 senior researchers from 1991 to 2000) indicated that scientometric indices correlate well with peer reviews. On the other hand, authors of [11] estimated linear correlation between peer reviews of project proposals in 2003, 2005 and 2008 and participants scientometric indices. They revealed that scientometric indices and peer review agree with each other only in 45–65% cases. Juvznivc et al. conclude that this may be due to shortcomings of both peer review and scientometric approaches. Therefore, scientometric indicators cannot replace peer review for decision making.

To sum up, drawbacks of the previous works can be grouped into two: (a) bias in feature importance estimations; (b) insufficient power of the applied techniques (e.g. linear correlation) to describe dependency between peer review, scientometric indicators and the outcome. The latter makes it difficult to find out, if scientometry can replace peer review or not. To overcome these issues, we suggested the following:

- Estimate feature importance using multiple techniques (we will refer to them as estimators), possibly based on different theories.

- Compare the feature rankings by various estimators.
- If it makes sense, compose an ensemble and use it to deduce the final rankings.
- Estimate predictive power of various feature sets (application-based, peer review-based and scientometric).

3 Experimental Design

3.1 Dataset

The dataset is a collection of project proposals (applications for grant funding), scientometric indicators of participants, review forms and final decisions (results of peer review), which were submitted to the Russian Foundation for Basic Research within the "A" 2013–2014 program. "A" is a RFBR program to support personal initiative of research groups without predefined topic restrictions. Each project proposal is assessed using three-step scheme: individual peer review, panel review and final approval or rejection by the foundation board. During "A" contest in 2013 and 2014 all projects were evaluated using the same set of criteria. Data for other years have different format (another set of criteria) and will be an object of the future work. Therefore, the dataset used in this work consists of only data from 2013 and 2014 "A" contests. It contains information about final decision regarding each project. There are three review forms per project proposal. More than 65% of projects were rejected (Fig. 1).

The dataset can be thought of as a big table with each row representing a project and each column representing a feature (some attribute of projects). There are six major groups of columns in the dataset:

- *App*—project proposal info, e.g. project type (experimental or theoretical), research group experience (how many Ph.D's, DSc's, students, publications, what is the scientific area of the leader's thesis), etc.

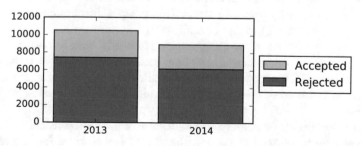

Fig. 1 Dataset size

Table 1 Structure of assessment form for individual peer review

Criterion (field of form)	Possible values
How accurate the proposed methods justified?	Precisely Vaguely Not justified at all
Is there a conflict of interests that prevents reviewer from assessing the project?	Yes No
Is the project fundamental?	Fundamental Partially fundamental Not fundamental
How significant are the previous results by applicants?	Results are very important Some results are interesting Results are ordinary Not evaluable
Why you cannot review this project?	Free-text field
Is the project feasible?	Most probably, the authors will succeed with the project There are some doubts Not evaluable
Are expected results important?	Fundamentally important Important for a particular area Somewhat useful Not important at all
Are problem and objectives clear?	Precise Vague Missing
Is the project experimental or theoretical?	Experimental and theoretical Purely experimental Purely theoretical
Overall rating	Number from 1 (strong reject) to 9 (strong accept)

- *Review*—project assessment info. These are the results of individual peer review (review forms filled by peer reviewers). This group of features contains all fields of review form (Table 1), except Overall rating.
- *Overall rating*—the overall rating, aggregated from all review forms for the project (Table 1). After a peer reviewer has completed reviewing a project, they assign overall rating to 1 if the proposal must be definitely rejected and 9 for strong accept.
- *Participants scientometry*—aggregated scientometric indicators of all project participants.
- *Leader scientometry*—Scientometric indicators of the project leader.
- *Outcome column*—final decision (whether the project received funding or not).

One could notice that the dataset does not contain information about projects completion. One peculiarity of funding program "A" is that only projects with high probability of successful completion are approved. Also, such information about previous experience of participants is implicitly encoded in "How significant are the previous results by applicants?" field of the review form. That is why we did not explicitly include such information into the dataset in this work. However, it will be interesting to investigate in the future research.

Scientometric indicators were collected manually from eLIBRARY,[2] which is the most complete source of scientometric data in Russian Federation. Due to difficulties of manual data collection, we collected it only for 1900 project proposals from "Math, Computer Science and Mechanics" scientific area (about 7800 people). Thus, experiments involving scientometry are executed only on this subset of the full dataset. These indicators include: the number of publications and citations, h-index, h-index without self-citations, mean impact factor of journals, where their publications are published. In this study we decided to use only those indicators, because most other ones were shown [4] to be highly correlated with them.

Each categorical field has a predefined and finite set of possible values without ordering. To represent categorical fields in a form suitable for machine learning, we employ binarization technique. So, the feature space can be formally described as $C = \cup_{\forall Cr \in CR}(Cr \times K_{Cr}) \cup NR$, where NR is a set of numeric criteria and Cr is a set of categorical criteria and K_{Cr} is a set of possible values of Cr. We will refer to elements of C as *features* and to the original fields of the review forms and project proposals as *criteria*. If an attribute had multiple values for a single project, we replaced the original column with three aggregated columns: min, max and mean. Publications number, citations number and h-indices were transformed using logarithm to make distribution more like Gaussian, which is more suitable for data mining techniques.

3.2 Methods

We conducted four experiments in this paper: (a) verified that the available criteria actually allow making decisions that correlate with final funding decisions made by the foundation board; (b) estimated the dependency of the overall project rating on other criteria; (c) estimated and compare importance of various criteria; (d) estimated the relationship between scientometric indicators and funding decisions.

Experiment 1—Project Proposals Classification During the first experiment we evaluated discriminative power of the review form elements. We built a set of models that predict project acceptance. Models were compared using traditional three-fold cross validation and macro-averaged recall, precision and F_1 scores. Each score achieves maximum of 1 and minimum of 0 (the more the better). Also, we conducted a "separability test" by evaluating classification metrics on the training dataset. We built separate models for each scientific area. Model hyper-parameters were optimized using randomized grid search and three-fold cross-validation [3]. The difference between "ideal model" (with all scores of 1), "separability test" and

---

[2]Scientific Electronic Library and Russian Science Citation Index http://elibrary.ru.

cross-validated "predictive power test" shows if models overfit or not and if we lack of additional external knowledge.

Experiment 2—Overall Rating Imputation The second experiment aimed on exploring the correlation between the overall rating and other criteria. This refers to a function approximation problem (or regression). We applied random forest and linear regression. The models were evaluated with mean absolute error, mean relative error and R^2-score (determination coefficient) macro-averaged over three-fold cross validation. Also, we analyzed how a criterion and each of its values affect the overall rating. It was done by comparing feature weights in linear regression. This became possible due to the fact that the input dataset was normalized before training.

Experiment 3—Criteria Importance Analysis This experiment consisted in estimation and comparison of criteria importance across various scientific areas. Overall rating was excluded from the feature set in this experiment (only *Review* feature subset is used). To estimate importance, we employed a number of various approaches:

- *Gini*—Random Forest internal importance measure (Gini Importance).
- *F1-RF*—F1-measure drop with random forest, when a feature is excluded from the input feature set.
- *LSVM*—feature weights in linear SVM with L2 norm.
- *F1-LSVM*—F1-measure drop with linear SVM, when a feature is excluded from the input feature set.
- *ReliefF*—ReliefF feature selection method.

Gini Importance $\Delta Gini(c)$ is a total Gini Impurity decrease among all nodes (only those using feature $c \in C$) in all trees in a random forest [5]. It is empirically shown that Gini Importance of a feature highly correlates with the real impact of this feature on the final decision. Resulting Gini Importance of a feature was obtained by summing over all (three) folds in cross-validation $W_{Gini}(c) = \sum_{i=1}^{k} \Delta Gini(c)$.

Feature importance according to *F1-RF* and *F1-LSVM* was calculated as drop of average F1-score in classification task (as in experiment 1, see Sect. 3.2), but with this feature excluded from the input feature set $W_{F1-\{RF,LSVM\}} = F_1(Model(C)) - F_1(Model(C \backslash c))$.

ReliefF method consists in (1) iterating over all samples from the training dataset; (2) searching the nearest neighbors from the same category and from others and (3) rising importance of features that have similar values for samples from the same category and lowering importance of others: $W_{ReliefF}(c, t) = W_{ReliefF}(c, t - 1) - dist(x_c, sameCategoryNN_c) + dist(x_c, anotherCategoryNN_c)$, where x_c is a column vector of feature c values from all samples from some category, $sameCategoryNN_c$ and $anotherCategoryNN_c$ are columns like x_c, but for the nearest neighbors of the corresponding samples from x_c.

Then, to calculate the importance of the criteria, we averaged importances of all features that represent this criteria (over all $Cr \times K_{Cr}$) pairs: $I(Cr) = \frac{1}{|Cr|} \sum_{\forall c \in Cr \times K_{Cr}} W(c)$.

After we computed importances, we built a criteria ranking, which is a sequence of criteria sorted by importance decrease.

$$Ranking(CR) = \left(Cr_1, Cr_2, ..., Cr_{|CR|}\right), I(Cr_i) > I(Cr_j), i < j$$

Rank of a criterion Cr is an index of the criterion in the ranking: $R(Cr) = i$: $Ranking(CR)_i = Cr, Cr \in CR$.

Final rank of a criterion is an average rank among all importance estimation techniques: $R(Cr) = \frac{1}{5}(R_{Gini}(Cr) + R_{F1-RF}(Cr) + R_{LSVM}(Cr) + R_{F1-LSVM}(Cr) + R_{ReliefF}(Cr))$.

Then, to investigate relationships of various importance estimation techniques, we compared rankings using Rank-Biased Overlap metric [30]. The feature selection techniques do not guarantee that feature importance will not degrade to zero (and if it does, the feature is excluded from the rankings). This renders traditional Spearman rank correlation unusable, because it requires equal contents of all rankings.

Rank-Biased Overlap is a relatively unknown metric, so we present a brief over-view on it. Let S and T be the rankings to compare. Let $S_{:d}$ be a d-prefix (first d elements) of S. $A(S, T, d) = \frac{|S_{:d} \cap T_{:d}|}{d}$ is relative overlap of d-prefixes. Thus, the final similarity is calculated as $RBO(S, T, p, k) = (1 - p) \sum_{d=1}^{k}(p^{d-1}A(S, T, d)) + RBO_{ext}(S, T, p, k)$, where $k = min(|S|, |T|)$, p is the weight decrease rate and $RBO_{ext}(S, T, d)$ is the upper bound of the remainder: $RBO_{ext}(S, T, p, k) = A(S, T, k)p^k + \frac{1-p}{p} \sum_{d-1}^{k} A(S, T, d)p^d$. $RBO(S, T, p)$ is normalized and equals to 0 when S and T consists of completely different elements and it equals to 1 when they are the same. We have to say that exact value of p does not matter, the only requirement is that p must be the same in all experiments. In this paper we chose p so that the longest possible prefix would have weight of 0.1 (the first prefix has weight of 1).

We used a combination of open implementations of aforementioned algorithms [20, 23] with home-brewed tweaks and glue code.

Experiment 4—Explanation Using Scientometric Indicators This experiment aimed on analyzing the role of scientometry in grant funding. To achieve this, we built a number of predictive models (random forest classifiers). Each classifier was trained on only a subset of all features. So, we could estimate what influence on the prediction each group of features has. The classifiers and feature sets were evaluated using three-fold cross validation and compared using F1-measure, precision and recall.

4 Results and Discussion

Table 2 presents *results of the first experiment* ("predictive power" test). We decided to omit results of "separability" test, because they are almost the same and only a bit better (< 0.04 F_1-score points). From this data, one can see that the classification

Table 2 F1-measure of funding decision prediction

Research area	Total	Accept rate (%)	RF	LSVM L2-norm	LogReg	Gaussian naive bayes	CART decision tree
Math, CS, mechanics	1859	31	**0.82 ± 0.02**	0.82 ± 0.03	0.82 ± 0.04	0.72 ± 0.03	0.81 ± 0.02
Physics, astronomy	2749	30	**0.86 ± 0.01**	0.85 ± 0.01	**0.86 ± 0.01**	0.69 ± 0.02	**0.86 ± 0.01**
Chemistry	2486	70	0.85 ± 0.01	0.86 ± 0.01	**0.87 ± 0.01**	0.75 ± 0.03	0.85 ± 0.01
Bio&Med	4097	30	**0.85 ± 0.02**	0.85 ± 0.03	0.85 ± 0.04	0.75 ± 0.04	**0.85 ± 0.02**
Earth science	2137	69	**0.83 ± 0.01**	0.83 ± 0.02	0.83 ± 0.01	0.71 ± 0.05	0.81 ± 0.02
Humanities	1367	23	**0.78 ± 0.03**	**0.78 ± 0.03**	0.78 ± 0.04	0.72 ± 0.01	0.74 ± 0.03
IT	1859	29	0.88 ± 0.01	**0.89 ± 0.01**	0.88 ± 0.01	0.80 ± 0.01	0.88 ± 0.02
Engineering	2738	30	**0.89 ± 0.02**	**0.89 ± 0.02**	**0.89 ± 0.02**	0.72 ± 0.03	0.87 ± 0.01

task is relatively easy, and because the difference between "separability test" and "predictive power test" is very small, we conclude that the models do not over- or under-fit and there are definitely some other factors that affect the final decision. We did not take into account such complex and potentially wide background knowledge as competition conditions etc. We leave this for the future work.

During the second experiment we tried to reconstruct overall project rating from other fields using RF and linear regression. Both models achieved good results with about 0.45 mean absolute error (5% relative error) and normalized $R^2 > 0.83$. This relation can be considered to be functional. Such a reconstruction model can be used to detect "unusual" review forms and double-check peer reviews. As a side effect, we identified criteria and the corresponding values that affect the overall rating most of all:

- Have the researches successfully completed many projects?

  + Yes, and the results are very important.
  − Maybe, but results are ordinary.

- Is the project fundamental?

  + Fundamental and interdisciplinary.
  − Not fundamental.

- Is the project feasible?

  + Most probably, the authors will successfully complete the project.

- Are proposed methods reasonable?

  + Reasonable.

- Are expected results important?

  − Not at all.

In the third experiment we compared various rankings using *RBO* similarity. We found that average similarity between rankings is 0.731 ± 0.153, average similarity

Table 3 Averaged rankings for all scientific areas

Criteria	Math, CS, mechanics	Physics	Chemistry	BioMed	Earth	Humanities	IT	Engineering
Are expected results important?	1	1	1	1	2	1	1	1
Have the researches successfully completed many projects?	2	2	2	2	1	2	2	2
Is the project feasible?	5	4	3	5	4	5	3	5
Are problem and objectives clear?	3	3	6	4	5	3	4	3
Are proposed methods reasonable?	4	5	4	3	3	8	6	4
Is the project experimental or theoretical?	6	7	5	6	8	6	5	7
Is the project fundamental?	7	6	7	7	6	4	7	6

Table 4 Comparison of rankings for different scientific areas

	Math, CS, mechanics	Physics	Chemistry	BioMed	Earth	Humanities	IT	Engineering
Math, CS, mechanics	1							
Physics	0.97	1						
Chemistry	0.91	0.90	1					
BioMed	0.95	0.92	0.91	1				
Earth	0.61	0.62	0.62	0.66	1			
Humanities	0.95	0.95	0.86	0.90	0.57	1		
IT	0.91	0.93	0.96	0.91	0.60	0.90	1	
Engineering	0.99	0.97	0.90	0.94	0.62	0.95	0.90	1
Random shuffle	0.44	0.44	0.44	0.43	0.43	0.43	0.43	0.44
Inverse	0.28	0.33	0.45	0.28	0.27	0.35	0.42	0.27

Table 5 Impact of scientometric indicators on Funding Decision Prediction

Configuration #	Features					Metrics		
	Application	Review	Overall rating	Participants scientometry	Leader scientometry	F_1	Precision	Recall
1	+	+	+	+	+	0.82 ± 0.03	0.85 ± 0.05	0.8 ± 0.02
2	+	+	+		+	0.82 ± 0.03	0.85 ± 0.06	0.79 ± 0.02
3	+	+	+	−		0.82 ± 0.03	0.85 ± 0.05	0.8 ± 0.02
4	+	+	+			0.82 ± 0.03	0.84 ± 0.05	0.81 ± 0.04
5			+			0.82 ± 0.03	0.83 ± 0.05	0.81 ± 0.05
6	+	+		+	+	0.75 ± 0.03	0.79 ± 0.04	0.72 ± 0.06
7	+	+				0.74 ± 0.03	0.77 ± 0.03	0.71 ± 0.06
8	+					0.62 ± 0.08	0.73 ± 0.04	0.55 ± 0.11
9	+				+	0.62 ± 0.07	0.71 ± 0.05	0.55 ± 0.1
10	+			+		0.61 ± 0.08	0.71 ± 0.05	0.53 ± 0.11
11					+	0.44 ± 0.03	0.61 ± 0.05	0.35 ± 0.05
12				+		0.42 ± 0.05	0.52 ± 0.07	0.36 ± 0.04

between meaningful rankings and randomly shuffled rankings is 0.444 ± 0.002 and average similarity between meaningful rankings and inverted meaningful ones is 0.389 ± 0.055. Thus, we conclude that all importance estimators produce rankings more similar to each other, than to random and inverse. However, they do not produce exactly the same results. This means that there is a lot of sense of combining all these estimators into an ensemble to reduce variance and bias.

Tables 3 and 4 present the comparison of rankings for different scientific areas (the third experiment). In Table 3 we omitted three criteria (overall project rating, conflict of interest, cause of denial to review), because the former is easily predictable and the latter two do not affect outcome at all. As one can see, the most significant factors are universal: (a) expected results are fundamentally important and (b) the applicants have already successfully completed some projects with significant scientific outcome. Also, even the two most different rankings (Humanities and Earth) are much more similar to each other, than to random ranking and inverse ones.

Table 5 presents the results of *the fourth experiment*: impact of scientometrics on prediction quality. From this table one can see that these indicators do not add much information to the base features of project proposals and review forms (either due to noise, or due to high correlation with these features). Hence, we conclude that scientometrics cannot be used to substitute peer review. The most significant improvement is achieved by the configuration presented in row 6 when compared to row 7 of Table 5 (review forms, proposals without overall rating). Finally, we conclude that peer review process cannot be greatly improved by incorporating scientometry.

5   Conclusion and Future Work

To sum up, in this work we described a machine learning approach to robustly estimate importance of criteria in science funding. The presented technique is able to deal with quantitative and qualitative criteria and suitable for large datasets.

The most significant criteria that affect decisions about research funding were identified. They include the importance of the expected results and quality of previous research conducted by applicants. We also showed that criteria importance is almost the same over all scientific areas. The overall project rating almost-functionally depends on other review form fields. The approximation of this dependency can be used for peer review quality check, for "too optimistic" or "too pessimistic" reviewers detection. Despite the outcome is quite accurately predictable, there are definitely some other "background" factors that affect decisions. The described technique can be used by scientific foundations to evaluate new criteria before they are included into review form (by simulation).

We also empirically showed that the examined set of scientometric indicators (h-index, number of publications and citations and mean impact factor of the journals in which the applicant has been published) is not sufficient to explain decisions. This complies with [11] findings. Moreover, due to high correlation with other features, these indicators do not allow to significantly improve the peer review procedure.

In fact, replacing the traditional peer review process with purely scientometry-based decision making is dangerous and may lead to unpredictable funding decisions.

However, we are not saying that scientometry is useless for examination of R&D results. Scientometric indicators together with peer review may be very useful to compare individual researchers, research groups, conferences within a particular scientific area.

Possible directions of the future research include incorporation of more background knowledge into the feature set, such as results of deep natural language processing [21]. Furthermore, we plan to analyze theoretical and experimental projects separately and apply the same pipeline to data from other scientific foundations and/or funding programs, which have more quantitative and free-text fields in their review forms. Another idea we are going to check is to analyze and predict final and intermediate indicators of project success.

We hope that all these additions together with traditional peer review will make the funding decisions more transparent and difficult for cheating. Moreover, we hope that our work will have a good impact on providing open access to scientific publications and reports.

Acknowledgements The research is supported by Russian Foundation for Basic Research (Grant 16-29-12881 ofi-m).

References

1. Atanassov, K.T.: Intuitionistic fuzzy sets. Fuzzy Sets Syst. **20**(1), 87–96 (1986)
2. Benner, M., Sandström, U.: Institutionalizing the triple helix: research funding and norms in the academic system. Res. Policy **29**(2), 291–301 (2000)
3. Bergstra, J., Bengio, Y.: Random search for hyper-parameter optimization. J. Mach. Learn. Res. **13**(Feb), 281–305 (2012)
4. Bornmann, L., Mutz, R., Hug, S.E., Daniel, H.D.: A multilevel meta-analysis of studies reporting correlations between the h index and 37 different h index variants. J. Informetr. **5**(3), 346–359 (2011)
5. Breiman, L.: Technical note: some properties of splitting criteria. Mach. Learn. **24**(1), 41–47 (1996)
6. Breiman, L.: Random forests. Mach. Learn. **45**(1), 5–32 (2001)
7. Colledge, L., James, C.: A basket of metricsthe best support for understanding journal merit
8. Geuna, A., Martin, B.R.: University research evaluation and funding: an international comparison. Minerva **41**(4), 277–304 (2003)
9. Green, J.T.: Evidence-based decision making in academic research: the snowball effect. Acad. Executive Br. **3**(1), 12–14 (2013)
10. Jousselme, A.L., Grenier, D., Bossé, É.: A new distance between two bodies of evidence. Inf. Fusion **2**(2), 91–101 (2001)
11. Južnič, P., Pečlin, S., Žaucer, M., Mandelj, T., Pušnik, M., Demšar, F.: Scientometric indicators: peer-review, bibliometric methods and conflict of interests. Scientometrics **85**(2), 429–441 (2010)
12. Kawakubo, H., Yoshida, H.: Rapid feature selection based on random forests for high-dimensional data. In: Proceedings of the International Conference on Parallel and Distributed Processing Techniques and Applications (PDPTA). p. 1. The Steering Committee of The World

Congress in Computer Science, Computer Engineering and Applied Computing (WorldComp) (2012)

13. Kira, K., Rendell, L.A.: The feature selection problem: traditional methods and a new algorithm. AAAI. **2**, 129–134 (1992)

14. Kononenko, I., Šimec, E., Robnik-Šikonja, M.: Overcoming the myopia of inductive learning algorithms with relieff. Appl. Intell. **7**(1), 39–55 (1997)

15. Larichev, O.: Science and art of decision making. Moscow Sci. 200 (1979)

16. Laudel, G.: The art of getting funded: how scientists adapt to their funding conditions. Sci. Public Policy **33**(7), 489–504 (2006)

17. Lee, C.J., Sugimoto, C.R., Zhang, G., Cronin, B.: Bias in peer review. J. Am. Soc. Inf. Sci. Technol. **64**(1), 2–17 (2013)

18. Mladenić, D., Brank, J., Grobelnik, M., Milic-Frayling, N.: Feature selection using linear classifier weights: interaction with classification models. In: Proceedings of the 27th Annual International ACM SIGIR Conference on Research and Development in Information Retrieval. pp. 234–241. ACM (2004)

19. Olson, K.: An examination of questionnaire evaluation by expert reviewers. Field Methods **22**(4), 295–318 (2010)

20. Olson, R.: ReliefF: First Release (2016). https://doi.org/10.5281/zenodo.47803

21. Osipov, G., Smirnov, I., Tikhomirov, I., Sochenkov, I., Shelmanov, A.: Exactus expert—search and analytical engine for research and development support. In: Novel Applications of Intelligent Systems, pp. 269–285. Springer (2016)

22. Oztaysi, B., Onar, S.C., Goztepe, K., Kahraman, C.: Evaluation of research proposals for grant funding using interval-valued intuitionistic fuzzy sets. Soft Comput. 1–16 (2015)

23. Pedregosa, F., Varoquaux, G., Gramfort, A., Michel, V., Thirion, B., Grisel, O., Blondel, M., Prettenhofer, P., Weiss, R., Dubourg, V., Vanderplas, J., Passos, A., Cournapeau, D., Brucher, M., Perrot, M., Duchesnay, E.: Scikit-learn: machine learning in Python. J. Mach. Learn. Res. **12**, 2825–2830 (2011)

24. Petrovsky, A., Roisesnson, G., Balyshev, A., Tikhonov, I.: Retrospective analysis of the research projects. In: Information Models & Analyses. p. 349. ITHEA (2012)

25. Pislyakov, V., Dyachenko, E.: Citation expectations: are they realized? study of the matthew index for russian papers published abroad. Scientometrics **83**(3), 739–749 (2010)

26. Saeys, Y., Abeel, T., Van de Peer, Y.: Robust feature selection using ensemble feature selection techniques. In: Joint European Conference on Machine Learning and Knowledge Discovery in Databases. pp. 313–325. Springer (2008)

27. Schilling, C., Mortimer, D., Dalziel, K.: Using cart to identify thresholds and hierarchies in the determinants of funding decisions. Med. Decis. Mak. 0272989X16638846 (2016)

28. Strobl, C., Boulesteix, A.L., Zeileis, A., Hothorn, T.: Bias in random forest variable importance measures: illustrations, sources and a solution. BMC Bioinform. **8**(1), 1 (2007)

29. Van Raan, A.F.: Comparison of the hirsch-index with standard bibliometric indicators and with peer judgment for 147 chemistry research groups. scientometrics **67**(3), 491–502 (2006)

30. Webber, W., Moffat, A., Zobel, J.: A similarity measure for indefinite rankings. ACM Trans. Inf. Syst. (TOIS) **28**(4), 20 (2010)

31. Weingart, P.: Impact of bibliometrics upon the science system: inadvertent consequences? Scientometrics **62**(1), 117–131 (2005)

32. Wessely, S.: Peer review of grant applications: what do we know? The Lancet **352**(9124), 301–305 (1998)

33. Zhu, J.J., Wang, H.H., Ye, C., Lang, Q.: Project evaluation method using non-formatted text information based on multi-granular linguistic labels. Inf. Fusion **24**, 93–107 (2015)

Comparing Robots with Different Levels of Autonomy in Educational Setting

Mirjam de Haas, Alexander Mois Aroyo, Pim Haselager, Iris Smeekens and Emilia Barakova

Abstract Robots' ability to learn and show autonomous/intelligent behavior is expected to bring a breakthrough in usage of robots in education and assistive technologies. We compared a fully remotely operated robot (e. g. a robot with low autonomy) with one that could recognize cards and develop a playing strategy (i.e. highly autonomous) in a quartet game. We tested whether children perceive the robot in both conditions differently. Using a within-subject design, fourteen typically developed children played with a robot with high or low autonomy. The results show that both robots were evaluated equally engaging for the children. However, the introduction of more autonomy in robot's behavior and interaction increased the time that the educator or therapist can pay attention to the child. Consequentially, the perceived usefulness of the robot and the triadic interaction between the robot, child and educator or therapist were considerably improved from the perspective of the educator.

M. de Haas (✉)
Department of Cognitive Science and Artificial Intelligence, Tilburg University, Tilburg, The Netherlands
e-mail: Mirjam.dehaas@uvt.nl

A. M. Aroyo
Italian Institute of Technology, Genova, Italy

P. Haselager
Donders Institute for Brain, Cognition and Behaviour, Radboud University, Nijmegen, The Netherlands

I. Smeekens
Karakter Centre for Child and Adolescent Psychiatry, Radboud University, Nijmegen, The Netherlands

E. Barakova (✉)
Department of Industrial Design, Technical University of Eindhoven, Eindhoven, The Netherlands
e-mail: e.i.barakova@tue.nl

© Springer International Publishing AG, part of Springer Nature 2018
V. Sgurev et al. (eds.), *Practical Issues of Intelligent Innovations*,
Studies in Systems, Decision and Control 140,
https://doi.org/10.1007/978-3-319-78437-3_13

1 Introduction

Over the past decades extensive research has been done on the intersection of intelligent systems and robotics. Although much of this research has led to the development of systems and solutions that have the potential to profoundly impact society, one of the remaining questions is how these systems will operate in real-life scenarios. In case the intelligent system is a robot, its embodiment creates richer opportunities to socially bond with the user. In recent research, more attention is given to the use of robots that interact with children, for example in educational settings [5, 34] and also in social and behavioral training for children with Autism Spectrum Disorder (ASD) [2, 3, 12, 13]. The use of robots within education and therapy is one of the most promising applications of robotics. Social education is not only important for all children, but is especially necessary for children with ASD.

Currently, one of the greatest challenges in social robotics is to increase the autonomy in an acceptable manner for the user. Interaction autonomy of robots would help to create richer interaction with people. Social robots evoke higher expectations, thus, their designers need to face additional challenges regarding their autonomy due to the need to comply with the rules set by humans and make decisions based on the conventions that are familiar for them. Therefore, in a human-robot interaction and, especially in child-robot interaction, the Wizard-of-Oz (WoZ) control, e.g. remote control, is used in most of the cases to meet these expectations [18]. The Wizard refers to a person that is remotely operating the robot without the participant's knowledge, making it seem fully autonomous. The WoZ can help researchers to tests their designs before creating the fully autonomous behavior, and to provide the possibility to understand the impact of the designed human-robot interactions when the implementation is under development [32].

This study focuses on the impact of robot autonomy in educational settings. First, we investigate whether typically developed children can notice the difference between an autonomous robot and a WoZ controlled one. Second, we measure the effect of the robot's autonomy on the educator's workload. We used a game-based interaction designed to increase the social skills of children and test this interaction with typically developing children. In this article we describe an experiment that is performed with typically developed children, to pretest how children perceive the social training with the robot, without focussing on the complications that the behavioral traits of children with ASD will bring. The results cannot be generalized for children with ASD, but the general interaction will be tested in a future for both groups of children.

There are two main reasons to use a social robot in the training of children with ASD. The first one is that research has employed social robots to enhance social interaction and communication in children with ASD [3, 12, 22, 26]. This has been attributed to the fact that social robots are easier for interaction, due to the simple facial and gestural features that do not overwhelm the child. In addition, it has been shown that question-asking training was as effective performed by a robot as if it is performed by a human therapist [22]. The other main reason is that a robot can reduce the workload of the teacher or the therapist, and can help to gather data about the quality of the interaction of the child. Moreover, robots can be re-used for different children, can help multiple children in parallel and can track the development of each child and can keep the therapist informed.

A recent investigation of Senft et al. [35] on the possibility of a robot to reduce the workload of therapists found that participants rated the workload lighter while interacting with a semi-autonomous robot (that is programmed to learn from the guidance of the supervisor), than with the remotely controlled robot. Furthermore, they found that the autonomous robot required fewer interventions by the therapist, which releases time for the therapist to focus on the child. The time required to learn how to operate the robot would also be reduced by a semi-autonomous robot.

When a robot is present in a therapy, the therapist's attention is divided between operating the robot and the treatment of the child. A therapist is still needed in the same room as the robot and child because he or she can help the child to develop a more detailed conversation or may need to explain the meaning to the robot's, otherwise, mechanical actions. Diehl [12] adds that the therapist is necessary during this therapy because the therapist can help to translate the children's individuals needs, behaviors and pronunciation for the robot [17, 33]. Moreover, the therapist can assist during games, handing the turn to the child or to the robot, or provide feedback to robot developers and improve robots or improve programs to control the robot.

The importance of the usage of an autonomous robot in therapy with children with ASD is determined by the way people comply with the robot. Adults did not obey an autonomous robot more than a controlled one [15]. However this can be different for children. Moreover, no children were included in the intervention in the experiment of Senft et al. [35], therefore, any effect on children could be measured. This also implies that the therapist was not in a realistic training environment, thus his/her training load and the perception of the therapy did not include paying attention to the children and their reactions. To compare the therapy with a robot with high autonomy and low autonomy (i.e. remotely/WoZ controlled), the robot behavior, and the children's interactions with the robot and their expectations of playing with the robot should be as similar as possible in both cases.

Although these interactions were designed to be as similar as possible, we first want to test whether the ability of the robot to autonomously recognize the cards, and take a decision which card to ask for is noticeable by the children and whether the children prefer to play with a robot that is highly autonomous. Moreover, we want to find out how the triadic interaction between a child, the robot and a therapist differs in the cases of autonomous or WoZ operated robot. With this study a game, designed to be used in behavioral training of children with ASD, with a therapist/teacher, is performed with a typically developing child, a robot and a teacher. Although in the low autonomy condition the experimenter will use exactly the same game strategy as the intelligent (highly autonomous) robot will do, the timing of the response of the experimenter will be different. In the high autonomy condition, only unplanned behaviors of the child need to be corrected by the therapist, while the other behaviors of the robot are fixed and repeatable.

In addition, the presence of the experimenter in the room, makes it possible for the children to notice that when the experimenter types, the robot reacts. Akerkar [1] argued that an autonomous machine will appear more intelligent, therefore, we hypothesized that the children will find the robot displaying the high autonomous behavior more intelligent than the robot with low autonomy (H1). We also expected that children will interact less with the experimenter in the high autonomy condition (H2) and that they find the game more enjoyable and are more motivated to play the game when playing with the robot with high autonomy than with the robot with low autonomy (H3).

2 Background

Social skill are important for everybody. People with ASD have problems with social interaction: they show less initiative to make contact with others, which decreases their opportunities for learning how to use language, to improve their conversation skills, and to learn the conversation rules etc. Especially, children with ASD, face significant consequences; lacking social skills in childhood correlates with lower peer acceptance, academic achievement, and mental health [20]. Children with ASD will not initiate actions and have difficulties cooperating with other children. The number of children with ASD is increasing (currently 1 in 68 children at age 8 have ASD according to the Centers for Disease Control and Prevention [8]), and, therefore, the therapist's workload also increases [16].

Recent research suggests that social robots can be as effective as a human therapist in social therapy for children with ASD that were trained with the Pivotal Responce Training [22]. Looije and colleagues showed that robot had positive effects on childs mood and openness by children with diabetes and sometimes they shared more with the robot than they will do with a human [27]. Although robots have been proven to be engaging for children with ASD, studies that integrate robots into an empirically supported treatment for ASD are surprisingly sparse. Without such information, clinics that treat ASD are unable to integrate robots within their therapies. In a previous investigation by [3, 22] a robot was used for an Applied Behavior Analysis intervention conducted by a robot to promote self-initiated question asking from children with ASD. The robot did not provide the therapy autonomously, an interactive script was combined with remote operation in cases of decision making by a trained therapist who was present in the same room as the robot and the children with ASD. Results showed that the therapy conducted by the robot was as effective as the same therapy conducted by a human trainer, and in that the robot improved the communication between two children with ASD. In both studies the therapists wanted to keep control on the interaction between the robot and the child.

In a later study Zubrycki and Granosik [36] reported that the therapists would like to see the robot as a support during their therapy and saw the value in adding features like automatic emotion recognition, and natural language processing. They especially stated in questionnaires that they see the added value of the robot in the long run, and that robots can reduce their workload and reduce burnouts. The experiments made by Zubrycki and Granosik [36] were performed with a robot that had very different level of agency—these were rater interactive toys that have no anthropomorphic or zoomorphic shape, and do not move proactively. It is likely that the therapists might have perceived them as device/toy rather than as a robot. The therapists also mentioned some design requirements that are at the moment unfeasible to implement (for example natural speech interaction) and in the short term, if using robots, the therapists expect that their workload will increase especially before the session. Prior to the sessions the therapist needs to learn how to operate the robot and create the scenarios with the robot. More surveys are conducted to study people's opinions about robots in therapy. Coeckelbergh et al. [11] asked potential stakeholders and parents. In the survey people agreed on that they would like to see the robot to have supervised autonomy, e.g. that the robot is able to communicate and monitor the child, and only provide help and assurance to the therapist, with the option that the therapist still can intervene when possible.

Another research [7] looked at how children respond to working with the robot alone versus working with the robot together with a human mediator. The children showed more physical activity with the autonomous robot than with the human-operated robot, which required human assistance. In this research they used speech recognition, but the robot could only understand Yes and No and they only used the children that were the easiest to understand by the robot in the autonomous condition. In the human-assisted condition, the child was not interacting alone with the robot, but together with the human.

Huinen and colleagues [21] have attempted to map the number of robots used in therapy and the educational objectives for children with ASD. They review the research performed with 14 different robots. The objectives that were most often targeted are reported to be: imitation and turn-taking behavior for the purpose of social/interpersonal interaction, and relations; imitation in playful interactions; collaboration/joint attention in to maintain social interaction and relations, and attention. Most of these behaviors are a subject to automation via learning algorithms, such as the ones proposed in [4, 6, 30].

In this experiment, we used a robot that can play the game fully autonomously, but the therapist still has the possibility to intervene and to prompt the child in the case of an unexpected action.

3   Materials and Methods

The experiment was designed to test how children would react to the robot with two different degrees of autonomy in a training scenario. The children played the interactive card game Quartet with the robot. During this game two players play against each other to collect four cards that belong to the same category and form a quartet (i.e. four cards from the same category). In each turn, a player will ask the opponent for specific card that would help to gather a Quartet. If the requested card is in the opponent's hand, he or she is obliged to hand over the card. The goal of the game is to collect as many Quartets as possible.

3.1 Participants

Twenty typically developed children, from an elementary school in the Netherlands took part in the experiment. The results from four children were excluded due to software difficulties during the first day of the tests and two other children were excluded because the voice of the robot changed during the experiments. The data of fourteen children is evaluated in this study. For all children both parents signed a consent form. The participants were between 7 and 8 years old (M = 7.75, SD = 0.65).

3.2 Procedure

Prior to the experiment all children were introduced to the robot in a demonstration. This demonstration was not limited to the children that would participate in this experiment, but for all the children of the school. On the day of the experiment the experimenter explained how the game works, and that the robot was going to play with each child, after the child entered the class room where the robot was located. After this, the robot greeted and asked the child to choose who should start the game. If the child responded, the experimenter entered the answer in the computer and the game continued with the turn of the child or the robot. In the cases when the robot started, it looked at its cards and requested a card from the child (depending on the experimental condition: either autonomously or with the help of the researcher). After the child showed the card to the robot and the robot or the researcher (depending on the condition) recognized it, the turn was handed to the child. During the childs turn he or she asked the robot for a card and the researcher pressed a button so the robot pointed at that card while asking the child to grab it. Each time the child had four cards of the same kind, the robot gave positive verbal feedback ("Well done!") and non-verbal feedback (cheering). At the end of the game, the robot congratulated the child or simply thanked the child for playing in the case that the child lost the game. After the game, the experimenter asked the child to fill in a questionnaire to reflect on the game and the child was brought back to his/her usual classroom.

3.3 Experimental Design

To explore the hypotheses as stated in the introduction, we used a within-subject design, consisting of two conditions: the robot with low autonomy that was completely remotely controlled by the experimenter, and the robot with high autonomy that uses pattern recognition based on visual information to recognize the cards and the employs a strategy to ask the child for the right missing card.

During the first (low autonomy) condition, the experimenter had to decide all the behaviors of the robot and send commands to the robot via notebook. The participant interacted with the robot, and because the experimenter was operating the robot in the same room, he or she was able to notice that the robot was remotely controlled by the experimenter.

In the second condition, the high autonomy condition, the experimenter only chose a behavior in case the child was talking (as a speech recognizer was not used, the card was selected by the experimenter). The robot used its sensory information to notice whether the child had finished his/her turn, to perceive which cards it has and to decide which card to ask for. Both conditions did not differ in the behavioral expression of the robot, only the input and operation of the behaviors was different. Robot was wearing different shirts for each condition to create an illusion of a different robot. The order of the robot behaviors and the shirt color were counter balanced.

3.4 Experimental Setting and Materials

The experiment was conducted in an empty classroom in the elementary school. Each participating child was called out twice of the class room to play with the robot. During those sessions, the child and the robot sat in front of each other at a table and the experimenter at another table but within sight of the child (see Fig. 1a, b). A digital video camera was placed between the robot and the experimenter, to record the childs behavior. A laptop was used to control the robot. After the child had interacted with the robot in both conditions the child filled in a questionnaire about the game and the robot.

The robotic platform used for this study is the NAO humanoid robot [19], with 25 degrees of freedom, 58 cm in height. NAO has simplistic facial features only with mouth and eyes, and its face resembles to the age of a child's face. Some of the robot behaviors that were used in this game include text to speech functions, hand gestures, and NAO LEDs. The robot had a female voice (Jasmine) and was speaking Dutch. In order to be able to differentiate the child's responses to the robot in the two conditions, the robot wore a different shirt (blue striped or white) in each one (see Fig. 1c). TiViPE, graphical programming environment was used to program the robot and to carry out the interaction during the experimental sessions [2, 23]. A specially designed interface of TiViPE was used for real-time interaction between robot and child and was connected to the previously programmed interaction scenario. The preprogrammed scenario consisted of a dynamical system of behaviors for complex interactions, emotional expressions and behaviors as the one of learning to recognize the cards.

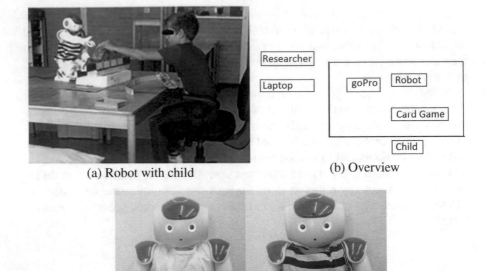

(a) Robot with child   (b) Overview

(c) Robot with the different shirts, that were weared
in low (left) and high (right) autonomy condition

Fig. 1 a A child interacting with the robot. **b** Experimental setup. **c** The robot wearing the different shirts in the different conditions

3.5 Intelligent Behaviors and Concepts for the Low and High Autonomy

The robot autonomy/intelligence was relying on the overall robot controller design, which is driven by a state machine that is embedded in a loop of constantly updating sensory and motor information of the robot. With this respect, the robot is a real-time dynamical system, including different learning behaviors in each state. The robot is aware of its own actuators (motors) position and movement, as proprioceptive sensory information defining the current 3D positioning (state) of the robot, letting the robot sense its own actions and the consequences of these actions. Providing much higher flexibility to the robot's behavior, as discussed in [4]. This implies that sensory information from the outside environment or the robot proprioceptive sensing can serve as an event that causes transition to a new state, i.e. new behavior. Once a task is accomplished, the scenario should proceed to the next state or set of states if no event occurs. The state concept makes possible to handle every event in a more flexible way than, for instance, if-than-else loops, since these only support applications with only one trigger and one action, lacking of real-time feedback from the outside world.

The behaviors that can be included in a scenario (i.e. in a state diagram) can be selected from approximately 500 existing module, also new ones could be developed as C++ modules for TiViPE [29]. The existing modules include biologically inspired filters similar to the ones existing in the human visual cortex, such as [28, 30].

In addition to the sensory-driven state concept, the robot is equipped with a very simple language consisting of 10 commands that can be used in parallel to the state concept, for example for creating new behaviors on a flow. The main merit of a textual robot language is that robot actions are described in terms such as move, LEDs on, etc., which are intuitive to human operator with no technical background. These actions can be executed in a mixture of parallel and serial actions or as a part of dynamic behaviors determined by environmental changes [29]. The complexity of a task such as designing parallel actions (such as speak and move accordingly) is reduced to basic logic using and character for sequential and "|" character for parallel actions. Square brackets are used to bind a set of these textual actions and give the priority of their execution where it is needed. One can easily understand how to schedule commands that come in parallel or the need to wait until the last of several parallel commands has been completed. A very important feature of this parallel processing is the synchronization between parallel processes (such as bodily expressions and speech). The advantage of describing actions as a text is that during the execution of the pre-programmed scenario, one can add new actions while training.

With this architecture we programmed the low and high autonomy robots. The robot with low autonomy went through a normal scenario in which the interaction was remotely controlled by an operator that through a keyboard and interface connected to the TiViPE controller can engage of speech interactions (through typing text), choose the card that the robot has to ask for or redirect the scenario.

Alternatively, we implemented an autonomous robot in the following way. The robot used its camera to detect the state of the game and the card while playing. Every card had an individual marker for the robot to identify the card. For the markers the ArUco markers were used [14]. The robot identifies the cards with the markers and decides which cards belong together (see Fig. 2).

Using these markers, the robot can identify the cards with higher reliability than with by recognizing the picture on the card, especially in real life conditions where there were severe illumination changes in the different days of testing. In addition detection of the shape and color of the animals that were pictured in the cards the robot can decide which cards belong together, and also be more specific when asking for a card. Therefore, a database was created with all the cards in the game with the following properties: the ArUco code, the category, and the color of the animal. With this information the robot was able to produce a variation of sentences, such as, "Can you give me the card with the purple fish?" or "I want the card with the fish that is purple", in these examples the category was fish, the color of the animal is

Fig. 2 The play cards with the ArUco markers

purple and the ArUco code corresponds to the one on the physical card. The robot also recognized whenever the child got four the same cards (Quartet) which means a milestone in the game and consequently, the robot could immediately and autonomously provide feedback to the child.

The robot's strategy was to let the children win. Each time it was the robot's turn, it would ask a random card from the cards that the robot did not possess. In order to let the children win, the robot avoid asking for the fourth card of the category and the robot had a bigger chance (twice as big) to ask a card from a category of which the robot had only two cards. This decreases the chance of the robot to complete a Quartet, and increases the chance that the child will assemble more Quartets.

The speech recognition of the robot is too unreliable to be used autonomously, this part of the scenario was remotely controlled by the experimenter. During the game the therapist could prompt the child when necessary, by using a text-to-speech interface [22]. This way, the experimenter could still intervene the interaction between the robot and child with possible prompts. The robot would sometimes ask on purpose for the wrong card in order for the child to protest. This was done randomly during the interaction and happened only once during each interaction, and was done by the experimenter (in the low autonomy condition) or the robot itself (in the high autonomy condition).

In the low autonomy condition, all possible robot behaviors could be activated by a simple keyboard press.

Robot: "Hi there! Can you give me the purple mermaid?"
Child: "Yes I have it here"

Child show the card to the robot.

Robot:"Yes that is right, can you place it in the card holder?"
Child: "Yes I can"

Child places the card in the card holder.

Robot: "Thank you!"

...After the child's turn...

Robot: "Hi there! I want the card with the red fish!"
Child: "Yes I have it here"

Child show the card to the robot and places it in the empty card holder.

Robot: "Thank you for placing the card in the card holder"

...After the child's turn...

Robot: "Hi there! Do you have the card with the blue seahorse?"
Child: "Yes I have it here"

Child does not give the robot the card.

Robot: "Can you please put the card in the card holder?"

The scenario above is an example of an interaction between the robot and a child in the beginning of the game. As shown, the robot gives two different responses to the behavior of the child. First, the child shows the robot a card and the robot asks to put the card in an empty place in the card holder. Second, the robot thanks the child for placing the card in the card holder. Every time the child says something, the experimenter can initiate a reaction by starting a robot behavior or typing a sentence that the robot will pronounce. For instance the experimenter used a text-to-speech option to make the robot ask the child again to place the card in the card holder since the child ignored its first question, which was a part of the preprogrammed scenario. The experimenter would only use prompts when the child does not react on the robot's actions and the robot should provide more guidance to the child.

3.6 Data Collection

All sessions were recorded using a video camera, the videos of the child-robot interaction were observed once the data-collection finished. All videos were random selections of one minute of the child interacting with the robot. For every condition three videos of one minute of the children were observed, therefore, the observers watched 6 videos of one minute per child, for 14 children. Three students and a researcher observed these videos; the students watched only half of the videos due to time restrictions, and consequently every video is observed at least two times. The observers were asked to count the times (frequency) that (1) the child looked at the robot, (2) the child looked at the experimenter, (3) the child said something to the robot that did not belong to the game (the observers did not count for the times the child asked for a card), (4) the child said something to the experimenter and (5) how much interest the child showed during this minute on a scale from 0 to 5 (The interest measurement was specified in a table which was adopted from [25], see Table 1).

The inter-observer agreement (IOA) was determined with the intraclass correlation coefficient (ICC). This statistic can be used to determine the degree that a group

Table 1 The explanation of the interest measurements as distributed for the observers

Low interest (0–1)	Neutral interest (2–3)	High interest (4–5)
Looks bored, uninvolved, and not curious about or eager to participate in the activity	Neither particularly interested nor disinterested	Readily attends to the activity
Yawns or tries to avoid the activity	Seems to passively accept situation	Is alert and involved in the task
Spend little time attending to the task	Does not rebel but is eager to continue	(Score as 4 or 5, depending on level of alertness and involvement)
A long response latency when there is a response (takes long for the child to respond)	(Score as 2 or 3, depending on extent of interest)	
(Score as 0, or 1, depending on extent of lack of interest)		

relates to each other. The coefficient varied between 0.57 and 0.94 with a mean of 0.79 (SD = 0.08), which indicates a substantial agreement between the observers.

3.7 Questionnaires

The questionnaires are based on the Immersive Experience Questionnaire [9] and indicate the immersion of the player. The questions were chosen to test the degree of attention of the child to the game and the degree of motivation to play the game and therefore relate to the player's overall experience of the game. Within this experiment, only thirteen questions were translated to Dutch and simplified for the level of understanding of the children. The questions that were explicitly about video games and duplicate questions were excluded. A pilot study performed by us indicated that children found it complicated to express themselves in degree of agreement (agree/slightly agree/slightly disagree/disagree). Therefore the children were asked to answer on a 4 point scale (yes!/slightly yes/slightly no/no!) or fill in that they did not know. The option for "Neutral" was excluded so the children had to choose. The questions were divided into four categories (questions about their enjoyableness (Q1–Q3), the robot (Q4–Q6), surroundings (Q7–Q9), the difficulty (Q10–Q13), see Table 2 for the questions).

After playing the game in both conditions, the children had to fill in a final questionnaire. They were asked which robot they preferred and with which robot they wanted to play the game again.

Table 2 The questionnaire (in English)

Question no.	Questions
1	I found it fun to play this game
2	I wanted to win
3	I did my best
4	I want to play the game again with the robot
5	I really wanted to help the robot
6	I felt sorry for the robot/me that he/I lost
7	I wanted to quit during the game
8	I was quickly distracted during the game
9	During the game I noticed nothing around me
10	I think the robot played really clever
11	I found this game challenging
12	I found the game really easy
13	The game was quickly finished

4 Results

This section shows the results of the observed behaviors of the children and the results of the questionnaires. The data analysis consists of five parts. The first part involved visual inspection of the data together with the calculation of the mean of all observers. The second part was a paired-samples t test with the data of the observers to see whether there was a significant difference between the two behaviors of the robot. In the third part, the two questionnaires about the robots after each condition were compared, for all questions a paired-samples t-test was performed. An ANOVA was used to check whether age and school year was of influence. Last, a chi-square goodness-of-fit test was performed for the final questionnaire.

4.1 Analysis of the Children's Behavior

Figure 3a summarizes the observed behaviors of the children in the two conditions: the robot with either high autonomy or low autonomy. It was expected that the children will look less at the experimenter when the children played with the highly autonomous robot. However, no significant differences were found between the two conditions. It is noteworthy that the observers rated the children speaking significantly more with the human than with the robot in both conditions (M = −1.31, SD = 1.33, p = 0.00 for the robot with low autonomy, M = −1.46, SD = 1.85, p = 0.00 for the robot with high autonomy). The children turned to the experimenter to express their excitement the robot.

Table 3 The results for the influence of the children's school level and their age on their answers for the questionnaire

	Grade (Groep)	Age
I wanted to quit during the game	$F(1,25) = 11.363^{**}$	$F(1,25) = 3.054$
During the game I did not notice anything around me	$F(1,25) = 0.128$	$F(1,25) = 0.000$
I was quickly distracted during the game	$F(1,25) = 8.703^{**}$	$F(1,25) = 3.405$
Multivariate	$F(3,23) = 4.837^{**}$	$F(3,23) = 1.493$

Note *p <0.05, **p <0.01

4.2 Questionnaires

The children filled in a questionnaire after playing the game with the robot in both conditions. It was expected that the children would like the robot with high autonomy more or find the robot smarter than the robot with low autonomy. As the results (see Fig. 3b, paired t-test) show, there are no significant differences between the conditions. The children's answers about the surroundings were positive for positive questions (I noticed nothing around me, M = 0.25, SD = 1.80) and negative for negative questions (I wanted to quit, M = −1.04, SD =1.71, I was quickly distracted during the game, M = −0.64, SD = 1.77). The influence of the children's age or grade on their answers was checked, because of the high standard deviation for some of the questions. It was expected that children of a younger age or lower grade, found the questions that had a longer length more complicated to understand and were less consistent with answering these questions. As the results in Table 3, the grade of the child significantly influences the answers of two of the questions. The children in a higher grade, showed less variation within the questions. In the final questionnaire the children were asked to choose between the two conditions which robot they liked more and with which robot they wanted to play again (see Table 4). There are no significant differences in the children's answers.

Table 4 The answers on the final questionnaire

	Low autonomy	High autonomy	Chi-Square	Asymp. Sig.
I liked the game more with	53.33	46.67	40	1
I would like to play again with	60	40	0.286	0.593

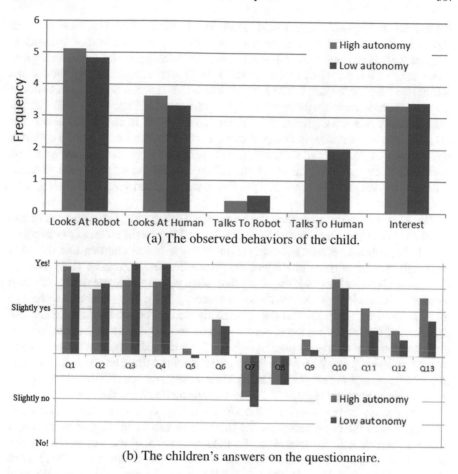

(a) The observed behaviors of the child.

(b) The children's answers on the questionnaire.

Fig. 3 **a** Results of the children's gaze and speech behavior rated by observers. The observers could rate the interest of the children between 0 and 5. **b** Results of the children's answers on the questionnaire

5 Discussion

The research presented in this paper explored the effects of an autonomous robot on childrens perception of the robot and the game. The main goal was to investigate whether children can perceive a difference between a robot that has a highly autonomous strategy and a robot with low autonomy and that was remotely operated. Fourteen children played a card game with the Nao robot displaying these different strategies in an experiment. The childrens gaze and speech behavior and the childrens subjective opinion were measured.

The children spoke more with the human than with the robot. It is possible that this is a consequence of the fact that the children understood that the robot could not process their speech, as they said to the experimenter. Therefore, more research should be done to further improve the flexibility of the human-robot interaction. A robot with high autonomy can easily adapt its strategy if the game appears to be too easy or too complicated for the child. With increased robot intelligence this can be done for more games. The robot can also detect when a child gets bored or uninterested and can adapt its behavior to that. Moreover, children with ASD interact more sociably when the preferred by them voice is used in the social interaction [24, 31]. The robot suddenly changed its voice during the game with two children, both of them noticed this and, therefore, were excluded from the data analysis. The possibility to change the voice of the robot allows create a more personalized robot. Changing the voice of the robot to a preferred, personalized voice can elicit greater interaction.

There are no significant differences between the robot with high autonomy or low autonomy in the gaze and speech behavior of the children or the children's answers on the questionnaire. This does not confirm our hypothesis that children find the robot with high autonomy more intelligent than the robot low autonomy (H1). Neither the children interact less with the researcher when playing with the robot with high autonomy than with the robot with low autonomy (H2). The observers were also asked to rate how much interest the children showed in the robot and the game. No significant differences were found between the two conditions. In both conditions the observers rated the children's interest slightly above neutral. A neutral interest means that the children were observed to be eager to continue the game, not to rebel and to passively accept the situation. The children were not more motivated to play the game in one of the two conditions (H3 is not confirmed either). These results suggest that the children did not experience the robot with high or low autonomy differently. This introduces the opportunity for the therapist to use an intelligent robot that is highly autonomous in the treatment. An autonomous robot reduces the therapists' energy for operating the robot and frees some time for the therapist, allowing her/his to focus on the child [35]. Because the children do not see a difference between the robot behaviors, we conclude that they can benefit from the more specialized attention of the therapist.

There are some limitations of this work. Frist, the novelty effect that might lead children to be focused mostly on the robot and not on the experimenter. The children, therefore, may not have noticed that the robot was controlled differently in the two conditions—either the experimenter, or the robot was autonomous. The answers on the questions about their surroundings support this suggestion. Second, there is a possibility that the children could not recall events that happened during the experiment. That would imply that the answers on the questions would be based on the last part of their interaction with the robot. Eight-year old children can relate better to events that occurred at some time in the past than seven-year old children [10]. The high variance of the questions about events that happened during the game underpins this possibility. The language development of children of the fifth and fourth grade is also different. Events that happened in the past are complicated because they ask the children to reflect on their behavior during the game. Children need more under-standing of language to be able to do so. Our results showed that the children's grade made a difference but not their age, and only for two of the three questions (I was quickly distracted during the game; during the game I noticed nothing around me). Children from a higher grade showed less variance in the two questions than the children from a lower grade. This means that children from a higher grade understood these complex questions better than children from a lower grade. For future experiments it will be important to reduce such interfering factors.

The third limitation of our experiment is that the game length was too short. During the short time that the children got to play with the robot, the children only wanted to focus on the robot and not on the surroundings. If they would have played with the robot for a longer time they might have noticed that the robot played autonomously and this might have had an influence on their behavior or opinion about the robot.

6 Conclusions

Extensive research is done towards interventions with children and robots in a Wizard of Oz setting with the researcher being invisible for the child. However, less research is done with the educator or the therapist in the same room, which is the way therapies and education with robots are likely to be used. The goal of this paper is to explore how the child and the teacher/therapist is influenced by robot that is either acting with high autonomy or low autonomy and is remotely controlled by the teacher/therapist located in the same room. The results of the study show that typically developing children did not respond differently to these two types of robot behavior. A clear advantage is that teacher/therapist can use a robot with increasing levels of autonomy instead of remotely controlled robot thereby reducing the workload of the therapist, and gradually finding ways of how the autonomy can gradually be increased. This, in turn, allows the therapists to focus more on the child and to even use the robot to record additional information for the child behavior and progress. More research should be done to measure different aspects of increased effectiveness of the therapy and whether our findings also apply to children with ASD, who might be more alert to the co-occurrence of the robot behavior and the actions of the therapists, and this to be differently affected by the different degrees of autonomy.

Acknowledgements We gratefully thank the Lindt elementary school in Helmond. Without the help of the school principal, all the involved teachers, children and parents, this study would not have been possible.

References

1. Akerkar, R.: Intelligent systems: perspectives and research challenges. CSI Commun. 5 (2012)
2. Barakova, E.I.: Robots for social training of autistic children. In: 2011 World Congress on Information and Communication Technologies (WICT), pp. 14–19. IEEE (2011)

3. Barakova, E.I., Bajracharya, P., Willemsen, M., Lourens, T., Huskens, B.: Long-term LEGO therapy with humanoid robot for children with ASD. Expert Syst. **32**(6), 698–709 (2015)
4. Barakova, E.I., Chonnaparamutt, W.: Timing sensory integration. IEEE Robot. Autom. Mag. **16**(3) (2009)
5. Belpaeme, T., Kennedy, J., Baxter, P., Vogt, P., Krahmer, E.E.J., Kopp, S., Bergmann, K., Leseman, P., Küntay, A.C., Göksun, T., Pandey, A.K., Gelin, R., Koudelkova, P., Deblieck, T.: L2TOR—Second language tutoring using social robots. In: L2TOR –Second Language Tutoring using Social Robots (2015)
6. Calinon, S., Billard, A.: Incremental learning of gestures by imitation in a humanoid robot. In: Proceedings of the ACM/IEEE International Conference on Human-robot Interaction, pp. 255–262. ACM (2007)
7. Cameron, D., Fernando, S., Collins, E.C., Millings, A., Moore, R.K., Sharkey, A.J., Prescott, T.J.: Impact of robot responsiveness and adult involvement on children's social behaviours in human-robot interaction. CoRR. abs/1606.06104 (2016)
8. Center for Disease Control and Prevention. Prevalence of autism spectrum disorder among children aged 8 years—autism and developmental disabilities monitoring network. http://www.cdc.gov/mmwr/preview/mmwrhtml/ss6302a1.htm?s_cid=ss6302a1_w (2010). Accessed 23 Dec 2015
9. Charlene, J., Cox, A.L., Cairns, P., Dhoparee, S., Epps, A., Tijs, T., Walton, A.: Measuring and defining the experience of immersion in games. Int. J. Hum. Comput. Stud. **66**(9), 641–661 (2008)
10. Child Development Institute. Language development in children. http://childdevelopmentinfo.com/child-development/language_development/ (2009). Accessed 23 Dec 2015
11. Coeckelbergh, M., Pop, C., Simut, R., Peca, A., Pintea, S., David, D., Vanderborght, B.: A survey of expectations about the role of robots in robot-assisted therapy for children with ASD: Ethical acceptability, trust, sociability, appearance, and attachment. Sci. Eng. Ethics **22**(1), 47–65 (2016)
12. Diehl, J.J., Crowell, C.R., Villano, M., Wier, K., Tang, K., Riek, L.D: Clinical applications of robots in autism spectrum disorder diagnosis and treatment. Comprehensive Guide to Autism, pp. 411–422. Springer (2014)
13. Feil-Seifer, D., Matarić, M.J.: Toward socially assistive robotics for augmenting interventions for children with autism spectrum disorders. Experimental Robotics, pp. 201–210. Springer (2009)
14. Garrido-Jurado, S., Muñoz-Salinas, R., Madrid-Cuevas, F.J., Manuel Jesús Marín-Jiménez, F.J.: Automatic generation and detection of highly reliable fiducial markers under occlusion. Pattern Recogn. **47**(6), 2280–2292 (2014)
15. Geiskkovitch, D.Y., Cormier, D., Seo, S.H., Young, J.E.: Please continue, we need more data: an exploration of obedience to robots. J. Hum. Robot. Interact. **5**(1), 82–99 (2015)
16. Gibson, J.A., Grey, I.M., Hastings, R.P.: Supervisor support as a predictor of burnout and therapeutic self-efficacy in therapists working in ABA schools. J. Autism Dev. Disord. **39**(7), 1024–1030 (2009)
17. Giullian, N., Ricks, D., Atherton, A., Colton, M., Goodrich, M., Brinton, B.: Detailed requirements for robots in autism therapy. In: 2010 IEEE International Conference on Systems Man and Cybernetics (SMC), pp. 2595–2602. IEEE (2010)
18. Goodrich, M.A., Crandall, J.W., Barakova, E.: Teleoperation and beyond for assistive humanoid robots. Rev. Hum. Factors Ergonomics **9**(1), 175–226 (2013)
19. Gouaillier, D., Hugel, V., Blazevic, P., Kilner, C., Monceaux, J., Lafourcade, P., Marnier, B., Serre, J., Maisonnier, B.: Mechatronic design of nao humanoid. In: IEEE International Conference on Robotics and Automation, 2009. ICRA'09, pp. 769–774. IEEE (2009)
20. Hartup, W.W.: Social relationships and their developmental significance. Am. Psychol. **44**(2), 120 (1989)
21. Huijnen, C.A.G.J., Lexis, M.A.S., Jansens, R., Witte, L.P.: Mapping robots to therapy and educational objectives for children with autism spectrum disorder. J. Autism Dev. Disord. **46**(6):2100–2114 (2016)

22. Huskens, B., Verschuur, R., Gillesen, J., Didden, R., Barakova, E.: Promoting question-asking in school-aged children with autism spectrum disorders: effectiveness of a robot intervention compared to a human-trainer intervention. Dev. Neurorehabilitation **16**(5), 345–356 (2013)
23. Kim, M.G., Oosterling, I., Lourens, T., Staal, W., Buitelaar, J., Glennon, J., Smeekens, I., Barakova, E.: Designing robot-assisted pivotal response training in game activity for children with autism. In: 2014 IEEE International Conference on Systems, Man, and Cybernetics (SMC), pp. 1101–1106. IEEE (2014)
24. Koegel, L.K., Koegel, R.L., Harrower, J.K., Carter, C.M.: Pivotal response intervention I: overview of approach. Res. Pract. Persons Severe Disabil. **24**(3):174–185 (1999)
25. Koegel, R.L., Egel, A.L.: Motivating autistic children. J. Abnorm. Psychol. **88**(4), 418 (1979)
26. Kose-Bagci, H., Ferrari, E., Dautenhahn, K., Syrdal, D.S., Nehaniv, C.L.: Effects of embodiment and gestures on social interaction in drumming games with a humanoid robot. Adv. Robot. **23**(14), 1951–1996 (2009)
27. Looije, R., Neerincx, M.A., Peters, J.K., Olivier, A.: Blanson Henkemans. Integrating robot support functions into varied activities at returning hospital visits. Int. J. Soc. Robot. **8**(4), 483–497 (2016)
28. Lourens, T., Barakova, E.: Orientation contrast sensitive cells in primate V1 a computational model. Nat. Comput. **6**(3), 241–252 (2007)
29. Lourens, T., Barakova, E.: User-friendly robot environment for creation of social scenarios. In the Proc. of International Work-Conference on the Interplay between Natural and Artificial Computation, pp. 212–221 (2011)
30. Lourens, T., Barakova, E., Okuno, H.G., Tsujino, H: A computational model of monkey cortical grating cells. Biol. Cybern. **92**(1), 61–70 (2005)
31. Paul, R.: Interventions to improve communication in autism. Child Adolesc. Psychiatr. Clin. N. Am. **17**(4), 835–856 (2008)
32. Riek, L.D.: Wizard of oz studies in hri: a systematic review and new reporting guidelines. J. Hum. Robot Interact. **1**(1) (2012)
33. Robins, B., Dautenhahn, K.: The role of the experimenter in hri research-a case study evaluation of children with autism interacting with a robotic toy. In: ROMAN 2006-The 15th IEEE International Symposium on Robot and Human Interactive Communication, pp. 646–651. IEEE (2006)
34. Sciutti, A., Schillingmann, L., Palinko, O., Nagai, Y., Sandini, G.: A gaze-contingent dictating robot to study turn-taking. In: Proceedings of the Tenth Annual ACM/IEEE International Conference on Human-Robot Interaction Extended Abstracts, pp. 137–138. ACM (2015)
35. Senft, E., Baxter, P., Kennedy, J., Belpaeme, T.: Sparc: supervised progressively autonomous robot competencies. In: International Conference on Social Robotics, pp. 603–612. Springer (2015)
36. Zubrycki, I., Granosik, G.: Novel haptic device using jamming principle for providing kinaesthetic feedback in glove-based control interface. J. Intell. Robot. Syst. 1–17 (2016)

Clustering Non-Gaussian Data Using Mixture Estimation with Uniform Components

Ivan Nagy and Evgenia Suzdaleva

Abstract This chapter considers the problem of clustering non-Gaussian data with fixed bounds via recursive mixture estimation under the Bayesian methodology. Here a mixture of uniform distributions is taken, where individual clusters are described by mixture components. For the on-line detection of data clusters, the paper proposes a mixture estimation algorithm based on (i) the update of reproducible statistics of uniform components; (ii) the heuristic initialization via the method of moments; (iii) the non-trivial adaptive forgetting technique; (iv) the data-dependent dynamic pointer model. Results of validation experiments are presented.

1 Introduction

The cluster analysis is a powerful tool of data processing solved by a great number of methods (e.g., well-known centroid, density based methods, etc.), see e.g., [1–3]. One of the cluster analysis domain is the mixture-based clustering, which is discussed in this paper.

The mixture components describe individual clusters in the data space. This means that the location, size and shape of components are important in the task of covering the data clusters. The location and the size are given by the expectation and the covariance matrix of the component, while the shape is defined by its distribution. Gaussian components are traditionally successful in detecting elliptic clusters [4–6]. However, clusters of a different shape require a solution with involved components of other distributions. The same situation occurs in the case of clustering non-negative,

I. Nagy · E. Suzdaleva (✉)
Department of Signal Processing, The Czech Academy of Sciences,
Institute of Information Theory and Automation, Pod vodárenskou věží 4,
18208 Prague, Czech Republic
e-mail: nagy@utia.cas.cz

I. Nagy
Faculty of Transportation Sciences, Czech Technical University, Na Florenci 25,
11000 Prague, Czech Republic
e-mail: suzdalev@utia.cas.cz

© Springer International Publishing AG, part of Springer Nature 2018
V. Sgurev et al. (eds.), *Practical Issues of Intelligent Innovations*,
Studies in Systems, Decision and Control 140,
https://doi.org/10.1007/978-3-319-78437-3_14

or somehow limited data (for instance, a vehicle speed under the speed limits). To take into account such a feature, components should have a limited support. Such a choice is e.g., the uniform distribution, which well covers clusters of the rectangle shape (for independent variables in the data space) or of the parallelogram shape in the case of dependent variables.

Estimation of data models with the bounded support including uniform ones was solved in various domains, e.g., clustering [7], individual state-space and regression models [8, 9] as well as mixture models [10], etc. In mixture-based clustering the bounded data the challenging task is the update of statistics of the uniform component parameters. Intuitively the prior chosen bounds of the uniform distribution are only expandable, but they are not floating in the data space to detect the centers of clusters. While estimating a uniform mixture, this feature is harmful and should be fixed. A solution to this task will allow to perform clustering on-line at each time instant using both the current and the previously available measurements. This task is highly desired in many application areas (fault detection, diagnostics, medicine, etc.).

This paper solves the problem based on the recursive Bayesian estimation algorithms proposed for normal regression components in [11–13] and applies them for derivation of the algorithm for the uniform mixture. The paper represents the extended version of the work [14]. The main contributions of the approach in addition to the recursive statistics update include also: (i) the initialization based on finding the initial centers of components with the help of the method of moments; (ii) the novel non-trivial adaptive technique of forgetting; (iii) using the categorical data-dependent dynamic model of switching, which assumes that the currently active component is modeled in dependence of the past active one and on discrete measurements too.

The proposed algorithm also enriches the clustering and classification tools developed within the current project under the adopted Bayesian recursive mixture estimation context. The systematic extension of the theory has already given algorithms for normal regression [15], state-space [16], mixed normal and categorical [17] and exponential components [18]. Here this line is continued by developing the systematic approach to uniform components and the data-dependent model of switching partially started in [17].

The present paper focuses on independent variables. Modeling dependent uniformly distributed variables with parallelogram-shaped clusters is definitely an important task that will be solved later. However, the main aim of this work is to cluster data with fixed lower and upper bounds, which is sufficiently covered by rectangle clusters provided by independent variables.

The paper is organized in the following way. Section 2 introduces models and formulates the problem. The preparative Sect. 3 recalls necessary basic facts about estimation of individual uniform and categorical models and explains the general idea of the approach. Section 4 presents a solution to the formulated task and the structural algorithm. Section 5 provides results of the experimental validation of the proposed algorithm. Conclusions and open problems are given in Sect. 6.

2 Models and Problem Formulation

Let's consider a multi-modal system, which at each discrete time instant $t = 1, 2, \ldots$ generates continuous data y_t, whose values are bounded by minimal and maximal bounds different within each working mode, and discrete data z_t with the set of its possible values $\{1, 2, \ldots, m_z\}$. It is assumed that the observed system works in m_c working modes indicated by values of the unmeasured dynamic discrete variable $c_t \in \{1, 2, \ldots, m_c\}$, which is called the pointer [11], and each of the pointer values also depends on values of the measured variable z_t.

The system is described by a mixture of uniform distributions presented by the probability density functions (pdfs)

$$f(y_t | \Theta, c_t = i), \ i \in \{1, 2, \ldots, m_c\}, \tag{1}$$

where $\Theta = \{\Theta_i\}_{i=1}^{m_c}$ is a collection of unknown parameters of all components, and $\Theta_i = \{L_i, R_i\}$ (for $c_t = i$) are parameters of the i-th component, where L_i is the minimum bound of the data y_t, and R_i is the maximum bound.

Switching the components describing the data is described by the following data-dependent dynamic pointer model:

$$f\left(c_t = i | \alpha, c_{t-1} = j, z_t = k\right) = \tag{2}$$

	$c_t = 1$	$c_t = 2$	$\cdots$	$c_t = m_c$
$c_{t-1} = 1$	$(\alpha_{1|1})_k$	$(\alpha_{2|1})_k$	$\cdots$	$(\alpha_{m_c|1})_k$
$c_{t-1} = 2$	$(\alpha_{1|2})_k$		$\cdots$	
$\cdots$	$\cdots$	$\cdots$	$\cdots$	$\cdots$
$c_{t-1} = m_c$	$(\alpha_{1|m_c})_k$		$\cdots$	$(\alpha_{m_c|m_c})_k$

where the unknown parameter α is the $(m_c \times m_c)$-dimensional matrix, which exists for each value $k \in \{1, 2, \ldots, m_z\}$ of z_t. Its entries $(\alpha_{i|j})_k$ are non-negative probabilities of the pointer $c_t = i$ under condition that the previous pointer $c_{t-1} = j$ with $i, j \in \{1, 2, \ldots, m_c\}$ and $z_t = k$.

The task is to cluster the data on-line at each time instant t according to the determined active component based on the available data collection and newly arriving data. Under Bayesian methodology adopted in [11–13] it leads to looking for a recursive algebraic computation of statistics of the involved distributions, which is obtained by substituting the prior pdf to be propagated into the Bayes rule [19, 20]:

$$f(\Theta | \Delta(t)) \propto f\left(y_t | \Theta\right) f(\Theta | \Delta(t - 1)), \tag{3}$$

where the denotation $\Delta(t) = \{\Delta_0, \Delta_1, \ldots, \Delta_t\}$ represents the collection of data available up to the time instant t; Δ_0 denotes the prior knowledge; the data item Δ_t includes the pair $\{y_t, z_t\}$; and $f(\Theta | \Delta(t - 1))$ is the prior pdf.

Within the considered context the clustering problem is specified as the recursive estimation of

- all component parameters Θ;
- the pointer model parameter α;
- the value of the pointer c_t expressing the active component at each time instant t.

3 Preliminaries

3.1 Individual Uniform Model Estimation

As it is known [21], description of the individual uniform pdf can be presented twofold: via minimal and maximal bounds or using the mid-point and the mid-range. The multivariate uniform pdf for independent variables will be the product of uni-variate marginal pdfs, and the distribution will have generally the rectangle support. The assumed independence of variables leads to a straightforward extension of the univariate case up to the multivariate one, which means that the whole estimation is performed independently over individual dimensions. Here, for simplicity omitting c_t from the condition of (1), the uniform pdf can be presented as

$$f(y_t|\Theta) = f\left(y_t|L,R\right) = \begin{cases} \frac{1}{R-L} & \text{for } y_t \in (L,R), \\ 0 & \text{otherwise,} \end{cases} \tag{4}$$

$$= f\left(y_t|S,h\right)$$

$$= \begin{cases} \frac{1}{2^K \prod_{l=1}^K h_l} & \text{for } y_t \in (S-h,\ S+h) \\ 0 & \text{otherwise,} \end{cases} \tag{5}$$

where K denotes the dimension of the vector y_t, $S = [S_1, \ldots, S_K]'$ is the vector of mid-points of the distribution support, and $h = [h_1, \ldots, h_K]'$ is the vector of mid-ranges.

For description (4) the maximum likelihood (ML) estimation leads to using the K-dimensional statistics $\mathcal{L}_t$ and $\mathcal{R}_t$ with their update for the new measurement y_t at time t in the following form [21] for each $l \in \{1, \ldots, K\}$

$$\text{if } y_{l;t} < \mathcal{L}_{l;t-1}, \text{then } \mathcal{L}_{l;t} = y_{l;t}, \tag{6}$$

$$\text{if } y_{l;t} > \mathcal{R}_{l;t-1}, \text{then } \mathcal{R}_{l;t} = y_{l;t}. \tag{7}$$

The point estimates of parameters L and R at time t are then obtained as

$$\hat{L}_t = \mathcal{L}_t, \quad \hat{R}_t = \mathcal{R}_t. \tag{8}$$

Notice here that in dependence of the initially chosen statistics $\mathcal{L}_0$ a $\mathcal{R}_0$, the point estimate $\hat{L}_t$ can be located to the left from $\mathcal{L}_0$, and then $\hat{R}_t$ to the right from $\mathcal{R}_0$. It means that the prior statistics can be only extended, which is problematic in the case of several components.

For description (5) the statistics for the parameter estimation are chosen based on the method of moments (MM) [22] as the sum $s_t = \left[s_{1;t}, \ldots, s_{K;t} \right]'$ and the sum of squares q_t, which is the matrix with the diagonal $\left[q_{1;t}, \ldots, q_{K;t} \right]$. Starting with the chosen initial statistics, their update for the actual data item y_t measured at the time instant t is

$$s_t = s_{t-1} + y_t, \tag{9}$$

$$q_t = q_{t-1} + y_t y_t'. \tag{10}$$

The point estimates of parameters S and h are obtained via expressing the covariance matrix of the uniform distribution with the help of the statistics s_t and q_t

$$D_t = \left(q_t - s_t s_t'/t \right)/t, \tag{11}$$

which gives

$$\hat{S}_t = s_t/t, \tag{12}$$

$$\hat{h}_t = \sqrt{3 \, \text{diag}(D_t)}, \tag{13}$$

where $\sqrt{3 \, \text{diag}(D_t)}$ denotes the square roots of entries of the vector $\text{diag}(D_t)$, which follows from the variance of the univariate uniform distribution.

The following remarks can be given regarding (9)–(10):

- The obtained statistics are similar to those used for the recursive estimation of parameters of the normal regression model [12]. This juncture between the uniform and the normal distribution gives a chance to apply the similar systematic approach.
- The first moment, which is the basis of the obtained statistics, has a property of "floating" in the data space. For instance, in the case of starting at zero and continuing at one hundred, the zero value will be forgotten and the pdf will be located around the hundred. It does not hold for the statistics $\mathcal{L}_t$ and $\mathcal{R}_t$, where the pdf would be expanded from zero to the hundred value. This difference is significant and can be used for detecting centers of components.
- However, the drawback of the statistics is a lack of the characteristics of limiting the data by the support. As a consequence, the component support can move to the prohibited data area. To avoid this, further restrictions should be used.

3.2   Individual Categorical Model Estimation

The individual categorical model (2) in the case of the measured values of c_t, c_{t-1} and z_t is estimated via (3) using the conjugate prior Dirichlet pdf according to [13] with the recomputable statistics $(v_{t-1})_k$, which is here the square m_c-dimensional matrix, existing for each value of z_t. Its entries for $c_t = i$, $c_{t-1} = j$ and $z_t = k$ would be updated for $i, j \in \{1, \dots, m_c\}$, $k \in \{1, \dots, m_z\}$ in the following way:

$$(v_{i|j;t})_k = (v_{i|j;t-1})_k + \delta(i, j, k; c_t, c_{t-1}, z_t), \tag{14}$$

where $\delta(i, j, k; c_t, c_{t-1}, z_t)$ is the Kronecker delta function, which is equal to 1, if $c_t = i$ and $c_{t-1} = j$ and $z_t = k$, and it is 0 otherwise. The point estimate of α is then obtained by

$$(\hat{\alpha}_{i|j;t})_k = \frac{(v_{i|j;t})_k}{\sum_{l=1}^{m_c} (v_{l|j;t})_k}, \tag{15}$$

The value of the pointer c_t at time t points to the active component. However, values of c_t and c_{t-1} are unavailable and should be estimated.

4   Mixture Estimation with Uniform Components

The discussed mixture-based clustering the bounded data is based on estimation of a mixture of uniform components and determination of the currently active one. Generally one of the key problems during the mixture estimation is initialization of the algorithm, i.e., the initial location of components in the data space. Difficulties with initialization of the mixture estimation algorithm still grow in the case of uniform components.

In order to estimate the mixture under the adopted theory [11, 12] (see the state of the art in Sect. 1), it is necessary to obtain the algebraic recursion of the update of the individual component and the pointer statistics in the form

actual statistics = previous statistics

+ weighted data-based statistics increment.

Another demand is the use of the statistics for effective computing the point estimates of parameters. If these two requirements are met, it is possible to use the following estimation scheme, sec, e.g., [18]:

- Measuring the new data item;
- Computing the proximity of the data item to individual components;

- Computing the probability of the activity of components (i.e., weights) using the proximity and the past activity, where the maximal probability declares the currently active component;
- Updating the statistics of all components and the pointer model;
- Re-computing the point estimates of parameters necessary for calculating the proximity.

Uniform components do not belong to the exponential family, and extension of the estimation approach to this class of components is not entirely trivial. The way how to utilize advantages of both the types of statistics during the recursive estimation and to keep the disjoint components is the algorithm of their combination along with the forgetting technique, which is proposed in this paper. This way is rather suitable, if positions of components are known before (at least, one data item from each component). Thus, it is suitable to initialize the estimation by using the statistics (9)–(10) for detecting the initial centers of components with the help of prior data, and then actualize (6)–(7) by new data for specifying the bounds.

4.1 Proximity as the Approximated Likelihood

The derivations of individual points of the above scheme are based on construction of the joint pdf of all variables to be estimated and application of the Bayes rule, which takes the form (under assumption of the mutual independence of Θ and α, and Δ_t and α, and c_t and Θ):

$$\underbrace{f(\Theta, c_t = i, c_{t-1} = j, \alpha | \Delta(t))}_{\text{joint posterior pdf}}$$

$$\propto \underbrace{f(y_t, \Theta, c_t = i, z_t = k, c_{t-1} = j, \alpha | \Delta(t-1))}_{\text{via chain rule and Bayes rule}}$$

$$= \underbrace{f\left(y_t | \Theta, c_t = i\right)}_{(1)} \underbrace{f(\Theta | \Delta(t-1))}_{\text{prior pdf of } \Theta}$$

$$\times \underbrace{f\left(c_t = i | \alpha, c_{t-1} = j, z_t = k\right)}_{(2)} \underbrace{f(\alpha | \Delta(t-1))}_{\text{prior pdf of } \alpha}$$

$$\times \underbrace{f(c_{t-1} = j | \Delta(t-1))}_{\text{prior pointer pdf}}, \tag{16}$$

$\forall i, j \in \{1, 2, \ldots, m_c\}$ and for $k \in \{1, 2, \ldots, m_z\}$. To obtain recursive formulas for estimation of c_t, Θ and α with the help of (16), it is necessary to marginalize it firstly over the parameters Θ and α.

The marginalization of (16) over parameters Θ provides the proximity, i.e., the closeness of the current data item y_t to individual components at each time instant t. It is evaluated in the same way as for the mixture estimation with normal regression components, i.e., as the approximated likelihood. It is the value of the normal pdf, which is the normal approximation of the uniform component, optimal in the sense of the Kullback-Leibler divergence, see, e.g., [13]. The proximity is obtained by substituting the point estimates of the expectation $(E_{t-1})_i$ and the covariance matrix $(D_{t-1})_i$ of each i-th uniform component from the previous time instant $t - 1$ and the currently measured y_t into the pdf

$$m_i = (2\pi)^{-K/2} |(D_{t-1})_i|^{-1/2}$$

$$\times \exp\left\{ -\frac{1}{2} \left(y_t - (E_{t-1})_i\right)' (D_{t-1}^{-1})_i \left(y_t - (E_{t-1})_i\right) \right\}, \tag{17}$$

where $(E_{t-1})_i$ and $(D_{t-1})_i$ are either (12) and (11) respectively obtained for the i-th component via the statistics (9)–(10), or the expectation $(E_{t-1})_i$ is the K-dimensional vector, each l-th entry of which is

$$(E_{l;t-1})_i = \frac{1}{2}((\hat{L}_{l;t-1})_i + (\hat{R}_{l;t-1})_i), \tag{18}$$

and the covariance matrix $(D_{t-1})_i$ contains on the diagonal

$$(D_{l;t-1})_i = \frac{1}{12}((\hat{R}_{l;t-1})_i - (\hat{L}_{l;t-1})_i)^2 \tag{19}$$

obtained via (8). The proximities from all m_c components form the m_c-dimensional vector m. Similarly, the integral of (16) over α provides the computation of its point estimate (15) using the previous-time statistics $(v_{t-1})_k$ for the actual value k of z_t.

4.2 Component Weights

In order to obtain the i-th component weight (a probability that the component is currently active) the proximities (17) are multiplied entry-wise by the previous-time point estimate of the parameter α (15) and the prior weighting m_c-dimensional vector w_{t-1}, whose entries are the prior (initially chosen) pointer pdfs $(c_{t-1} = j|\Delta(t-1))$, i.e.,

$$W_t \propto \left(w_{t-1}m'\right) .* (\hat{\alpha}_{t-1})_k \tag{20}$$

where W_t denotes the square m_c-dimensional matrix comprised from pdfs $f(c_t = i, c_{t-1} = j|\Delta(t))$ joint for c_t and c_{t-1}, and $.*$ is a "dot product" that multiplies the matrices entry by entry. The matrix W_t is normalized so that the overall sum of all its entries is equal to 1, and subsequently it is summed up over rows, which allows to obtain the vector w_t with updated component weights $w_{i;t}$ for all components. The maximal $w_{i;t}$ defines the currently active component, i.e., the point estimate of the pointer c_t at time t.

4.3 The Component Statistics Update

Using the obtained weights $w_{i;t}$ at time t, the component statistics are updated as follows. The updates (9)–(10) for the i-th component takes the form

$$(s_{l;t})_i = (s_{l;t-1})_i + w_{i;t}y_{l;t}, \tag{21}$$

$$(q_{l;t})_i = (q_{l;t-1})_i + w_{i;t}y_{l;t}^2, \tag{22}$$

$\forall i \in \{1, 2, \ldots, m_c\}$ and $\forall l = \{1, \ldots, K\}$.

The update of the statistics (6)–(7) is performed with the adaptive forgetting technique. The principle of this forgetting is as follows. If the larger value of the entry $y_{l;t}$ is measured, the corresponding l-th maximum bound $(\mathcal{R}_{l;t})_i$ of the i-th component moves on it. Otherwise it is a bit narrowed. If the larger values of $y_{l;t}$ do not arrive for some number of time instants, the maximum bound will decrease until it meets the active area where the data are measured. The identical principle is for the minimum bound on the opposite direction. The question is a reasonable start of forgetting. The idea is very simple: forgetting should not be performed as soon as the bound does not move, but after some period of time. Therefore, it is necessary to estimate how often the bounds should be updated, and start forgetting when they do not move too long.

If the maximum statistics $(\mathcal{R}_{l;t-1})_i$ is assumed to lie in e.g., 80% of the interval of the i-th uniform component (from the minimum to the maximum), the probability of measuring the value of $y_{l;t} > (\mathcal{R}_{l;t-1})_i$, which leads to its update (7), is 0.2. The number n of the time instants, when $(\mathcal{R}_{l;t-1})_i$ was not updated (due to $y_{l;t} < (\mathcal{R}_{l;t-1})_i$) is described by the geometrical distribution with the distribution function

$$F(n) = 1 - (1 - p)^n, \tag{23}$$

here with $p = 0.2$. Taking the confidence level, for instance, 0.05, it is possible to compute the number n, for which the following relation holds:

$$1 - F(n) = 0.05. \tag{24}$$

It is

$$n = \frac{\ln(0.05)}{\ln(0.8)} \doteq 13. \tag{25}$$

It means that if the statistics was not updated during $n = 13$ time instants, and the update follows, then with the probability 0.95 the current point estimate (according to (8)) lies in 20% from the population maximum border. This can occur either (i) if it really lies in 20% from the right bound—then shifting the bound to the left leads to the higher frequency of updating, or (ii) the point estimate of the right bound is caused by some outlier—then the shifting remains until the bound estimate reaches the corresponding data cluster. This will enable to get rid of inaccuracies brought by the prior statistics.

In this way, the update (6)–(7) takes the following form $\forall i \in \{1, 2, \ldots, m_c\}$ and $\forall l = \{1, \ldots, K\}$. For the minimum bound, the counter of non-updates is set as

$$(\lambda^L_{l;t-1})_i = 0, \tag{26}$$

and then

$$\delta_L = y_{l;t} - (\mathcal{L}_{l;t-1})_i, \tag{27}$$

$$\text{if } \delta_L < 0, \quad (\mathcal{L}_{l;t})_i = (\mathcal{L}_{l;t-1})_i - w_{i;t}\delta_L, \tag{28}$$

$$(\lambda^L_{l;t})_i = 0, \tag{29}$$

$$\text{else } (\lambda^L_{l;t})_i = (\lambda^L_{l;t-1})_i + 1, \tag{30}$$

$$\text{if } (\lambda^L_{l;t})_i > \underbrace{n}_{(25)}, \quad (\mathcal{L}_{l;t})_i = (\mathcal{L}_{l;t-1})_i + \phi w_{i;t}, \tag{31}$$

where ϕ is the forgetting factor, often set as 0.01. Similarly the update is performed for the maximum bound with the counter of non-updates set as $(\lambda^R_{l;t-1})_i = 0$, i.e., and

$$\delta_R = y_{l;t} - (\mathcal{R}_{l;t-1})_i, \tag{32}$$

$$\text{if } \delta_R > 0, \quad (\mathcal{R}_{l;t})_i = (\mathcal{R}_{l;t-1})_i + w_{i;t}\delta_R, \tag{33}$$

$$(\lambda^R_{l;t})_i = 0, \tag{34}$$

$$\text{else } (\lambda^R_{l;t})_i = (\lambda^R_{l;t-1})_i + 1, \tag{35}$$

$$\text{if } (\lambda^R_{l;t})_i > n, \quad (\mathcal{R}_{l;t})_i = (\mathcal{R}_{l;t-1})_i - \phi w_{i;t}. \tag{36}$$

4.4 The Pointer Update

The statistics of the pointer model is updated similarly to the update of the individual categorical model and based on [11, 13], however, with the joint weights $W_{i,j;t}$ [18] from the matrix (20), where the row j corresponds to the value of c_{t-1}, and the column i to the current pointer c_t

$$(v_{i|j;t})_k = (v_{i|j;t-1})_k + \delta(k; z_t)W_{j,i;t},\tag{37}$$

and the Kronecker delta function $\delta(k; z_t) = 1$ for $z_t = k$ and 0 otherwise.

4.5 Algorithm

The following algorithm specifies the estimation scheme with the above relations.
Initialization (for t = 0)

1. Set the number of components m_c.
2. Set the initial (expert-based or random) values of all component statistics $(s_{l;0})_i$, $(q_{l;0})_i$ and the pointer statistics $(v_0)_k$ $\forall k \in \{1, 2, \ldots, m_z\}$.
3. Using the initial statistics, compute the point estimates (12), (13), (11) and (15).
4. Set the initial weighting vector w_0.

Initialization of component centers (for t = 1, …, T)

1. Load the prior data item y_t, z_t.
2. Substitute (12), (11) and y_t into (17) to obtain the proximities.
3. Using (15) for the actual value k of z_t, compute the weighting vector w_t via (20), its normalization and summation over rows.
4. Update the statistics (21), (22) and (37).
5. Re-compute the point estimates (12), (13), (11) and (15) and go to Step 1 of the initialization of component centers.
6. For $t = T$ the result is $(\hat{S}_{l;T})_i$, which is the center of the i-th component for the l-th entry of y_t.

On-line bound estimation (for t = T+1, T+2, …)

1. Use the obtained centers to set the initial bounds $(\hat{L}_{l;T})_i = (\hat{S}_{l;T})_i - \varepsilon$, and $(\hat{R}_{l;T})_i = (\hat{S}_{l;T})_i + \varepsilon$ with small ε.
2. Measure the data item y_t, z_t.
3. For all components, compute the expectations and the covariance matrices via (18) and (19).
4. Substitute (18), (19) and the current y_t into (17).
5. Using (15) for the actual value k of z_t, compute the weighting vector w_t via (20), its normalization and summation over rows.
6. Declare the active component according the biggest entry of the vector w_t, which is the point estimate of the pointer c_t at time t.
7. Update the component statistics according to (26)–(36) for both the bounds.
8. Update the pointer statistics (37).
9. Re-compute the point estimates (18), (19) and (15) and go to Step 2 of the on-line bound estimation.

5 Results

5.1 Illustrative Example

A simple example of the algorithm application can be done by using simulated data with three uniform components. For programming the open-source software Scilab (www.scilab.org) was used.

Advantages of the update (26)–(36) with forgetting can be demonstrated as follows. 500 simulated data items of two-dimensional vector y_t are taken. Figure 1 (top) shows clusters detected by using the update with forgetting and compares them with clustering, where the update was taken without forgetting (bottom). It can be seen that even in the simple case of well-distinguishable components the clustering without forgetting in the bottom plot is not successful.

Figure 2 compares results for a more complicated situation, where components are located close to each other. Mixture-based clustering with forgetting coped with this task, which is demonstrated in Fig. 2 (top). Clustering without forgetting gave similar results as in Fig. 1 (bottom).

These results only verify correctness of programming. The next section provides validation experiments with realistic traffic simulations, which is a much more complicated task.

5.2 Validation Experiments

Realistic simulations from the transportation microscopic simulator Aimsun (www. aimsun.com) were used for testing the proposed algorithm. A series of validation experiments was performed. Here typical results are shown.

The aim of clustering was to detect different working modes in the traffic flow data.

5.2.1 Data

The data vector y_t contained:

- $y_{1;t}$—the non-negative traffic flow intensity [vehicle/period] bounded by the saturated flow on the considered intersection on the maximal value 32;
- $y_{2;t}$—the non-negative occupancy of the measuring detector [%] with the maximal value 100.

The discrete variable z_t was the measured indicator of the vehicle queue existence such that $z_t \in \{0, 1\}$, where 1 denotes that the queue is observed, and 0—there is no queue. The data were measured each 90 s.

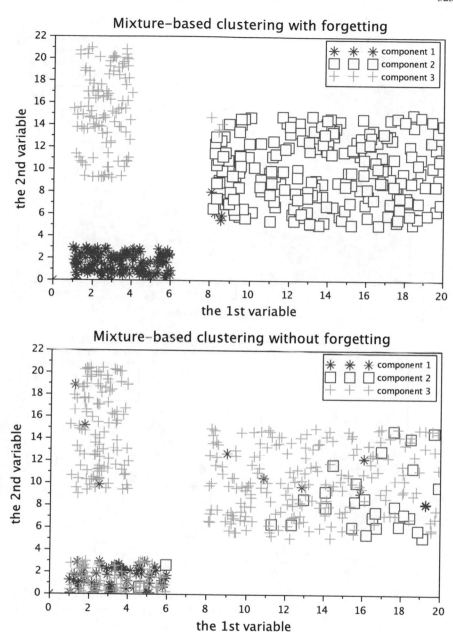

Fig. 1 Comparison of clustering with and without forgetting with well-distinguishable components. The top figure compares results of clustering with forgetting (top) and with- out forgetting (bottom). Notice that three well-distinguishable components are correctly detected in the top plot, but they are incorrectly determined in the bottom figure

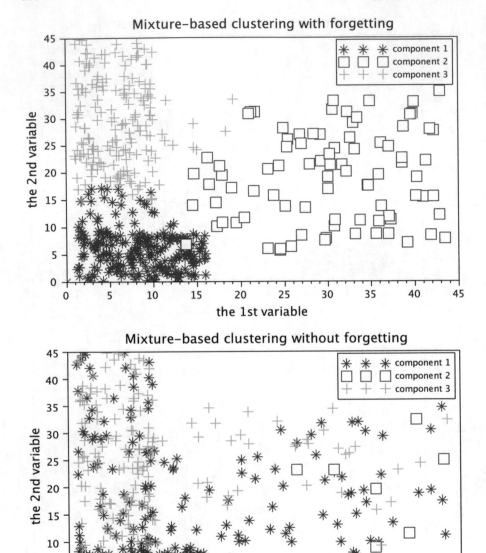

Fig. 2 Comparison of clustering with and without forgetting with closely located components. Notice that three components are sharply visible in the top figure, and they are incorrectly detected in the bottom figure

Table 1 Initial centers of three components

	$(\hat{S}_{T=400})_i$
$i = 1$	[16.1 85.4]′
$i = 2$	[16.3 33.5]′
$i = 3$	[8.6 2.4]′

5.2.2   Initialization

400 prior data items were used for the initial detection of component centers. The assumed number of components expressing the traffic flow modes was $m_c = 3$. The initial centers of components obtained according to the corresponding part of the algorithm are given in Table 1.

5.2.3   Results

1000 data items were used for the bound estimation. Figure 3 demonstrates results of the proposed mixture-based clustering (top) and compares them with the k-means clustering [2] (bottom).

The proposed algorithm detected three clusters on-line by actualizing the statistics by each arriving data item, see Fig. 3 (top).

The upper cluster with the center [16.1   85.4]′ (according to $i = 1$ in Table 1) corresponds to the approaching unstable traffic flow. This explains why the intensity is almost the same as for the middle cluster (corresponding to $i = 2$ in Table 1, i.e., with the initial value 16.3), however, the detector occupancy reports a high degree of workload for the upper cluster. The middle cluster can be interpreted as the stable flow. The bottom cluster in Fig. 3 (top) with the lower intensity and lower occupancy and the initial centers [8.6 2.4]′ corresponds to the free traffic flow.

Figure 3 (bottom) shows three clusters detected by iterative processing of the whole data sample by the k-means algorithm. The results differ from those obtained in Fig. 3 (top): the cluster with the free flow is partitioned in two clusters with two centers. The middle cluster is not found. The upper cluster is the same as in Fig. 3 (top).

5.2.4   Discussion

The obtained results look promising. Surely, the question can arise of application of the approach in the case of lack of prior data. The estimation with both the types of statistics can be performed also independently. When using the algorithm only with the statistics (9)–(10), the bounds can be probably exceeded, which can be fixed by additional restrictions. For the algorithm only with the statistics (6)–(7) the initialization with random centers can be applied in the absence of prior data. In this case

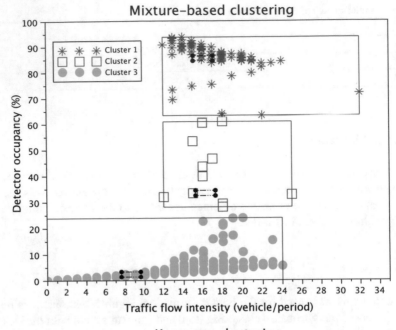

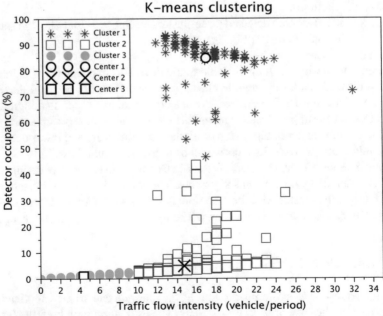

Fig. 3 Comparison of mixture-based (MB) (top) and k-means (KM) (bottom) clustering the traffic flow data The top figure shows three MB clusters: the upper one corresponds to the almost unstable traffic flow, the middle one to the stable flow, and the bottom cluster is the free traffic flow. Initial bounds of clusters are shown as internal rectangles indicated by the dashed line. The estimated bounds of clusters are plotted as external rectangles. The bottom figure plots three KM clusters and their final centers. The upper cluster is similar to the MB results, however, the clusters of stable and free traffic flow are different

setting the adaptive forgetting technique is expected to considerably help in finding the bounds. Analyzing the series of performed experiments, it can be said that the most suitable application of the approach is the combination of statistics given in Algorithm 4.5.

One of the advantages of the proposed algorithm is also a possibility of its on-line running with real-time measurements and the gradual update of component bounds, while iterative algorithms in this area (e.g., k-means) focus on off-line processing the whole available data sample at once. The unified systematic Bayesian approach used for other types of components (see Sect. 1), where each of models requires a specific update of statistics, is here presented for uniform components. A combination with other distributions for covering different shapes of clusters can be also a further extension of the approach.

6 Conclusion

This paper proposes the algorithm of mixture-based clustering the non-Gaussian data with fixed bounds. The specific solutions include (i) the recursive Bayesian estimation of uniform components and the switching model; (ii) the initialization via the moment method; (iii) the forgetting technique. The results of testing the algorithm are provided.

However, there still exists a series of open problems in the discussed area, where the first of them is modeling dependent uniformly distributed variables with parallelogram-shaped clusters. Further, extension of the clustering and classification tools for other distributions is planned within the future work on the present project.

Acknowledgements The research was supported by project GAČR GA15-03564S.

References

1. Berkhin, P.: A survey of clustering data mining techniques. In: Kogan, J., Nicholas, C., Teboulle, M. (eds.) Grouping Multidimensional Data, pp. 25–71. Springer, Berlin Heidelberg (2006)
2. Jain, A.K.: Data clustering: 50 years beyond K-means. Pattern Recogn. Lett. **31**(8), 651–666 (2010)
3. Bouveyron, C., Hammer B., Villmann T.: Recent developments in clustering algorithms. In: Verleysen, M. (ed.) ESANN, pp. 447–458 (2012)
4. Fraley, C., Raftery, A.E.: How many clusters? which clustering method? answers via model-based cluster analysis. Comput. J. **41**(8), 578–588 (1998)
5. Fraley, C., Raftery, A.E.: Bayesian regularization for normal mixture estimation and model-based clustering. J. Classif. **24**(2), 155–181 (2007)
6. Bouveyron, C., Brunet-Saumard, C.: Model-based clustering of high-dimensional data: a review. Comput. Stat. Data Anal. **71**, 52–78 (2014)
7. Banfield, J.D., Raftery, A.E.: Model-based gaussian and non-gaussian clustering. Biometrics **49**(3), 803–821 (1993)

8. Pavelková, L., Kárný, M.: State and parameter estimation of state-space model with entry-wise correlated uniform noise. Int. J. Adapt. Control Signal Process. **28**(11), 1189–1205 (2014)
9. Jirsa, L., Pavelková, L.: Estimation of uniform static regression model with abruptly varying parameters. In: Filipe J., Madani, K., Gusikhin, O., Sasiadek, J. (eds.) Proceedings of the 12th International Conference on Informatics in Control, Automation and Robotics ICINCO 2015, Colmar, France, 21–23 July 2015, pp. 603–607
10. McLachlan, G., Peel, D.: Finite Mixture Models. Wiley-Interscience (2000)
11. Kárný, M., Kadlec, J., Sutanto, E.L.: Quasi-Bayes estimation applied to normal mixture. In: Rojíček, J., Valečková, M., Kárný, M., Warwick, K. (eds.) Preprints of the 3rd European IEEE Workshop on Computer-Intensive Methods in Control and Data Processing, CMP'98 /3./, Prague, CZ, 07-09 sept 1998, pp. 77–82
12. Peterka, V.: Bayesian system identification. In: Eykhoff, P. (eds.) Trends and Progress in System Identification, Oxford, Pergamon Press, 1981, pp. 239–304
13. Kárný, M., Böhm, J., Guy, T.V., Jirsa, L., Nagy, I., Nedoma, P., Tesař, L.: Optimized Bayesian Dynamic Advising: Theory and Algorithms. Springer, London (2006)
14. Nagy, I., Suzdaleva, E., Mlynářová, T.: Mixture-based clustering non-gaussian data with fixed bounds. In: Proceedings of 2016 IEEE 8th International Conference on Intelligent Systems IS'2016, Sofia, Bulgaria, 4–6 Sept 2016, pp. 265–271 (2016)
15. Nagy, I., Suzdaleva, E., Kárný, M., Mlynářová, T.: Bayesian estimation of dynamic finite mixtures. Int. J. Adapt. Control Signal Process **25**(9), 765–787 (2011)
16. Nagy, I., Suzdaleva, E.: Mixture estimation with state-space components and Markov model of switching. Appl. Math. Model. **37**(24), 9970–9984 (2013)
17. Suzdaleva, E., Nagy, I.: Recursive mixture estimation with data dependent dynamic model of switching. J. Comput. Graph. Stat. (submitted)
18. Suzdaleva, E., Nagy, I., Mlynářová, T.: Recursive estimation of mixtures of exponential and normal distributions. In: Proceedings of the 8th IEEE International Conference on Intelligent Data Acquisition and Advanced Computing Systems: Technology and Applications, Warsaw, Poland, 24–26 Sept 2015, pp. 137–142
19. Gelman, A., Carlin, J.B., Stern, H.S., Dunson, D.B., Vehtari, A., Rubin, D.B.: Bayesian Data Analysis, 3rd edn. Chapman & Hall/CRC Texts in Statistical Science, Chapman and Hall/CRC (2013)
20. Lee, P.M.: Bayesian Statistics: An Introduction, 4th edn. Wiley (2012)
21. Casella, G., Berger R.L.: Statistical Inference, 2nd edn. Duxbury Press (2001)
22. Bowman, K.O., Shenton, L.R.: Estimator: method of moments. Encyclopedia of Statistical Sciences, pp. 2092–2098. Wiley (1998)

Printed in the United States
By Bookmasters